T0243104

Using and Developing Measurement Instruments in Science Education

A volume in
Science & Engineering Education Sources
Calvin S. Kalman, *Series Editor*

Using and Developing Measurement Instruments in Science Education

A Rasch Modeling Approach
(Second edition)

Xiufeng Liu
State University of New York at Buffalo

INFORMATION AGE PUBLISHING, INC.
Charlotte, NC • www.infoagepub.com

Library of Congress Cataloging-in-Publication Data

A CIP record for this book is available from the Library of Congress
http://www.loc.gov

ISBN: 978-1-64113-934-2 (Paperback)
978-1-64113-935-9 (Hardcover)
978-1-64113-936-6 (E-Book)

Printed in the United States of America

Contents

List of Tables

Using and Developing Measurement Instruments in Science Education, pages ix–x

List of Figures

Using and Developing Measurement Instruments in Science Education, pages xi–xvi

Preface

Since the beginning of the 21st century, the landscape of research methodology in science education has changed. While qualitative research methods remain popular among many science education researchers, more and more are turning to quantitative research methods, and the demand for measurement instruments is growing. For example, there are frequent postings on the NARST (National Association for Research in Science Teaching) LISTSERV looking for measurement instruments for specific research questions. In fact, in her presidential address to the 2009 NARST annual conference, Charlene Czerniak identified measurement as one of the top 10 grand challenges in science education research based on a survey of NARST members. Quantitative research using measurement instruments is gaining momentum.

No one can dispute the importance of measurement instruments in scientific research. Becoming familiar with standardized measurement instruments and developing new measurement instruments are integral to scientific inquiry and thus an essential component of graduate research training in natural sciences; this has not been the case in science education, regrettably. Few current doctoral programs in science education offer a course specifically on measurement instruments in science education. As a result, many science education researchers are not knowledgeable in using and developing measurement instruments; often they find themselves to be at a loss. There is a need to improve the status of using and developing measurement instruments in science education. Unfortunately, there is

Using and Developing Measurement Instruments in Science Education, pages xvii–xx

currently not a single book devoted to measurement instruments in science education. This book, *Using and Developing Measurement Instruments in Science Education: A Rasch Modeling Approach*, should fill this void. It intends to be a comprehensive and introductory measurement book in science education.

This book is for anyone who is interested in knowing what measurement instruments are available and how to develop measurement instruments for science education research. For example, science education researchers may use this book as a reference for locating measurement instruments when designing a research study. This book can be a main text for a course related to measurement in science education or science education research methods at the doctoral level. Although the book contains some statistics, no prerequisites in mathematics beyond the high school level are assumed.

This book contains seven chapters. Chapter 1 provides an overview of the historical evolution of measurement instruments in science education, validity and reliability of measurement, and current standards of measurement; it also reviews fundamental skills in writing questions for developing measurement instruments. Chapter 2 introduces approaches to developing measurement instruments; the focus is on Rasch modeling to develop measurement instruments. Rasch modeling applies a statistical model to raw data (no matter if they are nominal or ordinal) to produce measures of item difficulties and student abilities; as an outcome of the measurement revolution since the 1960s, it is considered to be one of the best approaches currently available for developing measurement instruments. Although applications of Rasch modeling in testing industries (e.g., Educational Testing Services) and large-scale international assessments (e.g., Trend in International Mathematics and Science Study) are routine, its application in science education is still preliminary. A focus on Rasch modeling in this book intends to change this situation by promoting a broader application in science education research.

Based on the conceptual framework and technical skills developed in Chapters 1 and 2, the subsequent chapters introduce measurement instruments published in refereed science education research journals over the last few decades related to common domains of measurement, and describe step-by-step procedures for developing measurement instruments. Specifically, Chapter 3 focuses on using and developing instruments for measuring conceptual understanding; Chapter 4 focuses on using and developing instruments for measuring affective variables; Chapter 5 focuses on using and developing instruments for measuring science inquiry; Chapter 6 focuses on using and developing instruments for measuring science learning progression; finally, Chapter 7 focuses on using and developing instruments for measuring science learning environments. Chapters 3 through 7 are independent from each other; they can be completed in any order.

It is my hope that this book will serve as a starting point for science education researchers to use and develop measurement instruments for science education research. Although measurement is a highly technical and well-developed field, this book emphasizes the applied side of measurement theories and techniques. This book is fully grounded in science education research literature. All the instruments are from refereed science education research journals; all the examples are specifically related to science education research problems. Readers should be able to develop a sound understanding of measurement theories and approaches, particularly Rasch modeling, in using and developing measurement instruments for science education research.

Acknowledgments

This book would not be possible without the support of many individuals. First and foremost, I thank my family (my wife, son, and daughter) for their unconditional support and love over the years. Dr. William Boone of Miami University (Ohio) has been inspirational in my pursuit of scientific measurement in science education; I thank you, Bill, for your shared passion in Rasch measurement and collaboration over the years.

I thank masters and doctoral students at the University at Buffalo, the State University of New York, for their interests in science assessment. This book is a result of courses related to science assessment, science education research designs, and measurement theories that I have taught over the past 7 years. The Department of Learning & Instruction at the University at Buffalo has been consistently supportive of my teaching and research related to science assessment; I feel fortunate to be a member of this academic home.

Last but not the least, I thank Information Age Publishing for bringing this book into print, and Prof. Calvin Kalman, editor-in-chief of the book series *Science & Engineering Education Sources*, for his support during the book preparation and final proofreading along with valuable suggestions. Anonymous reviewers provided valuable suggestions to strengthen the book, I thank you. Ms. Nancy Wojcik edited the final version of the manuscript; I hope readers will find the book to be highly readable.

—Xiufeng Liu, PhD
Buffalo, NY
December 2009

Preface to the Second Edition

It has been almost ten years since the publication of this book. Much progress has been made in science education research in terms of developing measurement instruments. If Rasch modeling was still a novel idea at the time of this book's first publication in 2010, it is now commonly used for developing measurement instruments in science education. There is a growing recognition within the science education community that application of Rasch modeling is necessary in order for strong validity claims to be made and for an instrument to be published in major science education research journals.

The increasing acceptance of Rasch modeling in science education is a testimony of powers of Rasch models. As always is the case with adoption of an innovation or a new method, increasing uses of Rasch modeling in science education research is accompanied with confusions and even misuses. While there have been exemplary applications of Rasch modeling in developing measurement instruments in science education, there have also been cases of casual applications and reporting of Rasch measurement. The science education community would benefit from a common framework in applying Rasch modeling for developing measurement instruments in order to advance from a stage of "storming" toward the stage of "norming." This book is intended to contribute to this development.

In keeping with the original intent of the book, that is, providing a comprehensive source for using published instruments and a conceptual introduction for using Rasch modeling to develop new measurement

---

Using and Developing Measurement Instruments in Science Education, pages xxi–xxii

instruments, this second edition maintains the original structure of the book. Major changes to the first edition of the book include (a) expanding the introduction of published instruments by including instruments published after 2010, (b) increasing clarity about the steps for developing measurement instruments by following a uniform structure, and (c) improving the writing and formatting. While the main ideas remain unchanged from the first edition, this updated edition is more current, comprehensive, coherent, and readable.

Writing this second edition has been a long journey. In fact, the attempt started almost four years ago. Due to my many other commitments, particularly taking on administrative responsibilities at the department, school, and university levels, the progress to complete this edition was slow. I thank Information Age Publishing and the book series editor, Prof. Calvin S. Kalman, for their patience and faith in me. I am also grateful for my home institution, University at Buffalo, State University of New York, for granting me a 6-month sabbatical in the Fall of 2018 during which the final dash to the finish line was made. Ultimately, I thank the readers including students who took my Rasch course over the years for their support and encouragement for a new edition. I hope this second edition meets your expectations!

—Xiufeng Liu, PhD
Buffalo, NY
April 2019

1

Essential Concepts and Skills for Using and Developing Measurement Instruments

Measurement is a process of quantifying observations and an essential component of scientific inquiry in science education. Measurement can be done in a variety of ways; using standardized instruments to conduct measurement is one way. A *measurement instrument* is a standardized tool with its associated procedures to quantify observations; it possesses empirical technical qualities. For example, in natural sciences a thermometer is a standardized measurement instrument; its making and use follow an international convention. As a result, temperature readings based on a thermometer are universally understandable and comparable. In science education, we also use measurement instruments to collect quantitative data on a wide variety of constructs. There are well-known measurement instruments in science education research, such as the TIMSS (Trends in International Math and Science Study) and PISA (Program for International Student Assessment) for measuring science literacy. Other measurement

Using and Developing Measurement Instruments in Science Education, pages 1–27

instruments, although less well-known, are routinely used and play important roles in science education research. Developing measurement instruments is a time-consuming and expensive process; it is important for science education researchers to become knowledgeable about a wide variety of measurement instruments currently available and be able to develop new measurement instruments for science education research when necessary. This knowledge begins with an appreciation of how measurement instruments have evolved throughout the history of science education.

Evolution of Measurement Instruments in Science Education

Using and developing measurement instruments are an integral part of the history of science education research. A brief overview of this history can help understand how using and developing measurement instruments closely relate to efforts to improve science education. Use and development of measurement instruments in science education began in the late 19th century and became popular in the early 20th century. According to Travers (1983), the Boston Survey conducted in 1845 was the first use of printed tests for large-scale assessment of student achievement in science. Soon afterwards, more school boards and states across the United States began to assess student achievement in science using standardized instruments. For example, in the early 1900s, the New York Board of Regents began to offer the Regents exams. According to Doran, Lawrenz, and Helgeson (1994), the second Curtis Digests, which reviewed science education research during the period from 1925–1930, showed that science educators began to use more "new" or standardized test questions such as completion, modified true–false, multiple-choice, and modified multiple-choice questions. Standardized achievement measurement instruments in science began to proliferate during the period from 1920s to 1930s. The purpose of introducing standardized measurement instruments into science achievement assessment was mainly to reduce errors in scoring, thus making assessment more scientific. The emergence of standardized measurement instruments was a cornerstone of the first generation of evaluation (Guba & Lincoln, 1989), which may be considered the first stage of using and developing measurement instruments in science education. One key characteristic of this first stage was to obtain objective measures about student science achievements.

The second stage of using and developing measurement instruments in science education was from the 1940s to the 1950s. This stage was characterized by expanding measurement targets beyond the cognitive domain and by explicitly tying measurement instruments to curriculum objectives, that is, ensuring the alignment between measurement instruments and curriculum objectives. As a result, paper-and-pencil based measurement instruments for evaluation of laboratory instruction became available in the 1940s. *The Forty-Sixth Yearbook* (National Society for the Study of Education, 1947) recommended evaluation to include skills, interests, appreciation, attitudes, as well as the functional understanding of facts, large conceptions, and basic principles of science. As a result, measurement instruments in all those domains became available.

The third stage of using and developing measurement instruments in science education took place in the 1960s and 1970s. After the former Soviet Union's Sputnik launch in 1957, the need to evaluate National Science Foundation funded curriculum efforts such as *BSCS Biology* (Biological Science Curriculum Study), Project CHEM, and *PSSC Physics* (Physical Science Study Committee) resulted in a new wave of using and developing measurement instruments in science education. Using measurement instruments to assess appropriate curriculum objectives continued. Measurement domains continued to expand to include measurement of science inquiry skills, attitude toward science, nature of science, as well as teachers' perceptions of science teaching and learning (Liu, 2008). For example, Tamir and Lunetta (1981) designed a comprehensive scheme for analyzing laboratory skills—the Laboratory Assessment Inventory (LAI). In addition, objectives-free measurement instruments were also advocated. Robert Stake's countenance of evaluation included explicit external standards or benchmarks for making judgments about the attainment of curriculum intents (Stake, 1967). In the past, evaluators were reluctant to judge that "gaining one skill is worth losing two understandings" (Stake, 1967, p. 95), or conclude that Curriculum I objectives were more worthy than Curriculum II objectives. Measurement instruments for externally formulated objectives were also valued in evaluation. Soon afterwards, large-scale national and international science assessments emerged. Congress mandated the NAEP (National Assessment of Educational Progress) to be given to Grades 4, 8, and 12 students every 4 years beginning 1969. The International Association for the Evaluation of Educational Achievement (IEA) coordinated its first international science achievement study during 1970–1971, and the second during 1983–1984.

The fourth stage of using and developing measurement instruments in science education occurred during the 1980s and the 1990s. During this period, qualitative approaches to science education research became dominant, and use and development of measurement instruments retreated to a less prominent role. However, due to an increasing interest in student alternative conceptions in science, use and development of diagnostic measurement instruments for assessing student alternative conceptions were a noticeable exception (Wandersee, Novak, & Mintz, 1994). A large number of assessment instruments for diagnosing student alternative conceptions were developed during this period. Also, non-paper-and-pencil based measurement instruments, that is, alternative science assessments, began to gain momentum (Mintzes, Wandersee, & Novak 1999). Alternative science assessment was a direct response to the limitations of paper-and-pencil based standardized measurement instruments using primarily selected response assessment methods, such as multiple-choice. Performance assessment in particular received wide acceptance as an alternative assessment, and gradually became standardized for large-scale administration. For example, performance assessment was a part of national and international science assessments, such as the NAEP, and the 1994–1995 Third International Mathematics and Science Study (Martin & Kelly, 1997). Performance assessment was perceived to be authentic because it approximated the scientific inquiry process.

Entering the 21st century, we are now in a new era of using and developing measurement instruments. This new era is characterized by a renewed interest in quantifying observations using measurement instruments. A number of factors have contributed to the revival of the use and development of measurement instruments in science education. The continuing interest in identifying student alternative conceptions has created a demand for more efficient and large-scale surveys of student alternative conceptions. Also, as standards-based science education reform is gaining momentum in the United States and around the world, new demands for standardized measurement instruments for school accountability are emerging. Just as important as measuring student attainment of learning standards is the measurement of opportunities to learn in the classroom, at home, and in school. "Students cannot be held accountable for achievement unless they are given adequate opportunity to learn science" (National Research Council [NRC], 1996, p. 83). Consequences of using standardized measurement instruments for making high-stake decisions also need to be explicitly evaluated. In addition, a growing recognition of limitations of qualitative research methods and a call for randomized experimentation in education research (NRC, 2002) have created a conducive context for using and developing measurement instruments in science education. Today, use and development of measurement instruments in science education is becoming increasingly routine in all fields of science education research, including measurement of conceptual understanding, affective variables (e.g., attitude toward science), science inquiry, learning progression, and classroom and laboratory learning environments. Importantly, as a result of new advances over the past few decades in our knowledge about how students learn as well as in measurement theories and in applications of technology to teaching and learning, a major change in approaches to developing measurement instruments is currently underway. One such approach is Rasch modeling—the focus of this book.

Fundamental Theories for Using and Developing Measurement Instruments

Underlying the evolution of using and developing measurement instruments in science education is change in conceptions of validity and reliability. *Validity* refers to the adequacy of claims made about a measurement process and outcome. The process of establishing validity of measures of a measurement instrument is called *validation.* Validation used to be only based on correlations, which results in *criterion-related evidence* of validity. In order to establish criterion-related evidence of validity, a criterion variable must be identified, and measures of the criterion variable must be available and assumed valid. The correlation between the two sets of measures of the same group of examinees, one based on an instrument of the criterion variable and another on the instrument under validation, is then computed. If the correlation is statistically significant, then measures of the construct of the current instrument under validation may be claimed to be valid, and vice versa. For example, if we want to establish evidence for the validity of measures of the test of students' conceptual understanding of force and motion, the criterion variable may be identified as the students' physics achievement. If we already know a group of students' valid physics achievement scores based on a test, then we can give the force and motion conceptual test to the same group of students, and compute the correlation coefficient between students' scores on the physics achievement test and on the force and motion conceptual test. If the correlation coefficient is statistically significant, then we can claim that the force and motion conceptual understanding measures are valid. If measures of the criterion variable are collected after measures based on the instrument under validation are also collected, we describe the criterion-related evidence as *predictive validity.* If measures of the criterion variable are collected about the same time as measures obtained from the instrument under validation, we describe the criterion-related evidence as *concurrent validity.* For example, if we want to validate the force and motion conceptual measures described above for ninth grade students, then students' 12th grade physics course grades can be used to establish predictive validity, and students' ninth grade physical science test grades can be used to establish concurrent validity.

Related to criterion-related evidence of validity are convergent validity and divergent/discriminant validity. *Convergent validity* is concurrent criterion-related validity when there is evidence to support the assumed statistically significant positive correlation between the measures based on the instrument under validation and measures based on an instrument of the criterion variable; *divergent/discriminant validity* is concurrent criterion-related validity when there is evidence to support an assumed statistically significant negative correlation between measures based on the instrument under validation and measures based on an instrument of the criterion variable. For example, if it is assumed that there should be a positive correlation between students' math achievement and science achievement, then a statistically positive correlation between students' science achievement test scores based on an instrument being developed and students' math achievement test scores based on an already validated instrument is considered as evidence for the convergent validity of the science achievement test. On the other hand, if it is assumed that there should be a negative correlation between students' test anxiety and science achievement, then a statistically negative correlation between students' science achievement test scores based on an instrument being developed and students' test anxiety based on an already validated instrument is considered as evidence for the divergent validity of the science achievement test.

Subsequently, evaluating the match between the intended content domain and the actual content domain of the measurement instrument became an acceptable approach to validation. This so-called *content-related validation* produces content-related evidence of validity. Content-related evidence may be established based on a table of test specification and the content coverage of test items. Typically, a panel of content experts reviews the table of test specification to decide if the test domain specified by the table of test specification provides adequate sampling of the content domain of the construct being measured. The expert panel may also review the test items of an instrument to see if they provide adequate coverage of the content domain of the construct being measured.

Since the 1950s, the focus of validation shifted from external, that is, criterion-related and content-related toward the internal structure of test items. This approach of validation relied on construct related evidence of validity. *Construct validity* refers to the adequacy of claims about the actual measurement of the intended construct. Evidence for construct validity can be established through multiple approaches. Typically, a hypothesis is derived from a theory that defines the construct of measurement. Because the validity of the theory is assumed, accepting or failing to reject a hypothesis derived from the theory is considered evidence of the validity of the measurement instrument. One approach to construct validation is examination of correlation patterns of items of the measurement instrument. Items measuring a same construct should correlate with each other more than items measuring a different construct. In order to examine correlation patterns among items, a statistical method called factor analysis is commonly used. There are two approaches to factor analysis. Principal component factor analysis is an exploratory approach to examining item correlation patterns by creating distinct groups of items that are thought to measure different latent traits or factors. Items within the same group correlate highly with one common latent factor, but lowly with other latent factors. The number of factors that provides maximal discrimination among different factors is considered evidence for the number of constructs the items measure. If the number of factors and their corresponding distinct item groupings are consistent with the hypothesis based on the defined construct and instrument development, then there is evidence for construct related validity. Confirmatory factor analysis is an explicit hypothesis testing approach to examining item correlation patterns. First, an explicit hypothesis on the number of factors and item groupings according to the factors is formed and expressed as a structural equation model. The structural equation model is then submitted to structural equation modeling (SEM), another statistical analysis approach, to see if data fit with the model. If there is a good model-data-fit, then there is evidence to support the hypothesis, thus construct related validity. If there is not a good model-data-fit, then there is not sufficient evidence to support the hypothesis, and thus there is lack of construct-related validity.

For many years, validity evidence of standardized measurement instruments was conceptualized to be three types: criterion-related, content, and construct. Since the late 1980s, a unified theory of validity based on construct validity has become widely accepted. Messick (1989), in a seminal review of validity theories, defines validity as "an integrated evaluative judgment of the degree to which empirical evidence and theoretical rationales support the adequacy and appropriateness of inference and actions based on test scores or other modes of assessment" (p. 13). Under this *unified conception of validity*, evidential bases of validation, that is, criterion-related validation (predictive or concurrent), and content validation are all special cases of construct validation. Further, the consequence of standardized measurement instrument uses, that is, the consequential base of validation, is also conceptualized as part of the construct validation. The *consequential basis of construct validation* attends to "the appraisal of both potential and actual social consequences of the applied testing" (Messick, 1989, p. 20). Thus, validation may be conducted through analyzing test content, response processes of examinees, internal structures of test items, relations of test scores to other variables, and consequences of testing (Joint committee of AERA, APA and NCME, 1991/2014).

One major limitation of the unified conception of construct validation is the lack of consideration of alternative hypotheses. Because a theory cannot be proven but can only be rejected, construct validation suffers from an apparent confirmatory bias of test developers (Kane, 2006). In order to address this limitation, current conceptions of validation, while still accepting the unified nature of validity, have expanded from a confirmatory hypothetical-deductive reasoning to a system of coherent arguments. That is, validation is now considered a process of establishing evidence to argue for the validity of the proposed interpretations or uses of test scores, which is called the *argument-based approach to validation* or argument-based validity. A key difference between construct-based validation and argument-based validation is that argument-based validation goes beyond confirming the hypothesis for the construct by evaluating the viability of alternative hypotheses. Thus, argument-based validation no longer relies solely on hypotheses derived from a formal theory; it considers any evidence that may help support or counter-argue for the appropriateness of claims made about the test scores and uses.

Kane (2006) states that there are two types of arguments in argument-based validity: interpretative argument and validity argument. "The *interpretative argument* specifies the proposed interpretations and uses of test results by laying out the network of inferences and assumptions leading from the observed performances to the conclusions and decisions based on the performances" (p. 23). Kane (2006) further suggests that interpretive argument may be based on the following four interpretations: (a) trait interpretation, (b) theory-based interpretation, (c) qualitative interpretation, and (d) decision procedures. With the exception of qualitative interpretation, the above interpretations may be roughly considered as corresponding to the previously conceived content validation, construct validation, and consequence validation. The *validity argument* provides an evaluation of the interpretative argument in order to ensure that the interpretative argument is coherent and reasonable, and its assumptions are plausible.

Reliability of measures of a standardized measurement instrument is concerned with the precision of test scores obtained from the instrument. The central concern of reliability is the consistency or replicability of scores across repeated applications of the measurement instrument. The classical test theory approach to reliability is to estimate the amount of errors due to various possible sources such as raters, different test forms, test-retest, and so forth. Accordingly, reliability may be defined as inter-rater reliability, equivalent form reliability, and test-retest reliability. Inter-rater reliability measures the consistency of scores given by two different scorers; equivalent form reliability measures the comparability of scores from two forms of the same test; test-retest reliability measures the stability of scores over time. The above three types of reliability can be quantified by a correlation coefficient. Specifically, *inter-rater reliability* can be established by calculating the correlation coefficient between two sets of scores given by two independent scorers to the same group of examinees, the *equivalent form reliability* can be established by calculating the correlation coefficient between two sets of scores obtained by the same group of examinees from two forms of the measurement instrument, and the *test–retest reliability* can be established by calculating the correlation coefficients between two sets of scores obtained by the same group of examinees from two successive administrations of the same measurement instrument.

Extending the concept of equivalent forms reliability by considering every item of a test as an equivalent form gives rise to the concept of internal consistency reliability. For internal consistency reliability, the most commonly used *Kuder-Richardson formula 20* (KR-20) provides an estimation of the degree of internal consistency among a set of dichotomously scored (e.g., multiple-choice) questions in measuring a group of examinees' performances. KR-20 can be calculated as follows:

$$\text{KR-20} = \frac{k}{k-1}\left(1 - \frac{\sum_{1}^{k} p_i q_i}{\sigma_x^2}\right)$$

where k is the number of items on the test, p_i is the percent of examinees answering item i correctly, q_i is the percent of examinees answering item i incorrectly, and σ_x^2 is the variance of test scores of examinees on the test. KR-20 measures the percent of variation in examinees' test scores that are due to the variation in examinees' true abilities. KR-20 ranges from 0 to 1; the higher the KR-20 is, the more reliable measures of the instrument are.

A more generalized reliability measure for internal consistency among a set of any items (multiple-choice, constructed-response, etc.) is Cronbach's alpha. An *alpha coefficient* is a generalized KR-20; its value also ranges from 0 to 1, indicating the percentage of variance in a sample of examinees' scores due to the covariation among items caused by examinees' true abilities. An alpha coefficient may also be interpreted as the averaged correlation coefficient among all possible pairs of items of a test; it is defined as

$$\alpha = \frac{k}{k-1}\left(1 - \frac{\sum_{1}^{k}\sigma_i^2}{\sigma_x^2}\right)$$

where k is the total number of questions on the test, σ_i^2 is the squared standard deviation (also called variance) of students' scores on item i, and σ_x^2 is the squared standard deviation of students' scores on the entire test, that is, variance.

A more recent and comprehensive approach to reliability is generalizability theory (Brennan, 2001; Haertel, 2006). While reliability is concerned with replicability of test scores across items, raters, times, and so forth, *generalizability* is concerned with inferring scores from one test design to a universe of test designs. A test design consists of facets, such as items, raters, settings (e.g., performance test, paper-and-pencil test), time (pretest, posttest), test forms, and so on. Each facet may further be conceptualized as a random sample or an exhaustive list of its universe. For example, a set of items may be considered as a random sample of items from a large item pool; two raters may be considered as a random selection from a large pool of potential raters or as a fixed selection not to change from test to test. The generalizability theory approaches reliability as two sequential studies: the generalizability study (or G study), and the decision study (or D study). The G study first estimates the variance components corresponding to different facets in a given test design. Each variance component represents the amount of variation contributed by the facet. Based on the variance components, the G study continues by calculating the degree of generalizability of test scores from the given test design to the universe of all possible test designs. This degree of generalizability is in the form of a coefficient ranging from 0 to 1 that describes the percentage of total variance in examinees' observed test scores explained by the facets of the test design. Following the G study, the D study investigates the effects of different test designs when varying the combination of facets on the generalizability coefficients, so that the most efficient test design (i.e., high generalizability but less costly) may be decided for future test administration.

In order to develop a standardized measurement instrument that meets the above validity and reliability requirements, researchers follow a systematic process to create the instrument and to establish evidence of validity and reliability. This systematic process typically consists of the following 10 steps (Crocker & Algina, 1986):

1. Identify the primary purpose(s) for which the test scores will be used.
2. Identify behaviors that represent the construct or define the domain.
3. Prepare a set of test specifications by delineating the proportion of items for each type of behavior identified in Step 2.
4. Construct an initial pool of items.
5. Have items reviewed (and revised, as necessary).
6. Hold preliminary item tryouts (and revise, as necessary).
7. Field-test the items on a large sample representative of the examinee population for whom the test is intended.
8. Determine statistical properties of item scores and, when appropriate, eliminate items that do not meet pre-established criteria.
9. Design and conduct reliability and validity studies for the final form of the test.
10. Develop guidelines for administration, scoring, and interpretation of the test scores.

The above steps progress generally in sequence, but looping between Steps 4 and 9 is also common. The ultimate goal of the above systematic process is to create a measurement instrument that meets expected standards of validity and reliability.

Measurement Standards

The most current measurement standards are the *Standards for Educational and Psychological Testing* developed by a joint committee of the American Educational Research Association (AERA), the National Council on Measurement in Education (NCME), and the American Psychological Association (APA; Joint Committee of the AERA, NCME, and APA, 2014). Published jointly by the three associations, the *standards* are updated regularly to maintain its relevance and reflect most recent measurement theories. The first version of standards was published in 1954. There have been six earlier versions of the standards before its current 2014 version. According to the *standards*, the purpose of the standards is to "provide criteria for the development and evaluation of tests and testing practices and to provide guidelines for assessing the validity of interpretations of test scores for the intended test uses" (Joint Committee of the AERA, NCME, and APA, 2014, p. 1).

The standards' document is organized into three parts. Part I deals with foundations of test development and evaluation. This part includes the following standards: (a) Standard 1: Validity; (b) Standard 2: Reliability/Precision and Errors of Measurement; and (c) Standard 3: Fairness in Testing. Part II deals with operations. This part includes the following standards: (a) Standard 4: Test Design and Development; (b) Standard 5: Scores, Scales, Norms, Score Linking, and Cut Scores; (c) Standard 6: Test Administration, Scoring, Reporting, and Interpretation; (d) Standard 7: Supporting Documentation for Tests; (e) Standard 8: The Rights and Responsibilities of Test Takers; and (f) Standard 9: The Rights and Responsibilities of Test Users. Part III deals with testing applications. This part includes the following standards: (a) Standard 10: Psychological Testing and Assessment; (b) Standard 11: Workforce Testing and Credentialing; (c) Standard 12: Educational Testing and Assessment; and (d) Standard 13: Uses of Tests for Program Evaluation, Policy Studies, and Accountability. Each standard is organized into two components; the first component is the standard itself, and the second component is a commentary providing elaboration of the standard.

Essential Skills for Developing Measurement Instruments

A measurement instrument consists of individual tasks or items. The items can be in many different formats: multiple-choice, checklist, rating scale, Likert-scale, performance task, to name a few. Given that the most common item formats for standardized measurement instruments in science education are multiple-choice questions, true–false questions, and Likert scale questions, this section provides an overview of basic skills for developing multiple-choice, true-and-false, and Likert scale questions.

Guidelines for Writing Multiple-Choice Questions

Multiple-choice (MC) questions are probably the most commonly used format for assessment in the cognitive domain. There are many advantages for using this format; Haladyna and Downing (1989a) summarize the following advantages:

1. Sampling of content can be very comprehensive, which generally leads to more content-valid test-score interpretation.
2. Reliability of test scores can be high when there are many high-quality MC items.
3. MC items are objective and efficient to score.
4. MC items are flexible to be used to assess a wide variety of learning outcomes and contents such as both lower-order and higher-order thinking skills.

Because of its popularity, writing MC questions has been a focus of scholarship for many years. Haladyna and Downing (1989a) reviewed 46 major measurement textbooks published since 1935 and summarized MC item writing guidelines contained in the textbooks into 43 rules. Although agreements among the textbooks on these 43 guidelines vary, with some unanimously recommended by all textbooks and others only recommended by a few, the 43 guidelines have been either assumed to be good practices or studied empirically for its validity (Haladyna & Downing, 1989b). In a follow-up review, Haladyna, Downing, and Rodriquez (2002) revised the original 43-guideline taxonomy based on textbooks and empirical studies published after the original review. The revised taxonomy now has 31 rules organized into the following five categories: content concerns, formatting concerns, style concerns, writing the stem, and writing the choices. The 31 guidelines are as follows:

Content Concerns

1. Every item should reflect specific content and a single mental behavior, as called for in test specifications (i.e., test blueprint).
2. Base each item on important content to learn; avoid trivial content.
3. Use novel material to test higher level thinking skills.
4. Keep the content of each item independent from content of other items within a same test.
5. Avoid assessing over-specific and over-general content.
6. Avoid opinion-based items.
7. Avoid trick items.
8. Keep vocabulary simple for the group of students being tested.

Formatting Concerns

9. Present MC items in a question, completion, true–false, or the best answer format; avoid the complex format involving various combinations of choices.
10. Format items vertically instead of horizontally.

Style Concerns

11. Edit and proof-read items.
12. Use correct grammar, punctuation, capitalization, and spelling.
13. Minimize the amount of reading in each item.

Writing the Stem

14. Ensure that the directions in the stem are very clear.
15. Include the central idea in the stem instead of the choices.
16. Avoid window dressing (i.e., excessive verbiage).

17. Word the stem positively. If negative words, such as NOT and EXCEPT, are used, use the word cautiously and always ensure that the word appears capitalized and boldface.

Writing the Choices

18. Develop as many effective choices as you can; in most situations three choices are adequate.
19. Make sure that only one of the choices is the right answer.
20. Vary the location of the right answer according to the number of choices.
21. Place choices in logical or numerical/alphabetical order.
22. Keep choices independent from each other; choices should not be overlapping.
23. Keep choices homogeneous in content and grammatical structure.
24. Keep the length of choices about equal.
25. *None-of-the-Above* choice should be used carefully.
26. Avoid the *All-of-the-Above* choice.
27. Phrase choices positively; avoid negatives such as NOT.
28. Avoid giving clues to the right answer, in such ways as
    a. specific determiners including *always, never, completely,* and *absolutely*;
    b. clang associations, choices identical to or resembling words in the stem;
    c. grammatical inconsistencies that cue the test-taker to the correct choice;
    d. conspicuous correct choice;
    e. pairs or triplets of options that clue the test-taker to the correct choice; and
    f. blatantly absurd, ridiculous options.
29. Make all distracters plausible.
30. Use typical errors of students to write distracters.
31. Use humor if it is compatible with the teacher and the learning environment.

Some of the above guidelines are universally or nearly universally promoted in measurement textbooks, such as including central ideas in the stem, avoiding clues to the correct answer, making all choices plausible, and making the choices relatively equal in length. Some of the above guidelines are not necessarily universally agreed on, such as presenting MC items in a question format. The following are examples to some of the above guidelines.

1. Include central ideas in the stem instead of choices.

Example:
Poor: A plant...
a. Absorbs water.
b. Takes in nutrients.
c. Makes its own food.
d. Reproduces asexually.

Better: How does a plant differ from an animal?
a. A plant absorbs water.
b. A plant takes in nutrients.
c. A plant makes its own food.*
d. A plant reproduces asexually.

2. Avoid window dressing.

Example:
Poor: In Florida, which season of the year in which records indicate the maximum statistical occurrence of hurricanes?
a. Fall*
b. Winter
c. Summer
d. Spring

Better: In Florida, which season has the most hurricanes?
a. Fall*
b. Winter
c. Spring
d. Summer

3. Make all choices plausible.

Example:

Poor: What are electrons?

a. Mechanical tools
b. Negative particles*
c. Neutral particles
d. Nuclei of atoms

Better: What are electrons?

a. Negative particles*
b. Neutral particles
c. Positive particles

4. Place choices in logical or numerical/alphabetical order.

Example:

Poor: How many bonding electrons does a chlorine atom have?

a. 2
b. 3
c. 1*
d. 4

Better: How many bonding electrons does a chlorine atom have?

a. 1*
b. 2
c. 3
d. 4

5. Avoid giving clues to the right answer.

Example:
Poor: Increasing the temperature will increase the pressure of a gas in a sealed container because
a. No container expansion.
b. Gas particles constantly move.
c. Gas particles collide with each other.
d. More gas particles collide with each other and with the container walls.*

Better: Increasing the temperature will increase the pressure of a gas in a sealed container because
a. Gas particles move more rapidly.
b. Gas particles expand bigger.
c. Gas particles collide more with each other.
d. Gas particles collide more with the container.*

6. Avoid the complex format involving various combinations of choices.

Example:
Poor: Which of the following is a renewable source of energy?
1. coal
2. hydro
3. wind
a. 1 & 2
b. 2 & 3*
c. 1 & 3
d. 3 only

Better: Which of the following is a renewable source of energy?
a. coal
b. hydro*
c. natural gas
d. oil

Guidelines for Writing True–False Questions

1. Avoid broad general statements if they are to be judged true or false.

Rationale: General statements are often not directly relevant to the target construct of assessment. A general statement often contains many aspects, and it is possible that a respondent may agree with some aspects but not others. Thus, it is impossible to tell what exactly a respondent agrees or does not agree with.

Example:		
Poor: Leaves are essential for plants.	T	F
Better: Photosynthesis takes place in leaves.	T	F

2. Avoid double negative statements.

Rationale: Double-negative statements are difficult to comprehend, and may easily confuse respondents.

Example:		
Poor: None of the senses is not useful in science experiments.	T	F
Better: All senses are useful in science experiments.	T	F

3. Avoid complex sentences.

Rationale: Complex sentences contain more than one idea, and it is possible that a respondent may agree with one idea but not others.

Example:		
Poor: Whales are mammals because they are large.	T	F
Better: Whales are mammals.	T	F

4. Avoid long sentences.

Rationale: Long sentences contain many ideas, and it is difficult for a respondent to judge all ideas to be true or false.

Example:
Poor: Despite the theoretical and experimental difficulties of determining the exact pH value of a solution, it is possible to determine whether a solution is acidic by the red color formed on litmus paper when it is inserted into the solution. T F

Better: It is possible to determine whether a solution is acidic by blue litmus paper. T F

5. Avoid extraneous clues in the form of specific determiners.

Rationale: Statements with such specific determiners as "always" and "absolutely" are usually considered unlikely to be true based on common sense statistics. Similarly, statements with such specific determiners as "under some circumstances" and "may" are usually considered likely to be true by test-wise respondents based on common sense reasoning.

Example:
Poor: Water always boils at 100 degrees Celsius. T F

Better: Water boils at 100 degrees Celsius under STP. T F

6. If an opinion is stated, attribute it to a source.

Rationale: Opinions cannot be judged to be true or false as every respondent is entitled to hold a particular opinion. However, if the source of an opinion is a fact, it is appropriate to ask a respondent to judge if the source is true or false.

Example:
Poor: Natural selection should be used to explain the evolution of living things. T F

Better: According to Charles Darwin, natural selection can explain the evolution of living things. T F

Guidelines for Writing Likert Scale Questions

Likert-scale questions are commonly used in measurement instruments of noncognitive domains pertaining to attitudes, interests, preferences, and so on. A typical *Likert-scale question* consists of a statement and a few categories of choices corresponding to different degrees of endorsement of the statements. Categories are symmetrical with equal numbers of positive and negative categories around the middle or neutral category representing degrees of agreement. An example of five categories are: *strongly agree* (SA), *agree* (A), *neutral* (N), *disagree* (D), and *strongly disagree* (SD). Similarly, an example of three categories of agreements are: *agree* (A), *neutral* (N), and *disagree* (D).

The Likert-scale question format was first proposed by Rensis Likert in 1932 (Likert, 1932). Over the past eight decades, many suggestions for writing Likert-scale questions, such as those in Likert (1967), have become conventions and are thus widely accepted. The following summarizes a few most common ones.

1. All statements should be expressions of desired behaviors, not statements of facts.

Rationale: Validity of facts should not be subject to personal judgment.

Example:
Poor: Science is an empirical approach to answering questions.
SA A U D SD
Better: Science should only use data to answer questions.
SA A U D SD

2. Each statement should be clear, concise, and straight-forward; it should involve only one aspect rather than more than one aspect.

Rationale: Higher than necessary reading comprehension can reduce the validity of responses. A respondent may have different opinions on different aspects, thus will have difficulty in stating one same opinion to all the aspects.

Example:
Poor: In order to improve student learning in science, the school should offer more elective courses in science, require more student labs in each course, and use only standardized assessment instruments.
SA A N D SD
Better: The school should offer more elective courses in science.
SA A N D SD

3. Present statements in the present tense.

Rationale: When a statement is in past tense, it presents more a fact than a value judgment.

Example:
Poor: Pollution was a major concern to me.
SA A N D SD

Better: Pollution is a major concern to me.
SA A N D SD

4. Have different statements worded so that about one half of them have one end of the attitude continuum corresponding to the left or upper part of the reaction alternatives and the other half have the same end of the attitude continuum corresponding to the right or lower part of the reaction alternatives.

Rationale: Even distribution of statements toward the two opposite ends of the attitude continuum can avoid any space error or tendency to a stereotyped response.

Example:
Positive statement: Science classes are interesting.
SA A N D SD

Negative statement: Science classes are boring.
SA A N D SD

5. Have statements worded so that the modal reaction to some is more toward one end of the attitude continuum and to others, more in the middle or toward the other end.

Rationale: Statements worded this way can produce a more evenly distributed response pattern across the entire attitude continuum.

6. Avoid using statements that contain absolute qualifiers (e.g., always, never) and indefinite qualifiers (e.g., only, many, often).

Rationale: Absolute qualifiers introduce extreme opinions, which decrease item discrimination. Similarly, indefinite qualifiers are ambiguous and can result in multiple interpretations, which also reduces item discrimination.

Example:
Poor: Technology design should never be part of a school science curriculum.
SA A U D SD

Better: Technology design should be part of a school science curriculum.
SA A U D SD

7. Avoid using too many choices.

Rationale: Too many choices are difficult to decide, which reduces validity, reliability, and discrimination.

Example:
Poor: Science laboratories are interesting.

Negative									Positive
1	2	3	4	5	6	7	8	9	10

Better: Science laboratories are interesting.
SA A U D SD

Locating Measurement Instruments

Because developing valid and reliable measurement instruments is time consuming and expensive, before developing a measurement instrument, science education researchers should first try to search if a suitable measurement instrument is available. There are various sources for locating standardized measurement instruments. The most comprehensive source is the *Mental Measurements Yearbooks* (MMY) published by the Buros Institute of Mental Measurement at the University of Nebraska, Lincoln. MMY contains a comprehensive collection and critical reviews of commercially available tests in many areas including achievement, personality, vocation, intelligence/scholastic aptitude, reading, mathematics, science, social studies, attitudes, and more. MMY was started by Oscar K. Buros in 1936; it has been periodically updated. The most recent version is the *20th Mental Measurements Yearbook* published in 2017. Although the actual copies of the tests are not included, pricing and ordering information is provided for every included test. Also, published periodically by the Buros Institute is *Tests in Print* (TIP). TIP provides a summary of tests reviewed in all the preceding MMY. It contains a comprehensive bibliography of all known published tests, and classifies the reviewed tests in the MMY by author, subject, publisher, and so on. Essentially, TIP is an indexing system for MMY. MMY is currently available in electronic databases searchable by authors, publishers, publication years, subjects, key words, and so on.

Another well-known source for locating standardized measurement instruments is the ETS (Education Testing Services) Test Collection Catalog (ETS-TCC; https://www.ets.org/test_link/about). Updated periodically, ETS-TCC contains information on both published and unpublished measurement instruments; it is published in six volumes: the achievement tests and measurements, vocational tests and measurements, tests for special populations, cognitive aptitude and intelligence tests, attitude tests, and affective measures and personality. Many tests are available for ordering from ETS. Also, the ERIC Clearinghouse on Assessment and Evaluation (http://ericae.net/) is a comprehensive online database searchable for assessment instruments.

As standardized measurement instruments are being continuously developed, many newer instruments may not be yet collected in MMY or ETC-TCC. Thus, it is always necessary to review major research journals in science education, such as the *Journal of Research in Science Teaching, Science Education, International Journal of Science Education,* to name a few. Sometimes measurement instruments are included in the appendices of journal articles; sometimes, only sample questions are provided. The authors can always be contacted for more information or the complete instrument when necessary.

Chapter Summary

A measurement instrument is a standardized tool along with the accompanying procedures to quantify observations; they possess demonstrated technical qualities. In science education, the use and development of measurement instruments have been ongoing for over a century with identifiable stages. The first stage took place during the late 19th and early 20th centuries; it marked the emergence of standardized measurement instruments for assessing science achievement. The second stage was during the 1940s and 1950s in which standardized measurement instruments were used to measure the attainment of curriculum objectives—the curriculum evaluation. The third stage was during the "curriculum reform" movement in the 1960s and 1970s in which measurement instruments were used to evaluate a wide variety of student and teacher outcomes associated with the new curriculums—the program evaluation. During the 1980s and 1990s, the use and development of measurement instruments in science education became less common due to the prevalence of qualitative approaches to science education research. Entering the 21st century, we are now in a new stage of use and development of measurement instruments. The current needs for using and developing measurement instruments are mainly due to our continuous interest in measuring student conceptual understanding, the worldwide standards-based movement toward science education, and a renewed call for scientific research in education.

Fundamental theories for using and developing measurement instruments are about validity and reliability. Theories of validity and reliability have evolved considerably. Validity used to be partitioned into various types, such as criterion-related, content, and construct validity. The current notion of validity is that validity is unitary, which may be called construct validity. Validity may be established through systematically collecting evidence and developing coherent arguments to support the intended interpretations and uses of test scores. Similarly, reliability used to be solely established by identifying sources of errors associated with items; our current notion of reliability is based on a systematic analysis of various sources of errors associated with not only items but also other facets of measurement (e.g., rater, testing setting) so that scores may be generalizable to all facets. This new notion of reliability is called the generalizability theory. As a result of our changing notions of validity and reliability, standards for educational and psychological testing have been updated. The most recent version of the standards was published in 2014 by the joint committee of AERA, APA, and NCME.

Development of measurement instruments uses a wide variety of item formats, including multiple-choice, true–false, constructed response, performance tasks, and the Likert scale. Writing high quality items is an essential skill science education researchers need to possess. Various guidelines are available for developing good multiple-choice, true–false, and Likert-scale questions. In addition to developing measurement instruments, there are also tools for locating previously developed measurement instruments. For example, the MMY is a continuous and periodic publication that catalogues and reviews published measurement instruments; a searchable computer database is also available.

Exercises

1. For each of the following scenarios, indicate what validity evidence is of importance for the indicated measurement instrument.
   a. A researcher is developing a standardized measurement instrument to measure the effectiveness of an experimental intervention for improving students' conceptual understanding of science concepts.
   b. A program evaluator is developing a standardized measurement instrument to measure science teachers' satisfaction with a summer teacher professional development workshop.
   c. A science teacher is developing an end-of-unit test to measure students' mastery of a science unit.
   d. A doctoral candidate is looking for a measurement instrument on high school students' attitudes toward technology to help answer research questions about the relationship between science and technology.
   e. A university placement test is used to measure students' readiness for taking calculus-based physics.
2. Indicate what type of reliability is needed for the measurement instrument used in each of the following scenarios.
   a. Two performance assessment tasks involving student hands-on problem-solving tasks that are part of a state standardized achievement test.
   b. A program evaluator is looking for a standardized measurement instrument for measuring students' mastery of a reform-based science curriculum.
   c. An attitudinal survey is used to measure elementary school students' attitudes toward science.
   d. A conceptual test will be used in both paper-and-pencil and online administrations.
   e. A high school biology conceptual test consisting of 30 multiple-choice questions is used to assess students' understanding of evolution.
3. Use the *Mental Measurements Yearbooks* database to identify a standardized measurement instrument on a construct of your interest. Then read the reviews included in the MMY and locate the actual measurement instrument from the source. Finally, apply the validity and reliability standards to evaluate the adequacy of the measurement instrument.
4. Search the ERIC Clearinghouse on Assessment and Evaluation (http://ericae.net/) for measurement instruments on a construct targeting a specific population.

References

Brennan, R. L. (2001). *Generalizability theory*. New York, NY: Springer-Verlag.

Crocker, L., & Algina, J. (1986). *Introduction to classical & modern test theory*. Orlando, FL: Holt, Rinehart & Winston.

Doran, R. L., Lawrenz, F., & Helgeson, S. (1994). Research on assessment in science. In D. L. Gabel (Ed.), *Handbook of research on science teaching and learning* (pp. 388–442). New York, NY: Macmillan.

Guba, E. G., & Lincoln, Y. S. (1989). *Fourth generation evaluation*. Newbury Park, CA: SAGE.

Haertel, E. (2006). Reliability. In R. L. Brennan (Ed.), *Educational measurement* (4th ed., pp. 65–110). Westport, CT: Praeger.

Haladyna, T. M., & Downing, S. M. (1989a). A taxonomy of multiple-choice item-writing rules. *Applied Measurement in Education, 2*(1), 37–50.

Haladyna, T. M., & Downing, S. M. (1989b). Validity of a taxonomy of multiple-choice item-writing rules. *Applied Measurement in Education, 2*(1), 51–78.

Haladyna, T. M., Downing, S. M., & Rodriguez, M. C. (2002). A review of multiple-choice item-writing guidelines for classroom assessment. *Applied Measurement in Education, 15*(3), 309–334.

Joint Committee of American Educational Research Association, American Psychological Association, & National Council on Measurement in Education. (2014). *Standards for educational and psychological testing*. Washington, DC: American Psychological Association. (Originally published in 1991)

Kane, M. T. (2006). Validation. In R. L. Brennan (Ed.), *Educational measurement* (4th ed., pp. 17–64). Westport, CT: Praeger.

Likert, R. (1932). A technique for the measurement of attitudes. *Archives of Psychology, 22*, 5–53.

Likert, R. (1967). The method of constructing an attitude scale. In M. Fishbein (Ed.), *Readings in attitude theory and measurement* (pp. 90–95). New York, NY: Wiley.

Liu, X. (2008). Standardized measurement instruments in science education. In M.-W. Roth & K. Tobin (Eds.), *World of science education: North America* (pp. 649–676). Rotterdam, The Netherlands: Sense.

Martin, M. O., & Kelly, D. L. (1997). *Third international mathematics and science study: Technical report (vol. II), implementation and analysis—Primary and middle school years*. Chestnut Hill, MA: Boston College.

Messick, S. (1989). Validity. In R. L. Linn (Ed.), *Educational measurement* (3rd ed., pp. 13–103). New York, NY: Macmillan.

Mintzes, J. J., Wandersee, J. H., & Nova, J. D. (999). *Assessing science understanding: A human constructivist view*. San Diego, CA: Academic Press.

National Research Council. (1996). *National science education standards*. Washington, DC: National Academy Press.

National Research Council. (2002). *Scientific research in education*. Washington, DC: National Academic Press.

National Society for the Study of Education. (1947). *The forty-sixth yearbook: Part I: Science education in American schools.* Chicago: University of Chicago Press.

Stake, R. (1967). The countenance of educational evaluation. *Teachers College Record, 68,* 523–540.

Tamir, P., & Lunetta, V. N. (1981). Inquiry related tasks in high school science laboratory handbooks. *Science Education, 65*(5), 477–484.

Travers, R. M. W. (1983). *How research changed American schools.* Kalamazoo, MI: Mythos Press.

Wandersee, J. H., Mintzes, J., & Novak, J. (1994). Research on alternative conceptions in science. In D. Gabel (Ed.), *Handbook of research on science teaching and learning* (pp.177–210). New York, NY: Macmillan.

2

Approaches to Developing Measurement Instruments

Although various measurement instruments in a wide variety of science education research areas are now available, the approaches to developing them have not varied much. Specifically, the approaches to developing measurement instruments in science education fall into two broad categories, one based on the classical test theory (CTT), and the other based on the Item response theory (IRT). Classical test theory has a long history, as long as that of developing measurement instruments. On the other hand, IRT is about five decades old, with routine applications of IRT for developing measurement instruments in science education only since the early 1990s. This chapter will first review the CTT; it will then discuss the IRT, specifically Rasch modeling, to develop measurement instruments.

Using and Developing Measurement Instruments in Science Education, pages 29–62

Classical Test Theory and Generalizability Theory

The CTT is an approach to modeling measurement data; it has been the dominant approach to developing measurement instruments in science education for a long time; it remains largely so even today. The popularity of CTT is attributable to its conceptual simplicity and relatively less technical complexity in application. Conceptually, CTT assumes a test score to be a random observation of a subject's true ability. If X represents the observed score, T represents the subject's true ability or true score, and e represents a random error, then for any person i, there exists a relationship as follows:

$$X_i = T_i + e \tag{2.1}$$

Since we can only observe X_i although our objective is to obtain T_i, the accuracy (validity) and precision (reliability) of using X_i to represent T_i depend on the estimation of the magnitude of e. One fundamental assumption when estimating e is that e is distributed randomly among all subjects within the population. There are many ways to minimize and estimate e for a target population, such as creating items based on a test blueprint, test–retest, internal consistency among items, and so forth. No matter how e is estimated, CTT has the following important characteristics:

1. X_i is a total score, that is, the sum over all individual item scores;
2. e is estimated for the entire population and is the same for all individuals of the population; and
3. T_i is a range, not a fixed score. That is, T_i can only be described with probability not certainty. For example, a 95% confidence interval of T_i can be expressed as:

$$T_i = X_i \pm 1.96\sigma_{xx} \tag{2.2}$$

where σ_{xx} is the standard error of measurement for the measurement instrument. Depending on sources of errors, the value of σ_{xx} can be estimated in various ways. For example, if the test could be unreliable due to scoring as represented by reliability coefficient r_{xx}, then the standard error of measurement can be estimated as follows:

$$\sigma_{xx} = \sigma_x \cdot \sqrt{1 - r_{xx}} \tag{2.3}$$

where σ_x is the standard deviation of the observed scores.

Two item properties affect validity and reliability measures; they are item difficulty and item discrimination. *Item difficulty* is the percent or proportion of students who have answered a question correctly. If 100 students answer a test question, and 80 students answer the item correctly, then the item difficulty is 80% or 0.80. *Item discrimination* refers to an item's capability to differentiate between students whose overall abilities are high and those whose overall abilities are low. There are many indices for measuring item discrimination. One index is based on the correlation between students' scores on the item and their overall performance on the test. That is, for each item, there is a set of students' item scores (e.g., 1s or 0s if the item is a multiple-choice question). Also, for the test, there is a set of students' test scores. A correlation between these two sets of scores can be used as an index of item discrimination. This correlation is also called biserial (if the item is scored on an ordinal scale based on a continuous construct) or point-biserial (if the item is scored dichotomously) correlation. Item discrimination index is typically expected to be above 0.3; the highest discrimination occurs when item difficulty is 0.5.

The reliability coefficient, r_{xx} can be estimated by various methods, such as internal consistency, test–retest, and equivalent forms. One key task when developing a measurement instrument following CTT is to estimate the reliability coefficient r_{xx} so that the standard error of the measurement instrument can be quantified.

One major extension to the CTT, specifically to the concept of reliability, is generalizability theory. Within the conceptual framework of generalizability, scores of subjects are generalizable when a large percentage of observed variance is due to true differences among subjects. In order to estimate the degree of generalizability, the observed variance can be partitioned into various components associated with differences among subjects, items, errors, and other measurement designs (e.g., multiple raters, multiple-forms, etc.). The conventional testing situation involving one sample of subjects answering all items of a test is called the one-facet measurement design; the measurement situation involving one sample of subjects answering all items of a test and multiple raters is called a two-facet measurement design. Facets may be fixed (i.e., representing themselves) or random (i.e., representing a larger population). For the one-facet measurement design, if the variance component among subjects or persons is σ^2_{p}, the residual component (i.e., errors) is σ^2_{res}, and n_j is the number of items on the test, then the generalizability coefficient (ρ^2_{p}) is:

$$\rho^2_{\mathrm{p}} = \sigma^2_{\mathrm{p}} / (\sigma^2_{\mathrm{p}} + \sigma^2_{\mathrm{res}} / n_{\mathrm{j}}) \qquad (2.4)$$

In the above one-facet measurement design, ρ_p^2 is identical to the reliability coefficient r_{xx}. For a two-facet measurement design involving items and raters, the coefficient of generalizability is:

$$\rho_p^2 = \sigma_p^2 / (\sigma_p^2 + \sigma_{pj}^2 / n_j + \sigma_{pr}^2 / n_r + \sigma_{res}^2 / n_j n_r) \qquad (2.5)$$

where σ_{pj}^2 is the variance component due to the interaction between persons and items, σ_{pr}^2 is the variance component associated with the interaction between persons and raters, and n_r is the number of raters.

Within the generalizability theory framework, there are two sequential studies: the generalizability study (G-study) to estimate the variance components associated with subjects and various measurement design facets, and the decision study (D-study) to estimate the generalizability coefficient and to identify the most efficient measurement design. The G-study is conducted through analysis of variance (ANOVA). The D-study is conducted by computing the coefficient of generalizability and studying the effect of varying measurement design facets, such as increasing the number of items or raters, on the coefficient of generalizability. The purpose of the D-study is to identify the best combination of facets, that is, the best measurement design, that results in an acceptable coefficient of generalizability with minimal cost (e.g., minimal number of raters, or shortest test).

While reliability describes the extent to which X_is are consistent, validity describes the relationship between X_is and T_is, that is, the extent to which X_is qualitatively represent T_is. For example, the relationship between X_is and T_is may be examined as the degree to which observed scores (X_is) represent subjects' performance in a content domain (T_is)—content validity, or as the extent to which the observed scores (X_is) describe a latent construct (T_is)—construct validity, or as the ability of observed scores (X_is) to predict subjects' performances on different measurement instruments because of (T_is)—criterion related validity. Fundamentally, CTT is a theory of reliability, but it has major implications for test validation, because one necessary condition for validity is that scores are reliable.

There are a number of fundamental limitations with the above approach, that is, CTT, to developing measurement instruments. They are:

1. T_is are dependent on X_is;
2. e is the same for every subject in the population; and
3. T_is and X_is are not interval.

There are many consequences of the above limitations. Specifically, Limitation 1 implies that T_is are always tied to the entire measurement instrument; changing items (e.g., removing or adding an item) to the instrument will result in changes in T_is. The reverse is also true. Thus, the true abilities of subjects are dependent on observed scores (X_is) based on an instrument. For example, an easier test will increase observed scores, thus true ability estimates, while a more difficult test will decrease observed scores of a sample, thus lowering their true ability estimates. Similarly, a sample of subjects with higher abilities produces higher observed scores on the instrument, which makes the instrument easier; and a sample of subjects with lower abilities produces lower observed scores on the instrument, which makes the instrument more difficult. Common sense tells us that a person's true ability should not change from test to test. Thus, CTT is, in fact, counterintuitive.

The mutual dependence between subject ability and item/test difficulty is a classic paradox that makes measurement in science education fundamentally different from measurement in sciences and engineering. In the mechanical world, a measurement instrument and the measurement target are always independent from each other. For example, a thermometer's degree scale does not change no matter what target the thermometer is used to measure; being water, alcohol, oil, and so forth. Similarly, the temperature of water is not going to change no matter what kind of thermometer, being alcohol or mercury, is used to measure. It is highly desirable to separate measures of targets from the instrument used to measure them if we want measures in science education to be more valid and reliable. Subject abilities at a given time should be considered constant no matter what instrument is used to measure them; and an instrument's difficulty or measures should be considered constant no matter what sample of subjects are being measured.

The consequence to Limitation 2 is that we cannot have precise measures for individual subjects. Common sense tells us that different subjects may respond to the same measurement instrument differently, thus should have different measurement errors. A same measurement error for all subjects is counterintutive, and undermines measurement precision for individual subjects. It is highly desirable to identify individualized measurement errors for different subjects if we want measures in science education to be more accurate and reliable.

The consequence to Limitation 3 is that theoretically we cannot apply inferential statistics such as *t*-test and *F*-test to measurement data, because inferential statistics assume that data to be analyzed are interval. Although in science education research we routinely treat measurement data based on

CTT as if they were interval, scores from a measurement instrument in the form of a total score or a percentage are in fact only ordinal. For example, a subject who scored 85 on an achievement test out of a total possible score of 100, thus 85%, should only be considered to have scored higher than a person who scored 80, because the difference score between these two persons, that is, 5, does not reflect the same amount of achievement difference as that between two other persons who scored 55 and 50. This inequality in difference can be appreciated from another perspective. That is, it is always more difficult to score even one more point for a person whose score is close to the highest possible score (e.g., 100) than a person whose score is close to the other end (e.g., 0) or to the middle point (e.g., 50). The floor (minimal possible score, e.g., 0) and ceiling (maximal possible score, e.g., 100) effects are one other indication that total scores or its percentage conversions are not truly interval, because there is no differentiation among subjects when their scores have reached the floor or ceiling. Pretending raw scores to be interval when applying inferential statistics would reduce the statistical power to reject null hypotheses because of higher error variance in raw scores. It is highly desirable to use interval scores if we want to have more power in statistical testing in science education research.

Rasch Models and Item Response Theory

Rasch models are a family of probabilistic models to describe response patterns of examinees to individual items; they originated from the pioneering work of the Danish mathematician Georg Rasch who developed a simple Rasch model, commonly called the Rasch model, for a test of items that can be scored correctly or incorrectly (Rasch, 1960). The Rasch model assumes that there exists a linear measure common to both items and examinees. For items, such a measure is the item *difficulty*; for examinees, such a measure is the *ability*. Both difficulty and ability measures are unidimensional, that is, only increasing and decreasing along one dimension. In a testing situation, the probability of a particular examinee answering a particular item correctly is solely determined by the difference between the ability and difficulty measures. The higher the ability is, the higher the likelihood for the examinee to answer the item correctly; similarly the higher the item difficulty is (i.e., more difficult), the lower the likelihood for the examinee to answer the item correctly. Applying the model to develop measurement instruments, that is, *Rasch modeling* is to create a set of items that uniquely define such a linear measure. This process is also called Rasch calibration, which is analogous to constructing a meter stick. In order to construct a meter stick from a blank wooden stick, we need to place marks along the stick so that together they form a linear measure. Once a set of items are calibrated to define a linear measure, the set of items can then form a measurement instrument and be applied to any sample of examinees to produce ability measures. What are unique about these ability measures is that they are truly interval, not dependent on the set of items used to measure them, and they have individualized standard errors of measurement, which have overcome the three limitations identified with CTT discussed earlier. Essentially, Rasch modeling is based on a statistical model applied to raw data (no matter if they are nominal or ordinal) to produce measures of item difficulties and student abilities in order to inform development of items of an instrument.

According to Rasch, for any item *i* with a difficulty D_i that can be scored as right ($X = 1$) or wrong ($X = 0$), the probability (P) of a person *n* with an ability B_n to answer the item correctly can be expressed as

$$P(X=1|B_n, D_i) = \frac{e^{(B_n - D_i)}}{1+e^{(B_n - D_i)}} \tag{2.6}$$

Equation 2.6 is the well-known Rasch model for dichotomously scored items. Because the likelihood or odds for an event is the ratio of the probability of happening over the probability of not happening, we can thus show that the likelihood or odds for the person to answer item *i* correctly is

$$\frac{P}{1-P} = \frac{e^{(B_n - D_i)}}{1+e^{(B_n - D_i)}} \Big/ \left(1 - \frac{e^{(B_n - D_i)}}{1+e^{(B_n - D_i)}}\right) \tag{2.7}$$
$$= e^{(B_n - D_i)}$$

If L is the natural logarithm of the likelihood or odds, which is commonly called log-likelihood or log-odds or simply logit, then

$$L = \ln\left(\frac{P}{1-P}\right) = B_n - D_i \tag{2.8}$$

Equation 2.8 states how the log-odds (logit) for a person to answer a question correctly is simply the difference between that person's latent ability and the item's difficulty. The bigger the difference, the more likely the person will answer the question correctly.

In the equations above B_n and D_i have some important properties. First and foremost, B_n and D_i are expressed on a true interval scale, thus have an important property called linearity. This linearity property can be shown as follows. If person n_1 has a higher ability (B_{n1}) than person n_2 (B_{n2}), then the difference in log-odds between person n_1 and person n_2 to answer the same item D_i correctly is

$$L_{n1} - L_{n2} = (B_{n1} - D_i) - (B_{n2} - D_i) = B_{n1} - B_{n2} \tag{2.9}$$

Therefore, only the difference in latent abilities determines the difference in log-odds, regardless of the difficulty of the item. Similarly, we can show that the difference in log-odds for the same person with an ability B_n to answer two questions with different difficulties (D_{i1} and D_{i2}) correctly is solely determined by the difference in item difficulties, regardless of the person's ability:

$$L_{n1} - L_{n2} = (B_n - D_{i1}) - (B_n - D_{i2}) = D_{i2} - D_{i1} \tag{2.10}$$

The linearity of B_n and D_i implies that ability and item difficulty measures are mutually independent, which is called item invariance property and person invariance property; they are unique properties of Rasch measures.

Another important property of B_n and D_i is that they are latent, that is, not direct observations or counts. Referring to Equation 2.6 above, we see that only *X*, which is the score of the person on an item, is observable from the test. B_n and D_i need to be derived from *X*s, or the response patterns of a test. Because of this, B_n is also called person or examinee ability parameter estimate and D_i item difficulty parameter estimate. Parameter estimates are calculated based on algorithms using computer programs described later.

Lastly, from Equation 2.8, we see that the difference between B_n and D_i are on a logarithmic scale of the likelihood or odds or logit. The values of logits vary from $-\infty$ to $+\infty$, thus there is no upper and lower limit, that is, no ceiling or flooring effect in measures. If the difference between B_n and D_i is 0, that is, the person's latent ability is the same as the item's difficulty, according to Equation 2.6, we obtain that the probability (*P*) for the person to answer the question correctly is 0.5 or 50%. Because the person has an equal chance of answering the question correctly to answering the question incorrectly, the odds are 1 to 1, and the log-odds is 0 (Equation 2.8). Similarly, if the difference between B_n and D_i is +3, then the person has a probability of 95% to answer the question correctly (Equation 2.6), and the odds are 20 to 1. On the other hand, if the difference between B_n and D_i is –3, then the person has a probability of 5% to answer the question correctly, the odds are 0.05 to 1, or 1 to 20.

Because of their unique properties above, B_n and D_i are also called measures. Almost 50 years ago, Thurstone (1959) considered linearity a fundamental property of measures. Linearity can be best understood by a meter stick's measurement scale (Liu & Boone, 2006). If Mike is 10 centimeters taller than Jane, and if Jane is 15 centimeters taller than Janet, then because the centimeter scale is a linear one, Mike should be 25 centimeters (10 cm + 15 cm) taller than Janet. In order to obtain linear measures, Thurstone had to use a large number of judges to sort a large number of items. Based on the grouping of items by the judges, each item was assigned a measure, and only the items that formed a linear progression by increasing measure of an equal distance were retained and used to form the measurement instrument. By applying Rasch modeling, however, fewer judges and items are needed to develop a measurement instrument, making the process more efficient and more accurate.

Measures are also unidimensional. *Unidimensionality* refers to the fact that measures describe only one attribute. For example, if we want to develop an instrument to measure high school students' orientation toward science careers, then the measures based on the instrument should only reflect students' science career orientation, not other constructs such as cognitive reasoning, academic achievement, and so on. Of course, unidimensionality does not mean that we cannot measure more than one construct; it means that different measures should be used to describe different constructs.

Last but not the least, measures are based on abstract units. Data directly resulting from observations, called raw scores, are not based on abstract units. Raw scores are counts—how many items have or have not been correctly answered on a test. Raw scores or counts are not measures. Consider how one compares apples at the market (Liu & Boone, 2006). If one wishes to compare apples, one does not do so by just counting apples. Some apples may be bigger than others. An improvement upon the concept of counts is to weigh apples, because weight is a measure of mass contained in the apple. Weight is a derived measure through the scale calibration and on an abstract unit. In science education research, if counts are utilized, and such counts are not measures, then analysis of such data is flawed. By utilizing Rasch modeling, measures can be created from raw scores or counts.

Wright (1999) succinctly summarized *measures* to be: (a) linear; (b) inferences by stochastic approximations, thus on abstract units; (c) of unidimensional quantities; and (d) impervious to extraneous factors. We have considered the first three of these, the 4th characteristic can be viewed as the goal of a measure not being compromised by factors unrelated to the core meaning of the measures. For example, a meter stick being used to measure height is not affected by the materials, legibility of numbers, or width of the stick. This issue will be further elaborated when we compare the item response theory to Rasch modeling later. In summary, Rasch modeling is an approach to analyzing data from observations in order to produce measures that demonstrate the above four characteristics.

Since the development of the Rasch model by Georg Rasch for dichotomously scored items (i.e., correct or incorrect), many extensions to the original model have been developed. Major extensions to the dichotomous Rasch model are: (a) Rasch rating scale model, (b) Rasch partial credit model, and (c) many-facet Rasch model. The Rasch rating scale model takes the following form (Andrich, 1978):

$$L = \ln\left(\frac{P}{1-P}\right) = B_n - (D_i + F_k) \qquad (2.11)$$

where D_i is the overall difficulty of an item, F_k is the threshold difficulty or transition point between two categories of an item, and P is the probability of a person with ability B_n to choose category k of an item.

For example, the most commonly used Likert scale has five categories: *strongly agree* (SA), *agree* (A), *neutral* (N), *disagree* (D), and *strongly disagree* (SD). Among the five categories of each item, there are four between-category transition points or thresholds, thus $k = 4$. For example, $k = 1$ refers to the threshold between SA and A. In a rating scale such as a Likert scale, because all items have the same category structure (i.e., number of categories), F_k is the same for all items.

Both the dichotomous and rating-scale Rasch models assume that all items of a measurement instrument have the same structure (e.g., all multiple-choice, or all Likert scale). However, in practice, it is more common to have a mixture of question formats within a same measurement instrument. For example, a conceptual understanding test may contain both multiple-choice questions and constructed-response questions using a scoring rubric. In this situation, it is necessary to have a model containing characteristics of both the dichotomous and rating-scale Rasch models. This mixture of Rasch models is the partial credit Rasch model (Wright & Masters, 1982). The partial credit Rasch model allows each item to have its own structure (e.g., dichotomous, rating, etc.). The partial credit Rasch model takes the following form:

$$L = \ln\left(\frac{P}{1-P}\right) = B_n - D_{ik} \qquad (2.12)$$

where D_{ik} is the difficulty of category k of item i. For a rating scale, k is equal to the number of thresholds; for a dichotomous item, $k = 1$.

In all the above Rasch models, only examinees and items are considered. In complex measurement designs, there may be more than one rater, or more than one setting involved. For measurement scenarios like this, the many-facet Rasch model (Linacre, 1989) has been developed. The many-facet Rasch model extends a two-facet Rasch model (e.g., the dichotomous and rating-scale models) into a many-facet Rasch model. The general form of a many-facet Rasch model is as follows:

$$L = \ln\left(\frac{P}{1-P}\right) = B_n - D_i - C_j \ldots \tag{2.13}$$

where C_j is the difficulty of facet C category j (e.g., rater 1).

In addition to the above extensions, multidimensional extensions of the Rasch models have also been made. As discussed earlier, one key characteristic of Rasch measures is that they describe only one construct—the requirement of unidimensionality. Although it is possible to administer an instrument to measure only one construct in order to meet the unidimensionality requirement, sometime it may be more efficient and even desirable to use complex items to simultaneously measure more than one construct, or to include different groups of items to measure different constructs concurrently in one test administration. Well-known examples of the above are the Trend in International Math and Science Study (TIMSS) and the National Achievement of Educational Progress (NAEP), in which multiple constructs are measured at the same time in one single test administration. When multiple constructs are measured simultaneously, multidimensional Rasch modeling is needed. Various multidimensional extensions of the Rasch model have been developed (Briggs & Wilson, 2004). One multidimensional Rasch model is the multidimensional random coefficients multinomial logit (MRCML) model (Wang, 1997). MRCML allows for the estimation of item difficulties and multidimensional abilities, and also calculates the correlations among the dimensions of abilities.

Multidimensional Rasch modeling is not contradictory to the unidimensionality requirement of Rasch modeling. This is because, for each dimension of abilities, measures are still unidimensional, that is, describing only one construct. Also it is possible for multiple dimensions and one single dimension to exist at the same time based on theoretical assumptions. For example, science achievement is both unidimensional and multidimensional. Depending on the number of content areas (e.g., biology, chemistry, physics, and earth science), student abilities can be described as both a unidimensional scale, that is, science achievement, and multidimensional scales: biology achievement, chemistry achievement, physics achievement, and earth science achievement. It is possible that a student is strong in one dimension (e.g., physics achievement), but weak in another (e.g., biology achievement). From a measurement point of view, differentiating the dimensions and calculating variance as well as covariance among them increase test reliability.

Concurrently, yet independently from the development of the Rasch model in the 1960s, American psychometrician Allan Birnbaum (1968) developed a set of logistic models to analyze item responses, the commonly-known one-parameter, two-parameter, and three-parameter logistic models. These logistic models form a family called item response theory (IRT) models. One-parameter models assume that only the item difficulty parameter is necessary and sufficient to describe items; they are identical to the Rasch models. Two-parameter models assume that both item difficulty and item discrimination parameters are necessary to describe items; and three-parameter models assume that, in addition to item difficulty and item discrimination parameters, the item guessing parameter is also necessary to fully describe items. The two-parameter and three-parameter logistic models are as follows:

$$L = \ln\left(\frac{P}{1-P}\right) = A_i(B_n - D_i) \tag{2.14}$$

$$L = \ln\left(\frac{P-C_i}{1-P}\right) = A_i(B_n - D_i) \tag{2.15}$$

It can be seen that the 2-parameter (i.e., difficulty D_i and discrimination A_i) and three-parameter (i.e., difficulty D_i, discrimination A_i, and guessing C_i) IRT models appear similar to the Rasch model. Mathematically, the Rasch model can be derived from the above 2- and 3-parameter models when A_i is assumed to be 1 (all items have the same discrimination power) and C_i is assumed to be 0 (no guessing involved). In fact, popular IRT books (e.g., Embretson & Reise, 2000; Hambleton, Swaminathan, & Rogers, 1991) refer Rasch models as 1-parameter logistic models, and consider the Rasch models to be a special case of the more complex IRT models.

However, there are fundamental differences between Rasch models and IRT models. In the literature particularly during the 1980s and early 1990s, there was intensive debate on the advantages and disadvantages of Rasch versus IRT models. As a result, two camps, the Rasch camp and the IRT camp, have formed. Proponents of Rasch measurement demonstrated that B_n and D_i estimates from IRT models are not interval measures. Wright (1999) argues that the Rasch model is the only parsimonious model to construct linear measures. Adding additional parameters to the Rasch model, such as item discrimination A_i and item guessing C_i parameter as in 2-parameter and 3-parameter IRT models will not result in linearity in B_n and

D_i, because additional parameters are always entangled in the estimates of B_n and D_i.

Wilson (2005) states that fundamental requirements of a measurement model are: (a) it must enable one to interpret the distance between an examinee and an item, and (b) it must enable one to interpret the difference between different items and the difference between different examinees. The above two requirements can be met by constructing a combined Examinee and Item Map—the Wright map if a Rasch model is used (Wilson, 2005). A *Wright map* is a combined item difficulty and examinee ability diagram showing the distribution of items and examinees along a same unidimensional logit scale. This combined diagram cannot be constructed if IRT models are used. Fischer and Molenaar (1995) claim that Rasch modeling is currently the only way to convert ordinal observations into linear measures.

In addition to the above technical differences, there is also an important conceptual difference between Rasch models and IRT models when they are applied to develop measurement instruments. When Rasch models are applied, the approach is to designing appropriate items as well as an instrument for data to fit a Rasch model so that the unique properties of B_n and D_i (e.g., linearity) can be obtained; When IRT models are applied, the approach is to finding the best model (e.g., 1-parameter, 2-parameter, 3-parameter, etc.) to fit the data so that variance in data can be maximally explained. Clearly, applications of Rasch models are top-down or measure-oriented. On the other hand, applications of IRT models are bottom-up or variance explanation oriented. In other words, Rasch modeling is a measurement approach, and IRT modeling is a statistical analysis approach. This fundamental conceptual difference has been called paradigm incompatibility (Andrich, 2004). Construction of measures must be based on theories, particularly cognitive theories (National Research Council Committee, 2001; Wilson, 2005). Rasch modeling is such a theory-based approach to developing measurement instruments. Given the above technical and conceptual differences, although 1-parameter IRT models are mathematically the same as Rasch models, the application of Rasch models to develop measurement instruments requires a quite different approach than that of 1-parameter logistic models; it is best not to equate the Rasch models with 1-parameter IRT models.

Using Rasch Modeling to Develop Measurement Instruments

Using Rasch models to develop measurement instruments, or Rasch modeling, is a systematic process in which items are purposefully constructed according to a theory and empirically tested through Rasch models in order to produce a set of items that define a linear measurement scale. Using Rasch modeling to develop measurement instruments consists of the following 10 steps:

1. State the purpose and intended population of measurement.
2. Define the construct.
3. Identify performances of the defined construct.
4. Pilot-test/field-test with a purposefully selected sample.
5. Conduct Rasch analysis.
6. Review item fit statistics and revise items if necessary.
7. Review the Wright map, unidimensionality as well as reliability, and add/delete items if necessary.
8. Repeat (4) to (7) until a set of items fit the Rasch model and define a scale/Examine invariance properties of item and person measures.
9. Establish validity claims.
10. Develop guidelines for using the measurement instrument.

The above 10-step process parallels the 10-step general test development process (Crocker & Algina, 1986) described in Chapter 1 and incorporates the four building blocks for constructing measures (Wilson, 2005). The 10-step process described in Chapter 1 is applicable to developing measurement instruments following CTT, while the four building blocks are applicable to developing measurement instruments using Rasch models. One key consideration in the development process based on Rasch modeling is the role of theories. The task for developing a measurement instrument is to create a set of items that produce data consistent with the theory. This process follows logical-hypothetical deductive reasoning commonly used in natural sciences. However, developing measurement instruments using Rasch modeling is not to reject a hypothesis; rather, it is to construct items that result in data to agree with the hypothesis. This approach is also used in natural sciences. For example, developing a temperature scale (e.g., thermometer) is based on the theory of thermodynamics. Developing a temperature scale has nothing to do with validating or rejecting the thermodynamic theory; rather its purpose is to construct a device to generate measures consistent with the theory of thermodynamics. In science education, we can follow the same rationale to develop measurement instruments using Rasch modeling. For example, in measuring students' understanding of the nature of science, initially we must have a good theory about the nature of science. Otherwise, the measures obtained may be irrelevant. This theory-based approach is the essence of construct validity; using Rasch modeling to develop measurement instruments ensures that the measures produced will have high construct validity.

Step 1: State the Purpose and Intended Population of Measurement

Before developing any measurement instrument, it is important to be clear about the purpose for which the measurement instrument will be used. Stating clearly intended uses of the standardized measurement instruments not only helps users decide appropriate uses of the instrument, but also guides the development of the standardized measurement instrument. One way to think about the intended uses of the standardized measurement instrument is the timing of its use during instruction. There are three possible uses of measurement instruments: diagnostic test, formative test, and summative test. A diagnostic test measures students' conceptual understanding of a specific concept or topic in order to better plan for instruction. For example, it is important to understand student alternative conceptions when planning effective science teaching. A diagnostic use of a measurement instrument requires the instrument to be sensitive to specific areas of student misunderstanding when planning for instruction. A formative use of a measurement instrument requires the instrument to be able to differentiate different stages of conceptual understanding. Finally, a summative use of a standardized measurement instrument requires the instrument to be directly relevant to a targeted set of learning objectives. Of course, it is possible that a measurement instrument may be used for two or more of the above three purposes. Generally speaking, the more diverse the intended uses are, the more complex the measurement instrument development process and the more challenging the validation will be.

Another way to think about the potential uses of a standardized measurement instrument is in terms of criterion-referenced score interpretation and norm-referenced score interpretation. A criterion-referenced use of a measurement instrument is to use a test score to infer how a student may have understood a particular concept or topic that is found in the curriculum. A norm-referenced use of a measurement instrument is to use a test score to infer how a student is compared to other students in terms of conceptual understanding. Although it is not impossible to develop a measurement instrument for both of the above purposes, a better measurement instrument is the one that is clearly intended for one of the purposes. This is because criterion-referenced measurement and norm-referenced measurement have different requirements in terms of item writing, test assembly, validation, and so on. As discussed later, sometimes the requirements from criterion-referenced and norm-referenced uses are incompatible.

When stating the intended uses of the measurement instrument, it is also important to clearly identify the target population. This is because different populations have different abilities and expectations in terms of conceptual understanding; an instrument intended for elementary school students is certainly not appropriate for secondary school students. Of course, it is possible that an instrument can be intended for both elementary and secondary school students, in which case the intended population is both elementary and secondary school students.

Step 2: Define the Construct

When defining the construct to be measured, one important consideration is that the construct has a theoretically unidimensional progression from a lower level to a higher level. For example, if the construct to be measured is attitude toward science laboratories, then there must be a good theory or research literature to suggest that there is a unidimensional progression in students' attitude toward science laboratories from less favorable to more favorable. It is possible that student attitude toward science laboratories may be multidimensional. But each measurement scale must be related to only one dimension, and a measurement instrument may contain more than one scale in administration. The theory is the starting point for developing a measurement instrument. If no such a theory is available, then consider postponing developing a measurement instrument. For example, if little had been known about the nature of temperature or there had been no commonly agreed-upon theories on temperature, there would have been no need to develop a measurement instrument to measure temperature, because any instrument developed would have been arbitrary and lacked credibility. In science education, the same should be expected when developing an instrument to measure a construct. If a construct is unknown, then developing an instrument to measure the construct is premature.

Step 3: Identify the Performances of the Defined Construct

Once the construct is defined in terms of a progression, the next step is to develop a test specification that defines the type of items to solicit examinees' responses or performances. Because the construct is unidimensional, the performances must also be hierarchical. For example, to measure students' conceptual understanding about the concept of energy, there must exist a hierarchy of understanding from a lower level to a higher level, such as from reasoning about energy phenomena/activities to energy sources and forms, energy transfer, energy degradation, and energy conservation. The test specification will directly inform the development of items.

Once a test specification is defined, an initial item pool is then created and item scoring keys and/or rubrics are developed. These items and their scoring keys/rubrics define a performance space. Typically, more items than needed in the final measurement instrument should be prepared in anticipation for deletion of non-fitting items (described later). Consider conducting small scale qualitative studies such as "think-aloud" to help develop items. The final initial items will then make a draft measurement instrument for pilot testing.

Step 4: Conduct Pilot-Testing/Field-Testing

Pilot-testing is to administer the draft measurement instrument to a purposefully selected sample so that data can be collected for Rasch analysis. Although a random sample from the population is ideal, what is important for Rasch modeling is the spread of examinees along the measured construct. That is, an important consideration is to ensure that the range of examinees' abilities within the sample represents the range of that in the intended population of the instrument.

Step 5: Conduct Rasch Analysis

After the pilot-testing, examinees' responses to items are entered into the computer, and Rasch analysis begins. There are a variety of commercial computer programs available that are specifically developed for Rasch analysis. These commonly used commercial Rasch analysis programs are: Winsteps, Facets, Quest/ConQuest, and RUMM.

Winsteps (Linacre, 2019) is a windows-based program and the most popular Rasch analysis program in the United States. It handles a variety of data types, including dichotomous, multiple-choice, rating, partial credit, rank order, and paired comparison. Different item types may also be combined in one analysis. The structure of the items and persons can be examined in depth, which facilitates exploration. For example, it identifies unexpected data points, as well as multidimensionality through principal component analysis of residuals. Measures can be fixed (anchored) at preset values. A free evaluation, student, trial, and demonstration version of Winsteps, capable of analyzing up to 25 items and 75 cases, can be downloaded from the website (http://www.winsteps.com). Its DOS-based predecessor, Bigsteps, with a capacity of 20,000 persons and 1,000 items, is also free. Also from the same developer, the Facet (Linacre, 1997) computer program extends the Rasch measurement into more complex measurement designs such as judged performances, and sub-tests of items replicated across tasks.

For example, Facet can analyze the ratings awarded by judges to examinees who are being tested on their performance on a number of different skill items for each of several tasks. All sample analysis results included in this book, unless specified otherwise, are based on Winsteps software.

Quest (Adams & Khoo, 1998) is a multi-platform program that runs on PC, Mac, and Vax/VMS. Quest offers a comprehensive analysis of data based on both the Rasch measurement theory and classical test theory. Quest can be used to construct and validate variables based on both dichotomous and polytomous observations from multiple-choice tests, Likert-type rating, short answer items, and partial credit items. The outputs from Quest include item estimates, person estimates, fit statistics, as well as traditional statistics such as counts, percentages, and point-biserials. A variety of reliability indices are also available. ConQuest is an expanded version of Quest (Wu, Adams, & Wilson, 1997). ConQuest does analysis based on a wider range of item response models than Quest. While Quest does not have a GUI (graphical user interface), ConQuest has a GUI for Windows and Windows NT.

RUMM, Rasch Unidimensional Measurement Model software (Sheridan, 1998) is a windows-based computer program. According to information posted on the website (http://www.rummlab.com.au/), RUMM provides a powerful and flexible means of Rasch item analysis. With each analysis, RUMM produces tables, displays, plots, and graphics to assist in the interpretation of items in relation to the latent trait under construction. All displays in RUMM can be saved to files for importing into a word processor (to enhance reports) and spreadsheet (for further analysis and plotting). RUMM implements user-friendly features, such as instant graphic displays, editing of invalid responses, data entry errors, rescoring of category responses after inspecting the category characteristic curves, multiple data sets, batch file creation, and anchoring item parameters from a previous analysis.

In addition to the above commercial software for Rasch analysis, there are also a few open source packages for conducting Rasch analysis. The R package, eRm (Mair & Hatzinger, 2016), allows for extended unidimensional Rasch modeling. It fits a variety of Rasch models including binary, linear logistic, rating scale, and partial credit models. In addition to item and person parameter estimates, eRm also produces various common item and person fit statistics (described later). The R package, TAM (Robitzsch, Kiefer & Wu, 2017), conducts marginal maximum likelihood estimation and joint maximum likelihood estimation for both unidimensional and multidimensional Rasch models.

The issue of sample size for Rasch modeling has been a focus of research for many years. There is no absolute minimal sample size required for Rasch modeling, it all depends on the quality of items, match between the item difficult range and the examinee ability range, and the tolerance level of measurement errors. As Wright and Tennant (1996) indicated, with a reasonably targeted sample of 50 persons, there is a 99% confidence that the estimated item difficulty is within ± 1 logit of its stable value, which is close enough for most practical purposes, especially when persons take 10 or more items. This is because Rasch models contain only one item parameter—difficulty, and the estimation of Rasch parameters does not depend on a sampling distribution. In response to common misconceptions about the minimal sample size for Rasch modeling, Wright and Tennant (1996) states that a large required minimal sample size for Rasch modeling (e.g., 200) is based on a misconception that Rasch parameter estimation requires a normally distributed sample. In fact, Rasch modeling escapes from such an awkward requirement by focusing on the separation between item parameters and person parameters.

The issue of required minimal sample size is essentially an issue of standard error of measures (Rasch person and item parameter estimates). Wright (1977) shows that for a typical test with a raw score between 20% and 80% correct, that is, a 5-logit difference range, the minimal sample size is

$$N = 6 / SE^2$$

where *SE* is the standard error of Rasch measures. For example, if we want our *SE* to be smaller than 0.35, which is typically adequate for pilot-testing, then the required minimal sample size is 50; for *SE* to be smaller than 0.25, which is typically considered acceptable for low-stakes testing situations, the required minimal sample size is 96; for *SE* to be smaller than 0.15, which is typically considered excellent for most testing situations, the required minimal sample size is 267.

Step 6: Review Item Fit Statistics

Although different computer programs may use different formats to report fit statistics, there are many commonalities in reported statistics on items. Common item fit statistics include the mean square residual (MNSQ) and the standardized mean square residual (ZSTD). Both MNSQ and ZSTD are based on the difference between what is observed and what is expected by the Rasch model. The MNSQ is a simple squared residual, while ZSTD is a normalized *Z* score of the residual. There are two ways to sum MNSQs and ZSTDs over all persons for each item, which produce four fit statistics. The infit statistics (infit MNSQs and infit ZSTDs) are weighted means by assigning more weights for those persons' responses close to the probability of 50/50, while outfit statistics (outfit MNSQs and outfit ZSTDs) are simple arithmetic means of MNSQs and ZSTDs over all persons. Thus, outfit statistics are more sensitive to extreme responses—outliers. The rule of thumb is that items with good model-data-fit have infit and outfit MNSQ within the range of 0.7–1.3, and infit and outfit ZSTD within the range of –2 to +2.

Keep in mind that MNSQ and ZSTD are sensitive to sample sizes. Increasing the sample size will change MNSQ toward the expected value of 1, which may result in under-detecting mis-fitting items; similarly, increasing the sample size will increase ZSTD, which may result in over-detecting mis-fitting items. This is a well-known dilemma in Rasch modeling. Through a simulation study, Smith, Schumacker, and Bush (1998) found that when sample size was over 500, corrections to commonly used mean square fit statistics criteria (e.g., MNSQ to be within 0.7–1.3) were needed.

Table 2.1 shows a sample fit statistics table for an elementary science concept test. The dichotomous Rasch model was applied because all the 15 items were multiple-choice questions.

In Table 2.1, measures are item difficulties in logits—the bigger the measure is, the more difficult the item is. The SE is the standard error of measurement for the item difficulty measure. Infit and outfit statistics are reported in both MNSQ and ZSTD. Finally, PTMEA is the point-measure

TABLE 2.1 Fit Statistics for an Elementary Science Concept Test

			INFIT		OUTFIT		
Item	Measure	SE	MNSQ	ZSTD	MNSQ	ZSTD	PTMEA
Q1	0.06	0.46	1.09	0.7	1.07	0.4	0.21
Q2	0.26	0.45	0.93	–0.5	0.90	–0.5	0.42
Q3	–2.01	0.75	0.91	0.0	0.80	0.0	0.29
Q4	0.87	0.46	1.26	1.6	1.38	1.8	–0.02
Q5	–0.61	0.50	0.87	–0.5	0.78	–0.6	0.45
Q6	–0.61	0.50	0.99	0.0	0.92	–0.1	0.30
Q7	–0.38	0.48	1.11	0.7	1.65	2.0	0.03
Q8	1.32	0.49	1.06	0.3	1.03	0.2	0.27
Q9	–0.15	0.47	0.99	0.0	1.06	0.3	0.29
Q10	0.06	0.46	1.09	0.7	1.04	0.3	0.21
Q11	3.01	0.76	0.80	–0.2	0.53	–0.4	0.50
Q12	1.56	0.51	0.73	–1.1	0.62	–1.3	0.68
Q13	–2.01	0.75	1.05	0.3	0.90	0.2	0.13
Q14	–2.01	0.75	1.0	0.2	0.78	0.0	0.21
Q15	0.67	0.45	0.91	–0.6	0.88	–0.7	0.45

correlation, indicating how the item contributes to ability measures. From Table 2.1, we can see that Q4, Q7, Q11, and Q12 may not fit the Rasch model well because some of their fit statistics are beyond the acceptable ranges. The negative point-measure correlation for Q4 also indicates that the question does not contribute to the measures.

In Figure 2.1, items are plotted based on their difficulty measures (*y*-axis) and their infit ZSTD statistics (*x*-axis), along with their standard errors of measurements (the size of circles or bubbles). Plots outside the range –2 to +2 indicate potential item mis-fitting, and the bigger the circles are, the more errors in their difficulty measures. From Figure 2.1, we can see that all items are within the –2 to +2 range, but Q11 has a big circle, indicating that Q11 may not be fitting the model well.

Another aspect related to model-data-fit for items is item categories. For multiple-choice questions, item categories are concerned with this question: are all choices functioning? For rating scales (e.g., Likert scale), item categories are concerned with the question: Do the choices progress in the expected order? Table 2.2 shows a partial output table for three multiple-choice questions. As we can see from the table, for Q8, no one selected choice D, and only one person selected B. For Q11, all choices were selected, although

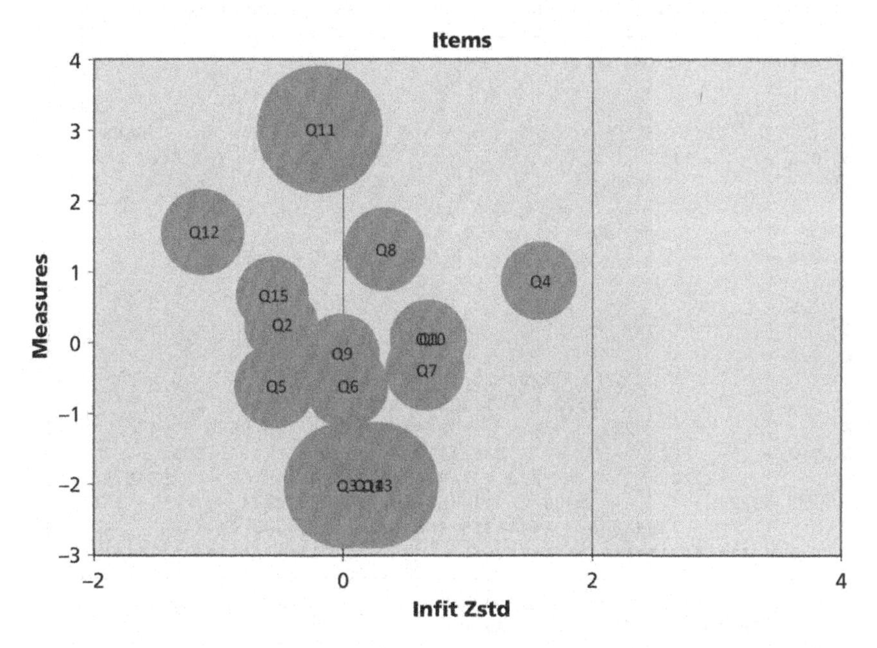

Figure 2.1 Bubble chart for an elementary science concept test. The size of the circle for an item is determined by its standard error of measurement; items outside the ±2 ZSTD range are potentially misfitting.

TABLE 2.2 Item Choice Statistics

Item	Choices	Score	Count (%)	Measure	SE	OUTFIT MNSQ	PTMEA
Q8	B	0	1 (5%)	–0.93	0.00	0.2	–0.41
	A	0	14 (64%)	0.44	0.20	1.2	–0.08
	C	1	7 (32%)	0.78	0.26	1.0	0.27
Q11	A	0	2 (9%)	–0.38	0.55	0.4	–0.36
	B	0	1 (5%)	–0.19	0.00	0.4	–0.19
	D	0	17 (77%)	0.49	0.15	1.0	0.01
	C	1	2 (9%)	1.67	0.76	0.5	0.50

most (77%) selected the wrong choice D. The PTMEA for incorrect choices should be negative and for correct choices positive. However, PTMEA for D of Q11, an incorrect choice, is positive. The above information suggests that the choice D for Q8 and Q11 may need some attention.

Figure 2.2 is the category probability curve for an attitude survey using the following categories: *strongly disagree* (1), *disagree* (2), *slightly disagree* (3), *slightly agree* (4), *agree* (5), and *strongly agree* (6). The *x*-axis represents subjects'

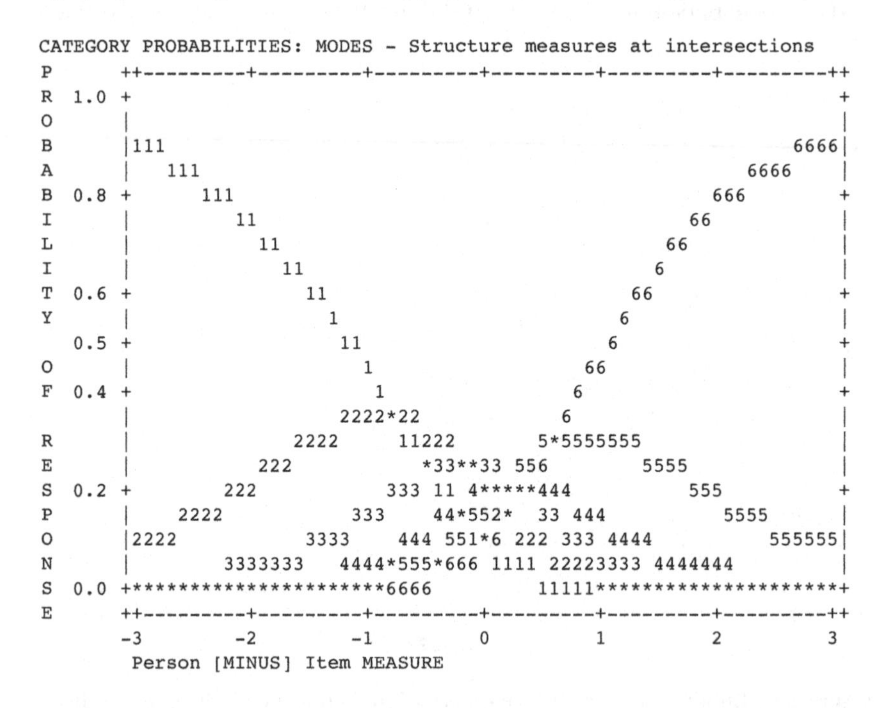

Figure 2.2 Sample category probability curves showing the probability associated with different differences between examinee abilities and item difficulties.

attitude measures in logits, and *y*-axis represents the probability of choosing a particular category. It shows that, as the difference between subjects' attitude and item difficulty increases, it becomes more likely for subjects to choose category 6 (*strongly agree*). However, categories 2, 3, 4, and 5 are buried under categories 1 and 6, indicating that the subjects are likely to respond with either category 1 (*strongly disagree*) or category 6 (*strongly agree*) depending on their overall attitude. These categories become redundant, thus may be revised. The revised survey may no longer need six categories; instead only *disagree, neutral,* and *agree* categories may be enough by combining categories 1 and 2, 3 and 4, and 5 and 6.

Step 7: Review the Wright Map, Dimensionality Plots, and Reliability

The Wright map shows how items target persons. Figure 2.3 is a Wright map for the elementary science concept test mentioned earlier.

In the Wright map shown in Figure 2.3, both item difficulties and person abilities are shown along the same linear scale in logit units (the middle line enumerated on the far left). A good measurement instrument should be able to target the intended population by matching its difficulty distribution with the sample's ability distribution. Any mismatch, that is, gap, indicates that subjects within that gap cannot be accurately differentiated because of the lack of items at that level. In Figure 2.3, a number of gaps exist: between Q11 and Q12, between Q8 and Q4, between Q2 and Q15, and between Q5/Q6 and Q13/Q14/Q3. Also keep in mind that this sample has only 22 students, which is quite small. Because this sample does have a good spread in the distribution of student abilities, further testing (i.e., field-testing) should involve a bigger sample with more broad ability variation.

Dimensionality is another important aspect to examine. A measurement instrument should be unidimensional. The unidimensionality requires that only one latent trait exists in item responses. This does not necessarily mean that only one latent trait exists among the examinees when they respond to the items; it is sufficient if one dominant factor exists in explaining the variances in item responses (Stout, 1990). There are various ways to examine dimensionality of data. First, fit statistics for items are examined and misfitting items could be measuring additional dimensions. Next, the dimensionality of Rasch residuals is examined. Rasch residuals are variances not explained by Rasch measures. If the items are unidimensional and fit the Rasch model well, then we should not expect a large variance of residuals left, and factor analysis, that is, principal component analysis, of residuals should result in no dominant factors. On the other hand, if factor analysis of Rasch residual

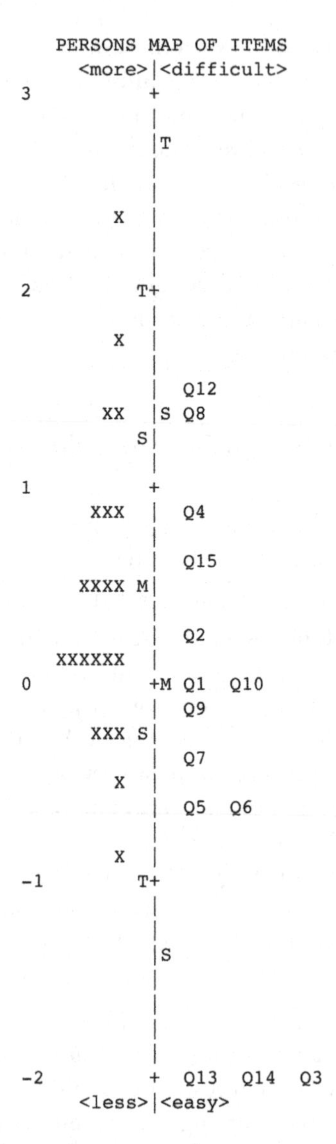

Figure 2.3 Wright map for an elementary science concept test. Both item difficulties and examinee abilities are arranged along a same logit scale shown on the far left.

variances results in one or more dominant factors, then some items may be measuring more than one construct, and the unidimensionality does not hold well. Figure 2.4 shows the plot of factor analysis of residuals for the elementary science concept test described earlier.

In Figure 2.4, the scatterplots show how items are correlated with a potential additional construct not modeled by the Rasch model. If the original

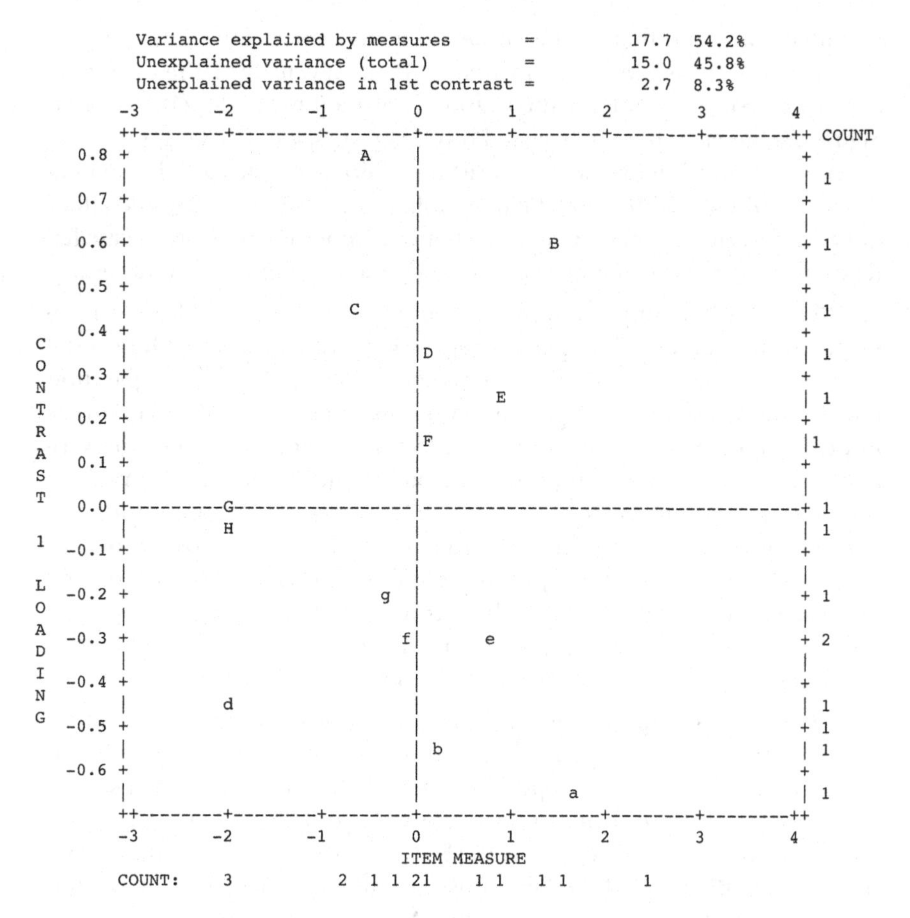

Figure 2.4 Factor loadings of rasch residual variance. Constrast loadings are correlation coefficients between an unintended construct and the items. Scatterplots should be randomly distributed within the ±0.4 loading range if items are unidimensional.

student response data are perfectly unidimensional, then the Rasch model would explain 100% variance. However, the Rasch model only explained 54% of total variance, leaving 46% variance unexplained—residuals. Among the 46% unexplained variance, it is possible that this variance is completely due to randomness, in which case no concern should be raised. However, it is also possible that an additional construct could exist among this unexplained variance. Typically, if the unexplained variance in the first contrast (i.e., the most dominant factor among the unexplained variance) is greater than 2.0 eigenvalue units (in this elementary science concept test, the size of the unexplained variance in the first contrast was 2.7), a second dimension may exist (Linacre, 2011). In this situation, items with high correlations with the potential construct, that is, the cluster of items with factor loadings < −0.4 and cluster of items with factor loadings > +0.4, may measure the additional construct besides the intended construct explained by the Rasch model. In Figure 2.4, items A, B, C, a, b, c, and d (Q2, Q3, Q5, Q6, Q8, Q11, Q12, and Q13) are potential items that measure an additional construct if the items at the top (represented by capital letters A, B, C, etc.) share common content or characteristics with the items at the bottom (represented by small case letters a, b, c, etc.). Those items need to be reviewed and modified if necessary.

Reliability is an important property of measures, essential for any measurement instrument. The person separation index indicates the overall precision in person measures as compared to errors. A person separation index is the ratio of true standard deviations to error standard deviations in person measures. Thus, a ratio greater than 1 indicates more true variance than error variance in person measures, and the bigger this ratio is, the more precise the person measures. This person separation index can also be converted to a Cronbach's alpha equivalent value from 0 to 1. In the elementary science concept test mentioned earlier, the person separation index was only 0.53, and alpha was only 0.22, indicating that person measures have poor reliability—it is unlikely to produce the same person measures if another equivalent sample retook the test.

Because Rasch modeling can also estimate a standard error of measurement (SEM) for every individual person and item, reliability in Rasch measurement becomes a property of individual persons and items, thus more precise. Different persons and items have different SEMs; persons and items with measures closer to the mean have smaller SEMs than those further from the mean. Based on individual SEMs for persons and items, it is also possible to calculate the overall SEMs for an entire instrument.

Step 8: Examine Invariance Properties

During Steps 6 and 7, if items are revised or new items are added, then the revised instrument needs to be administered to a new sample, that is, field-testing, and Steps 4–7 will repeat. Typically, two or more iterations are needed to result in an adequate measurement instrument. When there is good model-data-fit, measures are unidimensional and person reliability is high, item difficulty measures should be invariant from the sample used to calibrate the item difficulties, and person ability measures should be invariant from the set of items used to produce the ability measures. These two characteristics are called *item measure* and *person measure invariance properties,* or simply, item and person invariance properties. Thus, one final quality control when developing a measurement instrument is to examine the item and invariance properties.

In order to examine the item invariance properties, it is necessary to have two subsamples so that two sets of item difficulties are produced and compared for differences. The two subsamples are usually based on subject characteristics, such as gender, race, ability, and so forth. Because we expect that there should be no significant difference in item difficulty measures obtained from the two samples, any difference found may indicate bias in items, which is called item differential functioning (DIF). Bias due to DIF for gender or any meaningful group characteristics should be avoided. Rasch modeling provides various statistics to help identify DIF items.

Similarly, when the same sample of subjects taking two different sets of items or forms of an instrument that measure the same construct, person measures obtained from the two different tests should not be significantly different. However, because both item and person measures are interval instead of ratio, the 0 point of measures from different calibrations may be different. Thus, unless measures are obtained from the same calibration, before examining item or person invariance, measures from different calibrations should be equated through linear transformation first. Also, because item and person measures contain errors, the invariance should also be interpreted within the possible random sampling distribution.

Figure 2.5 shows the scatterplots between two sets of item difficulty measures based on two subsamples. Both x and y axes are in logits. The

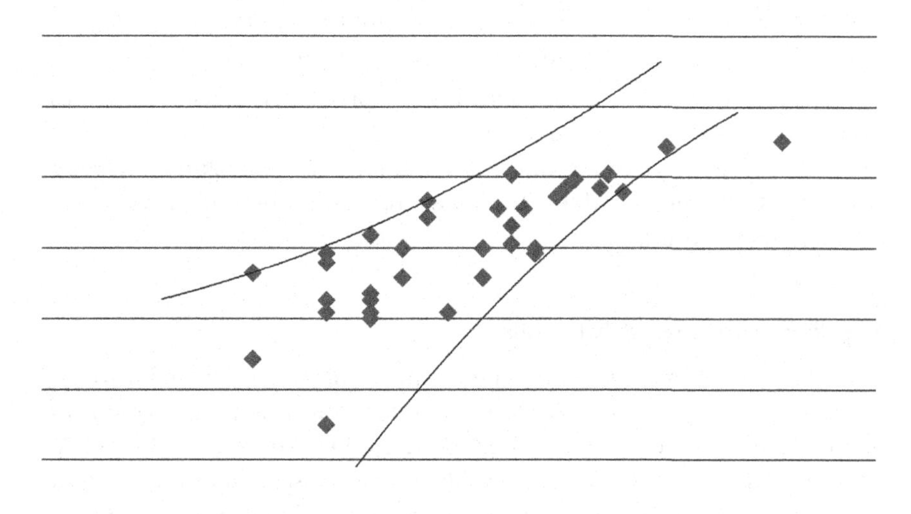

Figure 2.5 Invariance of item difficulty measures obtained from two samples of examinees. The smaller the band is, the better the invariance. Data are from bond and fox (2007, p. 74).

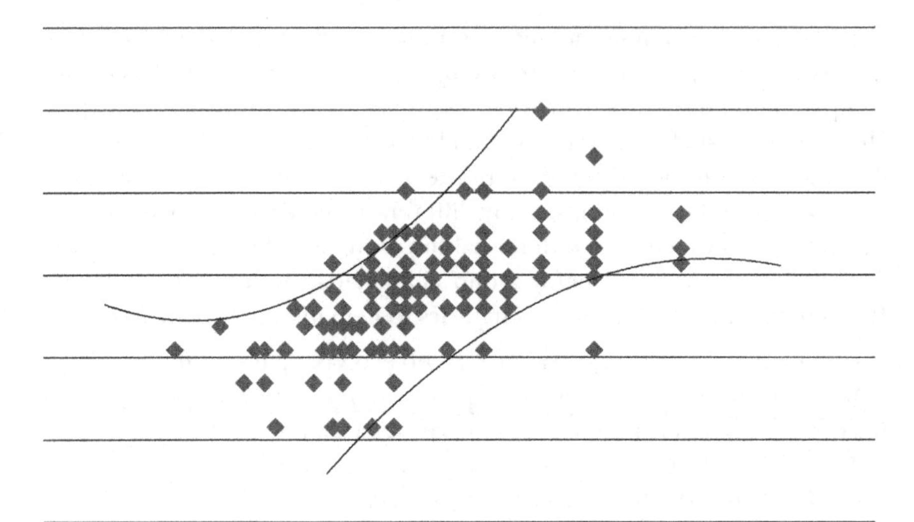

Figure 2.6 Invariance of examinee ability measures obtained from two sets of items or tests. The smaller the band is, the better the invariance. Data are from Bond and Fox (2007, pp. 96–99).

scatterplots are pairs of difficulty measures from two subsamples, and the two trend lines define a 95% confidence interval around the mean and combined standard error of the two sets of difficulty measures. We can see that most scatterplots are within the band, indicating that the two sets of item difficulty measures are, in general, invariant from samples.

Similarly, Figure 2.6 shows the invariance property of person measures from two forms of a measurement instrument. Once again, we see that most subjects' measures are within the 95% confidence interval band, indicating that, overall, the person ability measures from two different forms of the instrument are invariant.

Step 9: Establish Validity Claims

Validation of the measurement instrument may be conducted through analyzing test content, response processes, internal structures, relations to other variables, and consequences of testing (Joint Committee of the AERA, APA, and MCME, 2014). All of the validation processes support a coherent argument for the construct validity of the measures. First, because the Rasch approach to developing measurement instruments is theory-based, items have been purposely developed according to the progression of the defined construct; they should represent the content domain of the construct. Also, the phase of identifying student behaviors typically involves content experts, thus content validity can be further ensured. Think-aloud may also be used during the measurement development process to ensure that response processes of examinees are consistent with expectations. Second, Rasch measurement can provide detailed analysis of individual response patterns that reflect the reasoning processes of individuals involved in answering each question. Third, when data fit the Rasch model and are unidimensional, then there is evidence that items measure the intended construct, thus person measures possess the construct validity. Fourth, because Rasch modeling produces interval measures of individual persons, using Rasch ability measures of individuals is likely to result in a stronger correlation with measures of other variables. This is due to reduced error of measurement and increased statistical power for rejecting null hypotheses, which results in greater criterion-related validity. Comparing the rank order of individuals and items based on the Rasch ability and difficulty measures to that based on theories or other measures can also help establish criterion-related validity. Finally, consequence related to validation during Rasch modeling can be facilitated by providing clear interpretations of Rasch scale scores. Also, absence of DIF can be considered evidence for absence-of-bias.

Step 10: Develop Guidelines for Test Use

The final stage in developing measurement instruments is documentation. The purpose of documentation is to provide information to facilitate users in their appropriate uses of the measurement instrument. Important information included in the documentation should cover such aspects as the intended uses of the measurement instrument, the definition of the construct, the process of developing the instrument including pilot-testing, field-testing, Rasch analysis, and guidelines for administering the measurement instrument and reporting individual scores. Because it is unrealistic to expect users to conduct Rasch analysis when using the measurement instrument, a raw score to Rasch scale score conversion table should be provided. By referencing the conversion table, users can find out the equivalent Rasch scale score for each raw score without conducting Rasch modeling. Users should only use Rasch scale scores in the subsequent statistical analyses. Table 2.3 is a sample raw score to a Rasch scale score conversion table for the elementary science concept test mentioned earlier.

Table 2.3 demonstrates how each raw score corresponds to a Rasch scale score. Also, each Rasch scale score is associated with a standard error of measurement. Note that standard errors of measurement are bigger at the two ends of the scale, and smaller around the middle. This is because

TABLE 2.3 Raw Score and Rasch Scale Score Conversion Table

Raw Score	Rasch Scale Score	SE
0	–4.65	1.87
1	–3.32	1.09
2	–2.44	0.83
3	–1.85	0.72
4	–1.37	0.67
5	–0.95	0.63
6	–0.56	0.61
7	–0.20	0.60
8	0.16	0.60
9	0.53	0.60
10	0.91	0.63
11	1.33	0.67
12	1.83	0.74
13	2.45	0.86
14	3.39	1.13
15	4.79	1.90

there are fewer subjects at the two ends of the scale and there are more subjects around the middle to provide information for Rasch calibration.

It has to be noted that Rasch scale scores of individuals do not have to be negative. Because Rasch scale scores are interval, in order to make the Rasch scale score interpretation more intuitive, Rasch computer programs allow users to specify any score range to report Rasch scale scores. For example, the score range can be from 0 to 100, or in the same range as the raw scores, or with a mean of any value (e.g., 500) and a standard deviation of any value (e.g., 100). The scaling process involves only simple linear transformation by including a few commands in Rasch calibration.

Chapter Summary

There are two main approaches to developing measurement instruments; one is the classical test theory, and the other Rasch modeling. The CTT has been the dominant approach to developing measurement instruments in science education. Within CTT, one main task for developing measurement instruments is to estimate and reduce measurement errors. Measurement errors may be due to inconsistency of items to detect subject abilities, or instability of scores over time, or incomparability of scores across different test forms. Different sources of measurement errors can be quantified as different types of reliabilities. The KR-20 and Cronbach's coefficient alpha are quantification of internal consistency among items of a test. Other reliabilities are inter-rater reliability and test–retest reliability. Once reliability of a test is estimated, the standard error of measurement of a test can be computed, and the true score of a subject can be estimated. The generalizability theory is an extension to the CTT, specifically to the reliability concept.

A more recent approach to developing measurement instruments is to apply the Rasch models. Although Rasch models are mathematically identical to the 1-parameter logistic item response theory (IRT) models, they are conceptually different from IRT models. Applying Rasch models to develop measurement instruments is a theory-based approach. The process starts with a theory that describes a progression of a construct along a dimension. The purpose of Rasch modeling is to develop a set of items that produce data consistent with a Rasch model. When there is a good model-data-fit and unidimensionality in subjects' responses, then there is evidence for the construct validity of the subjects' measures. Rasch modeling also produces standard errors for individual items and subjects, in addition to measures of reliability for the entire instrument, such as the separation index and reliability coefficient.

Exercises

1. A research report includes the following information about a measurement instrument for measuring elementary school students' conceptions of scientists:

   > The measurement instrument has 12 multiple-choice questions; it measures elementary school students' conceptions of scientists. The development of the instrument involved: (a) definition the construct—conceptions of scientists, (b) development of multiple-choice questions according to the definition, (c) review of items for its adequacy by a panel of science educators, (d) pilot test of the instrument with 25 fifth grade students in one school, and (e) field test of the instrument with 250 students of Grades 3–6 from one school district. A principal component analysis of field-testing data indicated that all 12 items loaded highly on a same factor. The Cronbach's coefficient alpha for persons was 0.82.

   Suppose that a researcher is looking for a measurement instrument on seventh grade students' conceptions of scientists, and the above reported instrument is the closest the researcher could locate from the literature. Should the researcher use this test?

2. Some people state that the classical test theory for developing a measurement instrument is a bottom-up approach, while Rasch modeling for developing a measurement instrument is a top–down approach. Discuss the possible meanings of the "bottom–up" and "top–down" approaches, and evaluate the appropriateness of this statement.
3. Read the following research report that involves both Rasch modeling and CTT (Nehm & Schonfeld, 2008):
   a. Why do three assessment methods produce different findings about the same construct?
   b. How are conclusions about the validity and reliability of CINS based on Rasch modeling and CTT similar and different?
   c. What can you identify to be the advantages of Rasch modeling over CTT in developing measurement instruments?
4. Sondergeld and Johnson (2014) followed the 10-step process described in the earlier version of this book (Liu, 2010, 1st ed.) to develop the STEM Awareness Community Survey (SACS). Read the research report and describe how the 10-step process was followed appropriately in developing the instrument.

References

Adams R. J., & Khoo S.-T. (1998). QUEST–Version 2.1 the interactive test analysis system. *Rasch Measurement Transactions, 11*(4), 598.

Andrich, D. A. (1978). A rating formulation for ordered response categories. *Psychometrika, 43*(4), 561–573. https://doi.org/10.1007/BF02293814

Andrich, D. (2004). Controversy and the Rasch model: A characteristic of incompatible paradigms? In E. V. Smith Jr., & R. M. Smith (Eds.), *Introduction to Rasch measurement* (pp. 143–166). Maple Grove, MN: JAP Press.

Birnbaum, A. (1968). Some latent trait models and their use in inferring an examinne's ability. In F. M. Lord & M. R. Novick (Eds.), *Statistical theories of mental test scores.* Reading, MA: Addison-Wesley.

Bond, T. G., & Fox, C. M. (2007). *Applying the Rasch model: Fundamental measurement in the human sciences.* Mahwah, NJ: Erlbaum.

Briggs, D. C., & Wilson, M. (2004). An introduction to multidimensional measurement using Rasch models. In E. V. Smith & R. M. Smith (Eds.), *Introduction to Rasch measurement* (pp. 322–341). Maple Grove, MN: JAM Press.

Crocker, L., & Algina, J. (1986). *Introduction to classical & modern test theory.* Chicago, IL: Holt, Rinehart, and Winston.

Embretson, S. E., & Reise, S. P. (2000). *Item response theory for psychologists.* Mahwah, NJ: Erlbaum.

Fischer, G. H., & Molenaar, I. W. (Eds.). (1995). *Rasch models: Foundations, recent developments, and applications.* New York, NY: Springer-Verlag.

Hambleton, R. K., Swaminathan, H., & Rogers, H. J. (1991). *Fundamentals of item response theory.* Thousand Oaks, CA: SAGE.

Joint Committee of American Educational Research Association, American Psychological Association, & National Council on Measurement in Education. (2014). *Standards for educational and psychological testing.* Washington, DC: American Psychological Association.

Linacre, J. M. (1989). *Many-faceted Rasch measurement.* Chicago, IL: MESA Press.

Linacre, J. M. (1997). *Facets* (version 3.08). Chicago, IL: MESA Press.

Linacre, J. M. (2011). *A user's guide to WINSTEPS/MINISTEP Rasch-model computer programs.* Chicago, IL: Winsteps.

Linacre, J. M. (2019). *Winsteps* (Version 4.3.3). Chicago, IL: Winsteps.

Liu, X., & Boone, W. (2006). Introduction to Rasch measurement in science education. In X. Liu & W. Boone (Eds.), *Applications of Rasch measurement in science education* (pp. 1–22). Maple Grove, MN: JAM Press.

Mair, P., & Hatzinger, R. (2016). *Package 'eRm' user manual.* Retrieved from https://cran.r-project.org/web/packages/eRm/eRm.pdf

National Research Council. (2001). *Knowing what students know: The science and design of educational assessment.* Washington, DC: National Academy Press. https://doi.org/10.17226/10019

Nehm, R. H., & Schonfeld, I. S. (2008). Measuring knowledge of natural selection: A comparison of the CINS, an open-response instrument, and an oral interview. *Journal of Research in Science Teaching, 45*(10), 1131–1160.

Rasch, G. (1960). *Probabilistic models for some intelligence and attainment tests.* Copenhagen, Denmark: Danmarks Paedogogiske Institut. Chicago: University of Chicago Press.

Robitzsch, A., Kiefer, T., & Wu, M. (2017). *Package "TAM."* https://cran.r-project.org/web/packages/TAM/TAM.pdf

Sheridan, B. (1998). RUMM item analysis package: Rasch unidimensional measurement model. *Rasch Measurement Transactions, 11*(4), p. 599.

Smith, R. M., Schumacker, R. E., & Bush, M. J. (1998). Using item mean squares to evaluate fit to the Rasch model. *Journal of Outcome Measurement, 2,* 66–78.

Sondergeld, T. A., & Johnson, C. C. (2014). Using Rasch measurement for the development and use of affective assessments in science education research. *Science Education, 98*(4), 581–613.

Stout, W. (1990). A new item response theory modeling approach with applications to unidimensional assessment and ability estimation. *Psychometrika, 55*(2), 293–326.

Thurstone, L. L. (1959). *The measurement of values.* Chicago, IL: University of Chicago Press.

Wang, W. (1997). The multidimensional random coefficients multinomial logit model. *Applied Psychological Measurement, 21*(1), 1–23.

Wilson, M. (2005). *Constructing measures: An item response modeling approach.* Hillsdale, NJ: Erlbaum.

Wright, B. D. (1977). Misunderstanding the Rasch model. *Journal of Educational Measurement, 14*(3), 219–225.

Wright, B. D. (1999). Fundamental measurement for psychology. In S. E. Embretson & S. L. Hershberger (Eds.), *The new rules of measurement: What every educator and psychologist should know.* Hillsdale, NJ: Erlbaum.

Wright, B. D., & Masters, G. N. (1982). *Rating scale analysis.* Chicago, IL: MESA Press.

Wright, B. D., & Tennant A. (1996). Sample size again. *Rasch Measurement Transactions, 9*(4), 468.

Wu, M. L., Adams, R. J., & Wilson, M. (1997). *Conquest: Generalized item response modeling software—Manual.* Melbourne, Australia: Australian Council for Educational Research.

3

Using and Developing Instruments for Measuring Conceptual Understanding

This chapter is concerned with conceptual understanding. Conceptual understanding is one of the primary objectives of science education, thus a common domain of measurement. Various theoretical frameworks about conceptual understanding are available. This chapter will first review these frameworks. It will then review standardized instruments for measuring conceptual understanding. Finally, this chapter will describe the process of developing new instruments for measuring conceptual understanding using the Rasch modeling approach.

What Is Conceptual Understanding?

The Webster's Encyclopedic Dictionary of the English language defines *understanding* as "the ability to understand; the act of one who understands." It further defines *understand* as

Using and Developing Measurement Instruments in Science Education, pages 63–120

> to seize the meaning of; to be thoroughly acquainted with; expert in the use of practice of; to form a reasoned judgment concerning (something); to possess a passive knowledge of (a language); to appreciate and sympathize with; to gather and infer; to interpret, attribute a specified meaning to; to accept as fact, believe; to supply mentally (a word, idea, etc.); to have the power of seizing meanings, forming reasoned judgments, etc; to feel and show sympathy, tolerance, etc."

Thus, understanding is primarily an act. In order to demonstrate understanding, an individual needs to perform or go through a process.

Wiggins and McTighe (2005) propose *six facets of understanding*: explanation, interpretation, application, perspective, empathy, and self-knowledge; all the six facets imply actions. According to Liu's (2009, pp. 22–23) elaboration of the six facets, *explanation* is a person's ability to provide knowledgeable and justifiable accounts of events, actions, and ideas. An example of explanation is that the earth's tilt on its axis causes the season change. *Interpretation* provides meanings of events or objects through narratives or translations. For example, an elementary student interprets the earth's season change by describing how hot the summer is and how cold the winter is. *Application* is the ability to use knowledge effectively in a new situation to solve a problem. For example, a middle school student uses his knowledge of simple circuits to find a faulty light bulb in a device. *Perspectives* are the ability to appreciate different points of views. *Empathy* is the ability to understand another person's feelings and worldviews. You may not agree with another person, but you respect the person's views and can feel how strongly the person holds the view. For example, you share the feeling of the environmental activists in their efforts to protect endangered animals, but may not necessarily agree with some of their actions. Finally, *self-knowledge* is a person's ability to identify his/her own weaknesses and to actively seek improvement. For example, some students are better than others in self-evaluating their own learning and better able to look for additional resources to improve their learning. The six facets of understanding may function independently or concurrently; the more facets a person's understanding involves, the better.

Besides the above conception of understanding, which is primarily process-oriented, researchers have also described understanding to be a mental state. For example, Mintzes and Wandersee (1998) define *understanding as meanings* that are (a) resonant with or shared with others, (b) without internal contradictions, (c) without extraneous or unnecessary propositions, and (d) justified by the conceptual and methodological standards of the prevailing scientific paradigm. Therefore, Mintzes and

Wandersee's notion of understanding focuses mainly on the epistemological status of knowledge.

A both process- and state-oriented view about understanding is proposed by White and Gunstone (1992). White and Gunstone suggest that there can be six elements in which understanding of concepts may be stored and retrieved from long-term memory; they are proposition, string, image, episode, intellectual skill, and motor skill. Liu (2009, p. 23) provides an elaboration of those six elements. *Propositions* are facts, opinions, and beliefs. For example, the earth is one of the planets of the solar system. Facts, opinions, and beliefs, although distinct, may not necessarily be differentiable in student alternative conceptions. For example, students may think that it is a fact that plants take food from the soil, but this statement is only a personal belief that is scientifically incorrect. *Strings* are fundamental statements or generalizations that do not vary from situation to situation. Strings are usually in the form of proverbs, laws, and rules. For example, that matter cannot be created or destroyed, thus is conserved, is a string; but that water is matter is a proposition. *Images* are mental representations of sensory perceptions. For example, a clear lake is an image of water. *Episodes* are memories of events experienced directly or vicariously. For example, swimming in a lake is an episode of the buoyancy force. *Intellectual skills* are mental processes performed to solve a problem or conduct a task. For example, comparing the differences between physical change and chemical change involves an intellectual skill of differentiation. Finally, *motor skills* are procedures followed to conduct a physical task. An example of motor skills is performing a measurement. Six elements of understanding are located in different regions of long-term memory; they may require difficult mental processes to access. A person may possess understanding as an episode, while another person may possess understanding as a proposition or string. The more elements of understanding a person possesses, the better the person's understanding is.

All the above notions of understanding are general; conceptual understanding in science is also domain specific. There are various ways of defining domains of science, which is the task of a science content standard. For example, the document of *National Science Education Standards* (National Research Council, 1996) divides science content into three domains: the physical science, life science, and earth and space science. For the domain of physical science, students from K–4 are expected to understand properties of objects and materials; position and motion of objects; and light, heat, electricity, and magnetism. Measuring students' understanding of the above topics requires first of all a clear definition of types and levels of understanding on each of them, and use of appropriate assessment formats or tasks to solicit student understanding.

Consistent with the above conceptualizations that highlight three characteristics of conceptual understanding, that is, process-oriented, state-oriented, and domain specific, the revised Bloom's taxonomy of learning outcomes in the cognitive domain (Anderson & Krathwohl, 2001) consists of two dimensions—one dimension for types of knowledge and another for cognitive process skills. The *types of knowledge* include factual knowledge, conceptual knowledge, procedural knowledge, and meta-cognitive knowledge. The *cognitive process skills* include: (a) remember, (b) understand, (c) apply, (d) analyze, (e) evaluate, and (f) create. "Knowledge" in the original Bloom's taxonomy developed in the 1950s is elaborated into "Factual knowledge" (content) and "Remembering" (cognitive skill). Further, "Synthesis" in the original taxonomy has now become "Create" and the highest cognitive skill in the revised taxonomy. Table 3.1 summarizes the meanings of the revised Bloom's taxonomy.

TABLE 3.1 Revised Bloom's Taxonomy: Cognitive Dimension
1. *Remembering:* Retrieve relevant knowledge from long-term memory. 1.1 Recognizing 1.2 Recalling
2. *Understand:* Construct meaning from instructional messages, including oral, written, and graphic communication. 2.1 Interpreting 2.2 Exemplifying 2.3 Classifying 2.4 Summarizing 2.5 Inferring 2.6 Comparing 2.7 Explaining
3. *Apply:* Carry out or use a procedure in a given situation 3.1 Executing 3.2 Implementing
4. *Analyze:* Break material into its constituent parts and determine how the parts related to one another and to an overall structure or purpose. 4.1 Differentiating 4.2 Organizing 4.3 Attributing
5. *Evaluate:* Make judgments based on criteria and standards 5.1 Checking 5.2 Critiquing
6. *Create:* Put elements together to form a coherent or functional whole; reorganize elements into a new pattern or structure. 6.1 Generating 6.2 Planning 6.3 Producing

Source: Anderson & Krathwohl, 2001

We can see that the revised Bloom's taxonomy defines "Understand" in a more specific or narrower way than that defined by Wiggins and McTighe (2005). It seems that the entire cognitive domain of the revised Bloom's taxonomy is conceptual understanding in the sense by Wiggins and McTighe.

Instruments for Measuring Conceptual Understanding

A variety of standardized measurement instruments related to conceptual understanding are available in science education. The following summative descriptions introduce instruments published over the past 50 years in major science education research journals, mostly in *Journal of Research in Science Teaching, Science Education,* and *International Journal of Science Education.* The majority of the instruments have been developed based on CTT. They are for various intended uses and based on various theoretical frameworks of conceptual understanding. Each instrument description is information only, not intended to be a critical review of its strength and weakness. Also, descriptions do not follow a same format or even in similar length; how and how much an instrument is described depends on what is reported in the publication. In general, when available each description contains information about the instrument's intended population, purpose, composition (e.g., number and type of items), validation process, and key indices of validity and reliability. Instruments are described in their publication years and in alphabetical order of authors' last names.

Descriptions

Misconceptions Test (Doran, 1972)

This instrument contains 77 multiple-choice items, covering eight types of misconceptions on the particle theory of matter. The target population is elementary students (i.e., Grades 2–6). Average item difficulty ranged from 0.39 to 0.89 across types of misconceptions. Mean point-biserial correlation as a measure of item discrimination ranged from 0.26 to 0.54. Pilot-testing found that there was a statistically significant correlation between misconception scores and science achievement scores. Internal consistency reliability coefficients for the eight subtests were from 0.39 to 0.89 (adjusted to the 50-item length); and the alpha coefficient was 0.49 for all items together.

Misconception Identification Test (Wheeler & Kass, 1978)

The misconception identification test (MIT) contains 30 multiple-choice items for identifying high school chemistry students' misconceptions on chemical equilibrium. Each item asks students to predict the effect of changing certain variables, for example, temperature, pressure, and concentration, on the equilibrium conditions of selected chemical systems. The responses for all the questions are uniform, which include: (a) greater than at the first equilibrium, (b) less than at the first equilibrium, (c) the same as at the first equilibrium, and (d) insufficient evidence provided to decide. The test also asks students to write free responses to account for their reasoning of the predictions for five randomly chosen items. Two scores are obtained: (a) a performance score based on correct responses and (b) a misconception score based on certain misconceptions. Misconception scores were found to be statistically significantly related to students' performance on Piagetian formal operational tasks. The KR-20 was found to be 0.57 for misconception scores.

What Do You Know About Photosynthesis and Respiration? (Haslam & Treagust, 1987)

Photosynthesis and respiration (P & R) is a 13-item, two-tiered, multiple-choice instrument for measuring secondary school (Grades 8 to 12) students' alternative conceptions on photosynthesis and respiration. The development of the instrument followed the following process: (a) description of science content in terms of propositional knowledge statements and concept maps; (b) development of items based on interviews, open-ended pencil-and-paper tests, and related literature; and (c) the development of two-tiered test items. A series of five pilot studies were conducted to validate the instrument. The Cronbach's coefficient alpha was 0.72. Item difficulties ranged from 0.12 to 0.78 with a mean of 0.38. Discrimination indices ranged from 0.36 to 0.60 with a mean of 0.48. There was a statistically significant difference among grade levels in student correct answers; there was no statistically significant gender difference and interaction effects between gender and grade level.

Physical Changes Concepts Test (Haidar & Abraham, 1991)

The physical changes concepts test (PCCT) measures high school students' conceptions about dissolution, diffusion, effusion, and states of matter, as well as students' use of the particulate theory in their responses to questions about these concepts. It consists of two forms: the application form (A form) and the theoretical form (T form). The A form tests students' ability to utilize the concepts in everyday-life situations using everyday language. The T form tests students' scientific knowledge about these concepts using scientific language. In the T form, items cue students to use such terms as atoms and/or molecules. Validation involved reviews of items by science education professors and chemistry professors. Pilot-testing found that students' formal reasoning ability and their preexisting knowledge were associated with their conceptions and use of the particulate theory. Also, there was a statistically significant difference between students' applied and theoretical knowledge.

Force Concept Inventory (Hestenes, Wells, & Swackmaher, 1992)

The Force Concept Inventory (FCI) contains 30 multiple-choice questions that require a forced choice between Newtonian concepts and common-sense alternatives. The FCI questions are based on analysis of interviews of students from ninth grade to university undergraduate and graduate physics majors. About half of the questions in the FCI came directly from another similar diagnostic test called Mechanics Diagnostic Test. The authors interviewed 20 students and found that students gave very similar responses to FCI questions as to previous diagnostic questions. The FCI has been administered at a large number of universities from Arizona State University to Harvard. Lasry, Rosenfield, Dedic, Dahan, and Reshef (2011) noted that individual FCI responses are not reliable, but the total score is highly reliable.

Test of Understanding Graphs in Kinematics (Beichner, 1994)

The Test of Understanding Graphs in Kinematics (TUG-K) is for assessing high school and university students' understanding of graphs related to kinematics. Expert panels reviewed the list of objectives to be assessed, such as "given position-time graph, students will determine velocity." Three items were written for each objective. Open-ended questions were given to a small group of students first; common errors were used to create distracters. The instrument was revised a number of times based on pilot tests with both high school and college students. Item discrimination index was > 0.30 for all items (average 0.36). The KR-20 reliability coefficient was 0.83.

Diffusion and Osmosis Test (Odom & Barrow, 1995)

The diffusion and osmosis test (DOT) is a 12-item two-tiered multiple-choice test for measuring college introductory biology students' alternative conceptions of diffusion and osmosis. Correct responses to both tiers of a question are necessary to score a credit for the question. Validation of the test involved 240 students in a freshman biology laboratory course. Discrimination indices of the 12 items ranged from 0.2 to 0.69 with an average discrimination index of 0.45. Item difficulty indices ranged from 0.20 to 0.99 with an average difficulty of 0.53. Split-half internal consistency after the Spearman-Brown correction was 0.74. Post instruction test results showed that biology majors scored much higher than non-biology majors.

Force and Motion Conceptual Evaluation (Thornton & Sokoloff, 1998)

The Force and Motion Conceptual Evaluation (FMCE) is a 43-item multiple-choice test for university introductory physics students' alternative conceptions. Content validity was claimed to be high based on the fact that a group of physics professors reviewed the initial questions and considered them to be appropriate. Pre- and post-test results showed that 70%–90% of students answered questions from the Newtonian perspective at the end of the term after completing a reformed physics course, while only less than 20% of students did so at the end of a traditional physics course, suggesting that FMCE was very sensitive to new approaches to teaching physics for understanding.

Astronomy Diagnostic Test (Hufnagel et al., 2000)

Astronomy Diagnostic Test (ADT), Version 2, is a 21-question multiple-choice test, plus 12 additional demographic questions. Questions for ADT were mainly from two sources: the Project START Astronomy Concept Inventory (Sadler, 1998) and the Misconceptions Measure developed by one of the authors (Zeilik). The development of ADT 2.0 went through multiple cycles of pilot-testing and statistical analysis through the Collaboration for Astronomy Education Research (CAER), a multi-institutional collaborative across the United States. Pilot-testing institutions included public and private 2-year and 4-year colleges, and public and private research universities. A database of the pre-course ADT scores from 22 classes is also available to download from the website for comparison studies.

The Conceptual Survey of Electricity and Magnetism (Maloney, O'Kuma, Kieggelke, & Heuvelen, 2001)

The Conceptual Survey of Electricity and Magnetism (CSEM) is a 32-item multiple-choice question test for 2-year college introductory physics students. The instrument went through three revisions based on trials and feedback. The final version included the questions related to the following topics: change distribution on conductors/insulators; Coulomb's force law; electric force and field superposition; force caused by an electric field; work, electric potential, field, and force; induced charge and electric field; magnetic force; magnetic field caused by a current; magnetic field superposition; Faraday's law; and Newton's third law. Item difficulties ranged from 0.1 to a little over 0.8; item discriminations ranged from 0.1 to 0.55. Factor analysis revealed 11 main factors. The KR-20 reliability coefficient was 0.75.

Thermal Concept Evaluation Questionnaire (Yeo & Zadnik, 2001)

The Thermal Concept Evaluation (TCE) questionnaire is a pencil-and-paper instrument measuring understanding of thermodynamic concepts of 15–18 years old in everyday contexts. The items cover four themes: heat transfer and temperature changes, boiling, heat conductivity and equilibrium, and freezing, and melting. The items were developed based on student alternative conceptions reported in the literature. The instrument was administered to 478 Australian students from Grade 10 general science to Grades 11–12 physics and first year university introductory physics. The split-half reliability was 0.81. Validation involved content validity, face validity, and construct validity. Content validity was established through adequate sample of the content covered by the items, face validity was established through judgement of experienced physics lecturers, finally, construct validity was established through interviews of students and analysis of item discrimination.

Conceptual Inventory of Natural Selection (Anderson, Fisher, & Norman, 2002)

Conceptual Inventory of Natural Selection (CINS) is a 20-item multiple-choice test for assessing pre- and post-instruction knowledge of university biology non-majors and pre-instruction knowledge of majors. The test items are based on actual scientific studies of natural selection. The difficulty of test items ranged from 0.15 to 0.81, with an average difficulty index of 0.46. The item discriminations indicated by point-biserial correlation ranged from 0.16 to 0.52. The principal component analysis suggested seven interpretable factors, with all the 20 items (except one) having a significant loading on one factor. The seven factors accounted for 53% of total variance. There was also a positive correlation between student scores on CINS and on interviews. The KR-20 was 0.58 and 0.64 based on two samples of students.

Nehm and Schonfeld (2008) conducted additional validation of CINS. They found that CINS scores were statistically significantly correlated with student scores on both an open-ended survey and an oral interview of natural selection, providing additional evidence for the convergent validity. Discriminant validity was claimed based on statistically nonsignificant correlation with student test scores on an earth science achievement. Rasch modeling on student responses to CINS items showed that all items had a good fit with the unidimensional Rasch model, although principal component factor analysis still found 8 factors with 10 items loaded on the first factor. The instrument was overall easier for the biology-major students because about 15% of the sample had abilities above the most difficult item.

They also found that although CINS contained significantly more alternative conceptions represented by the distracters, CINS could only identify 75% of the alternative conceptions noted by the open-ended survey, while the open-ended survey could reveal a similar variety of student alternative conceptions.

Symbolic, Application, Particulate Test (Bunce & Gabel, 2002)

Symbolic, Application, Particulate (SAP) test is a 30-question multiple-choice test measuring students' understanding of major chemistry topics/concepts including states of matter, density, mixture/substance, conservation of mass, reaction type, moles, chemical reaction, solution, neutralization, and pH. There are three questions for each concept topic, with one related to the macroscopic representation, one related to symbolic representation, and one related to submicroscopic/particulate representation. The KR-20 for SAP was 0.76.

Chemistry Concept Inventory (Mulford & Robinson, 2002)

Chemistry Concept Inventory (CCI) includes 22 multiple-choice questions. It covers all key concepts typically taught in a first-semester college chemistry course. Graduate chemistry students scored almost perfect on the test. Chemical education researchers determined the questions to be appropriate for first-year chemistry students. Interviews with a limited number of students also confirmed that they interpreted and responded to the questions as intended. The Cronbach's alpha based on the pretest results of 928 students was 0.704, and 0.716 on the posttest.

Determining and Interpreting Resistive Electric Circuit Concepts Test (Engelhardt & Beichner, 2004)

Determining and Interpreting Resistive Electric Circuit Concepts Test (DIRECT) is a 29-item multiple-choice question test for both high school and university introductory physics students on the understanding of direct current electric circuits. The multiple choices for each item are based on students' open-ended answers. Multiple cycles of expert review, pilot-testing with a large number of students across the country, and revision took place over the years. The discrimination indices ranged from 0 to 0.43 with an average of 0.24, item-total correlations ranged from 0.07 to 0.46 with an average of 0.33, and difficulties ranged from 0.15 to 0.89 with an average of 0.49. No statistically significant difference in scores was found between university and high school students, but there was a statistically significant difference between male and female students. Factor analysis revealed 11 factors. The KR-20 reliability coefficient was 0.70.

Testing Students' Use of the Particulate Theory (Williamson, Huffman, & Peck, 2004)

Testing students' use of the particulate theory (TSUPT) is a 36-item primarily constructed-response question test for university introductory chemistry courses. The questions followed the same format and content scope as PCCT (Haidar & Abraham, 1991). Six questions test each of the following concepts: (a) dissolution, (b) insolubility, (c) saturation, (d) diffusion, (e) states of matter, and (f) effusion. Two questions come directly from PCCT and four are new. The six questions form a progression from everyday knowledge to more scientific knowledge by using scientific terms (molecules, atoms, particles, etc.). An independent evaluator randomly scored selected responses; inter-rater reliability was found to be 90%.

Brief Electricity and Magnetism Assessment (Ding, Chabay, Sherwood, & Beichner, 2006)

Brief Electricity and Magnetism Assessment (BEMA) is a 30-item multiple-choice test for college-level calculus-based introductory physics courses. Selected college faculty members reviewed the initial questions. A small number of students answered the questions with both the multiple-choice and short-answer formats. Students' common errors were used to construct the next version of the multiple-choice test. The draft instrument was then pilot-tested and revised. Item difficulty averaged 0.37, item discrimination averaged 0.34, and point biserial coefficient averaged 0.45. The KR-20 reliability coefficient was 0.85.

The Geoscience Concept Inventory (Libarkin & Anderson, 2006)

The Geoscience Concept Inventory (GCI) is a multiple-choice test developed for use in entry-level college earth science courses. It contains 73 questions. These questions cover topics related to general physical geology concepts, as well as underlying fundamental ideas in physics and chemistry, such as gravity and radioactivity. All questions are based on interviews of students' alternative conceptions. Data used to validate the inventory came from a wide range of institutions representing entry-level college students nationwide. Rasch measurement was used to construct the inventory and to establish indices of validity and reliability. Although there are 73 questions in the inventory, it is not necessary to use all of them; instead one set of 15 questions may be used by following the suggestions to ensure appropriate composition of questions in order to provide equivalent content coverage and difficulty distribution to that using the entire inventory.

Environment Literacy Instrument for Korean Children (Chu et al., 2007)

Environment Literacy Instrument for Korean Children (ELIKC) measures third-grade (8–9 years old) Korean students' environmental literacy. It includes the following scales: Environmental Knowledge (24 items), Environmental Attitude (22 items), Behaviors to Solve Environmental Problems (16 items), and Skills to Solve Environmental Problems (7 items). The knowledge and skill scales use multiple-choice questions, while attitude and behavior scales use the five category Likert-scale questions. The draft instrument including both multiple-choice and open-ended questions was pilot-tested with 250 second grade students at the end of the school year. Items were reviewed by four science education professors and three environmental education researchers for content validity. The final version was administered to 969 third-grade students. The Cronbach's alphas for the four scales were 0.65 (knowledge), 0.81 (attitude), 0.74 (behaviors), and 0.46 (skills).

The Measure of Understanding of Macroevolution (Nadelson & Southerland, 2009)

The Measure of Understanding of Macroevolution (MUM) is an instrument for measuring college undergraduate understanding of the scientific portrayal of macroevolution. The MUM comprises 27 multiple-choice items and one free-response item. The items cover the following five concepts of macroevolution: deep time, phylogenetics, speciation, fossils, and the nature of science. Content validation was conducted through content reviews of textbooks and by university biology faculty. Although conceptual understanding was a focus, some items also assessed specific factual knowledge. All items were based on scenarios. Pilot-testing was conducted with four undergraduate education students; field-testing was conducted with three cohorts of undergraduate students: first semester introductory biology course students (667), students at the beginning of an evolution biology course (67), and students nearing completion of an evolution biology course (54). Item analysis based on percent correct and point-biserial correlation was conducted. Cronbach's alpha based on Cohort 1 was 0.86, on Cohort 2 was 0.57, and on Cohort 3 was 0.65. The mean number of correct items was 13.33 (SD = 6.03). Student responses to the open-ended item (Q28) served to confirm the quantitative aspect of the MUM as a valid assessment of macroevolution knowledge and understanding.

Romine and Walter (2014) used MUM with a sample of undergraduate students taking a general education biology course with an emphasis on macroevolution. Three hundred and fifteen students completed the pretest of MUM, 291 completed the posttest of MUM, and 270 completed both the pre- and post-test of MUM. Rasch analysis was conducted on the pre- and post-test data. Results suggest that MUM was approximately unidimensional, not five-dimensional. Personal reliability was 0.75 and 0.78 for the pre- and post-test samples. Most items fit the model well, with the exception of three most difficult items (V3, V6, and V27). There was also a good stability of measures over time and absence of DIF from pretest to post-test. Overall, MUM provides a valid, reliable, and unidimensional scale for measuring knowledge of macroevolution in introductory non-science majors.

Mechanical Waves Conceptual Survey (Tongchai, Sharma, Johnston, Arayathanitkul, & Soankwan, 2009)

The Mechanical Waves Conceptual Survey (MWCS) is evolved from the open-ended Wave Diagnostic Test (WDT; Wittmann, 1998), specifically designed and shown to diagnose students' conceptions in wave mechanics. The MWCS consists of 22 multiple-choice questions, with eight questions on propagation, four questions on superposition, four questions on reflection, and six questions on standing waves. It was administered to 632 Australian students from high school to second-year university and 270 Thai high school students.

To develop the survey questions, 16 Thai students were individually interviewed first, with four from each subtopic. Then a pilot study was conducted for both Australian and Thai students. The difficulty range for different student groups was from 0.22 to 0.70 with an average of 0.38. The mean discrimination index based on top and bottom scored 25% students was acceptable with a value of 0.48, and the mean value of point-biserial correlation coefficients for all questions was 0.42. The Cronbach's alpha reliability coefficients for different groups of students ranged from 0.52 to 0.78, and the alpha for all students was 0.78. Ferguson's delta measuring the distribution of student scores over the possible score range based on the sample of all students was calculated as 0.97. The standard statistical analyses on the entire data showed that the survey could reveal differences among different groups of students on their conceptions of mechanical waves.

Quantum Physics Conceptual Survey (Wuttiprom, Sharma, Johnston, Chitaree, & Soankwan, 2009)

Quantum Physics Conceptual Survey (QPCS) is a 25-item multiple-choice test on university students' understanding of quantum physics. The initial development of QPCS included analysis of the syllabi from eight universities in Thailand, consultation with physicists, and analysis of a group of postgraduate physics education students' responses to a physics exam. Four themes were found to be fundamental in underlying quantum physics:

waves and particles, de Broglie wavelength, double slit interference, and the uncertainty principle. Wave function was dropped from the QPCS because introductory physics course students would not have studied this area yet. Further consultation with physics education researchers resulted in the addition of one more theme—the photoelectric effect.

The initial draft survey was administered to various groups of students, including experts at a university in Australia. Validation and revision of the items involved a Delphi study. The experts in the Delphi study were a group of five physicists who had more than 20 years of experience in teaching physics. They met as a group each week for several weeks until they found no inconsistencies or unwarranted omissions left in the proposed survey questions. The experts agreed that the questions were all on topics important to introductory quantum physics, and that the wording was unambiguous, with little cause for misunderstanding.

The main validation study took place in Australia. Item difficulties for the items ranged from slightly above 0.2 to slightly above 0.9, with most items between 0.3–0.7. The majority of the items (20 questions) had a discrimination index of 0.3 or greater and the average discrimination index was 0.35. The KR-20 test reliability index for QPCS was 0.97.

Wave Diagnostic Instrument (Caleon & Subramaniam, 2010)

The Wave Diagnostic Instrument (WADI) is a 14-item three-tier multiple-choice diagnostic test to investigate students' alternative conceptions on the nature and propagation of waves. Eight free-response questions were developed to formulate the first two tiers of the test on waves, and it was administered to 39 students; 10 of them were then interviewed individually. The list of propositional statements and concept map were sent for content validation to a panel of content experts, including four university physics lecturers and four secondary school physics teachers. The three-tier test was first pilot-tested on 78 Grade 10 students, and seven students were then individually interviewed. The final version of WADI was constructed and administered to 243 Grade 10 students in Singapore. The final version was administered twice to the same sample with 3 weeks in-between to enable calculations of reliability indicators by an interval of 3 weeks.

The internal consistency, in terms of Cronbach's alpha, for all three tires (content, total score for both content and reasoning, and confidence) was from 0.58 to 0.93. To test the consistency of decision for classifying students as passing or not passing, both proportion of agreement (P0) and Cohen's Kappa between Test 1 and Test 2 were used. The P0 was from 0.78 to 0.96 for three tiers; Cohen's Kappa was ranging from 0.24 to 0.55 for three tiers, and all were significant. Pearson correlations for test–retest were from 0.63 to 0.73 for three tires. The discriminating power of the WADI questions had a wide range of values, with a mean value of 0.39 for the content tier and 0.42 for both tiers. The difficulty level of the test items was from moderate (0.42) to high (0.12).

MalariaTT2 (Cheong, Treagust, Kyeleve, & Oh, 2010)

MalariaTT2 is an 18-item two-tiered multiple-choice diagnostic instrument to identify the common alternative conceptions about malaria. It may be used with Years 10–12 students, those in nursing education programs, as well as health workers in professional development programs involving the study of malaria and its control or prevention. The concepts covered by the test can be grouped into six main categories: symptoms, parasitism, transmission, prevention and control, global importance and distribution patterns, and immunity and vaccine.

To investigate possible alternative conceptions, interviews for the second tier (reasoning part) of the items were obtained from randomly selected students. Three hundred fourteen students in Year 12 in Brunei Darussalam was administered the test.

The internal reliability by Cronbach's alpha for the instrument was 0.65. Item difficulty ranged from 0.21 to 0.45. The discrimination index obtained for the items ranged from 0.07 to 0.84 with a mean of 0.46. The Fry readability test of the MalariaTT2 showed that 13 items were at reading levels below Grade 12, while five items were at the reading level of Grade 12 and beyond. These five items could be simplified further for students whose first language is not English.

Nature of Solutions and Solubility–Diagnostic Instrument (Adadan & Savasci, 2012)

Nature of Solutions and Solubility–Diagnostic Instrument (NSS–DI; Adadan & Savasci, 2012) is a two-tier multiple-choice diagnostic instrument to assess high school students' understanding of solution chemistry. It includes 13 items that address six aspects of solution chemistry: nature of solutions and dissolving, factors affecting the solubility of solids, factors affecting the solubility of gases, types of solutions relative to the solubility of a solute, the concentration of solutions, and the electrical conductivity of solutions.

The first version of the NSS–DI, which included 18 multiple-choice items with free response, was pilot-tested with 430 Grade 11 science-track students in 16 classes from four different randomly selected public schools in Istanbul, Turkey. Eleven (11) students who took the test were individually interviewed for the purpose of validating test items. A panel of experts validated the content of the propositional statements and a concept map, and the second version of the NSS–DI was content validated by four science education professors. The second version of the NSS–DI was tested with 154 Grade 11 science-track students in six classes from the same public high school. After some additional revisions, the final version of the NSS–DI was created, and a total of 756 Grade 11 students from 14 public high schools in Istanbul, Turkey participated in the final validation study. The reliability coefficients, in terms of the Cronbach alpha coefficient, for the content tier and both tiers were found to be 0.697 and 0.748. Cohen's Kappa (k) for the agreement between student interviews and responses to the multiple-choice questions was found to be 0.61 for the content tier and 0.54 for the reasoning tier. The item difficulty of the test ranged from 0.11 to 0.69, and the item discrimination indices ranged from 0.22 to 0.77.

Atmosphere-Related Environmental Problems Diagnostic Test (Arslan, Cigdemoglu, & Moseley, 2012)

The 13-item Atmosphere-Related Environmental Problems Diagnostic Test (AREPDiT) is a three-tier multiple-choice diagnostic test with the third tier to be certain or uncertain to responses to the first two tiers. It was designed to reveal common misconceptions of global warming (GW), greenhouse effect (GE), ozone layer depletion (OLD), and acid rain (AR). The AREPDiT was constructed and administered to 256 preservice teachers in the United States. The Cronbach alpha coefficient of the final version was 0.74. Content and face validations were conducted by senior experts. The mean difficulty indices decreased as the number of tiers increased, for first tiers: 0.42; first two tiers: 0.28; all three tiers: 0.19, respectively. Most of the point-biserial correlation coefficients were above 0.30, which means that the items function satisfactorily. The correlation between both tier scores and certainty scores was statistically significant ($r = 0.40$) at $p < 0.01$, and this moderate positive correlation indicates evidence for construct validity.

Star Properties Concept Inventory (Bailey, Johnson, Prather, & Slater, 2012)

The Star Properties Concept Inventory (SPCI) covers mass, temperature, luminosity, lifetime, nuclear fusion, and star formation typically taught in an introductory undergraduate astronomy course. It includes 23 multiple-choice questions intended to measure student learning gains over instruction. Development of items followed an iterative process using different item formats (i.e., open-ended, multiple-choice + explain, and multiple-choice) through three versions. Pilot-testing involved students taking general education natural science courses including Astronomy 101. Validation was conducted through content review by astronomy instructors, interview of select students, and pre- and post-course administration of the instrument. The pretest administration of version 3 yielded a Cronbach's alpha of 0.470, and the post-test alpha was 0.763 based on 417 students. Pretest item difficulty values ranged from 0.067 to 0.814, while the post-test difficulty values ranged from 0.161 to 0.918. Point-biserial correlation for items based on post-test ranged from 0.110–0.551. The average score on the pretest was 7.12 ($SD = 2.78$), and the average score on the post-test was 10.81 ($SD = 4.23$).

Acid Strength Concept Inventory (McClary & Bretz, 2012)

The Acid Strength Concept Inventory (ACID I) is a nine-item multiple-tier, multiple-choice concept inventory to identify alternative conceptions that second year undergraduate organic chemistry students hold about acid strength. The ACID I identified two significant alternative conceptions that students held about acid strength, that is, functional group determines acid strength and stability determines acid strength, and used them to make inferences for seven specific cases involving three distinct sets of acids. The instrument was developed by the data from three deep structure prediction tasks and student interviews. A sample of 104 students was administered the final version test. A confidence mean quotient (CDQ) was calculated for each item to determine if students were actually aware of what they did and did not know. Almost all items on ACID I had a negative CDQ, except for Item 4 and Item 8 having positive quotients (a negative CDQ value indicates that students who answer incorrectly are more confident with their responses than students who answer correctly, suggesting that students do not know what they do not know). The difficulty of the nine items varied widely, though no item was answered correctly by more than 70% of the students. The mean of the point-biserial coefficients was 0.41, suggesting acceptable item discrimination. The reliability of ACID I, calculated as a Cronbach alpha, was 0.41, which is below the standard value of 0.50.

Inventory of Student Evolution Acceptance (Nadelson & Southerland, 2012)

Inventory of Student Evolution Acceptance (I-SEA) is a finer-grained instrument that assesses acceptance on three evolution subscales: microevolution, macroevolution, and human evolution. The final instrument contains 24 Likert-scale items and was administered to 404 high school biology and 397 university students. The internal consistency reliability of Cronbach's alpha was 0.95 for the entire item set, and alphas for the three subscales were

0.90, 0.90, and 0.94. Exploratory and confirmatory factor analyses showed that the items loaded onto three components that reflect documented evolution acceptance conditions. The items were also reviewed by college biology faculty for their alignment with the intended three subscales.

Sbeglia and Nehm (2019) used Rasch modeling to further validate I-SEA. The I-SEA was given to over 2,000 undergraduate students taking an evolution-focused biology course. Dimensionality analysis supported a three-dimension (i.e., microevolution, macroevolution, and human evolution) structure. The three dimensions were generally in the expected order of difficulty, that is, the microevolution items were significantly more difficult than human evolution items, macroevolution items were more difficult than human evolution items although this difference was not statistically significant. Difficulties of the items did not align with the abilities of most of the participants, and many items within a dimension had similar item difficulties, indicating possible redundancy in trait measurement.

Assessing Contextual Reasoning About Natural Selection (Opfer, Nehm, & Ha, 2012)

Assessing Contextual Reasoning About Natural Selection (ACORNS) is an instrument that measures students' use of the core concepts of natural selection when explaining evolutionary change. The ACORNS items are short-answer questions asking students to explain evolutionary change in taxa and traits likely to be familiar as well as unfamiliar to them. A scoring rubric based on a number of key concepts and naïve ideas was developed to grade student responses. Two expert scorers independently coded student explanations for the presence or absence of the key and core concepts (KC) and cognitive biases (CB). Independent scoring of all KCs for the four new items exceeded inter-rater agreement scores of 0.81 (kappa), whereas independent scoring of CBs revealed more variable scores (kappas from 0.52 to 0.80). In the present study, the reliability values for the four new ACORNS items were 0.76 for KCs and 0.68 for CBs, which was similar to the past work. The Cronbach's alpha in prior study showed the reliability of ACORNS items was typically higher for KCs than for CBs, with 0.77 and 0.67, respectively. Prior studies also showed that students' scores on ACORNS were statistically positively correlated with students' scores on CINS (Anderson et al., 2002) and with scores based on interviews.

Assessment of Knowledge of Influenza (Romine, Barrow, & Folk, 2013)

The Assessment of Knowledge of Influenza (AKI) instrument is to evaluate high school students' knowledge of influenza, with a focus on misconceptions. In the first phrase of instrument development, items were written to address seven topic subthemes: the flu vaccine, flu transmission, nature of influenza, symptoms, complications, treatment, and prevention. A convenience sample of 205 students enrolled in Grades 9–12 at a rural, predominantly White school was used in the pilot study. Then, a refined version of the assessment containing 15 out of the original 38 items was administered to a more robust convenience sample of 410 students enrolled in Grades 9–12 from 6 schools.

Content validity was established through two review panels composed of experts in medicine and high-school pedagogy, respectively. Structural equation modeling suggested that the items fit a two-factor structure well. Overall, items fit well with the Rasch model; mean-square infit values ranged from 0.73 to 1.30; the range of mean-square outfit values was from 0.61 to 1.66. Principal component analysis of Rasch residuals showed that the eigenvalue for the largest residual component was 1.64, suggesting a unidimensional structure of items. Chronbach's alpha for the eight-item transmission and seven-item management subscales was measured at 0.68 and 0.76, respectively, while Rasch person reliability measures were 0.58 and 0.49, and the person reliability for the 15 items together was measured at 0.77.

Scientific Literacy Assessment (Fives, Huebner, Birnbaum, & Nicolich, 2014)

Scientific Literacy Assessment (SLA) measures middle school students' scientific literacy. The SLA has two parts: the SLA-D assesses five components of demonstrated scientific literacy and the SLA-MB, which was adapted from published instruments, measures motivation and beliefs. There are two versions of the SLA-D, each includes 26 multiple-choice items (11 shared and 15 unique items on each version). The SLA-MB is composed of three subscales for a total of 25 Likert items measuring three motivation and belief scales, value of science, scientific literacy self-efficacy, and personal epistemology.

Three pilot studies were conducted to validate the instrument. In Pilot Study 1, think aloud interviews were performed for six middle school students to gather evidence for the response-process related validity of SLA-D. The KR-20 reliability for the two versions of the SLA-D was 0.83 and 0.82, respectively. The discrimination indices for the 41 items that make up both versions of the SLA-D ranged from 0.30 to 0.85 with a mean and median of 0.58. The highest percent correct for any one item was 89% and the lowest was 22%. For SLA-MB, exploratory principal component factor analysis on the 25 items indicated that items formed three well-defined and unique components; each scale demonstrated sound reliability: value of science:

$\alpha = 0.80$; self-efficacy for science literacy: $\alpha = 0.72$; source and certainty of scientific knowledge: $\alpha = 0.88$. In addition, the correlation between self-efficacy for science literacy and value of science was 0.53 ($p < 0.001$), between self-efficacy for science literacy and SLA-D was 0.40 ($p < 0.01$), and between personal epistemology and SLA-D was 0.37 ($p < 0.01$). No significant correlation was found between other scales.

Visual-Perceptual Chemistry Specific (Oliver-Hoyo & Sloan, 2014)

The Visual-Perceptual Chemistry Specific (VPCS) assessment instrument consists of 33 multiple-choice questions aligned to eight visual-perceptual skills considered as needed by undergraduate chemistry students: association, constancy, discrimination, figure ground, form perception, memory, orientation, and sequencing. The first version of the instrument consisted of 47 questions and was administered to 266 chemistry students. Version I was submitted to peer review from professors in various fields to ask for comments and suggestions. Students were also interviewed to determine if questions appear to the participants to be aligned to the different categories as described, and 26 interviews were conducted. Principal Components Analysis (PCA) was used to eliminate items from the 47 in Version I to the 33 questions in Versions II and III. The final version was taken by 978 students.

Cronbach's alpha and Kuder-Richardson 20 (KR-20) were used to determine the internal consistency for the final version. The Cronbach's alpha for Version III was 0.66; the KR-20 value obtained was 0.63. Two-parameter item response models was applied to the data; the difficulties for all items ranged from −12.91 to 6.08, and eight items were identified as "difficult." The discrimination coefficients ranged from 0.13 to 2.03, with 15 items greater than 0.60. Confirmatory factor analysis was also applied to the data with good model-data fit for the eight factor solution.

Assessing Interdisciplinary Understanding of Osmosis (Shen, Liu, & Sung, 2014)

The knowledge integration framework informed the development of the instrument. First, a set of key concepts related to osmosis from different disciplines of natural sciences was identified, and interdisciplinary assessment items required students to integrate those disciplinary concepts. The instrument includes two types of items. The disciplinary items focus on relevant, individual science concepts that form building blocks needed to explain osmosis. The interdisciplinary items present scenarios that involve concepts from multiple disciplines. The initial version included 12 multiple-choice disciplinary items, and 14 multiple-choice and

7 constructed-response interdisciplinary items. After the pilot study, there were 15 multiple-choice disciplinary items, and 25 interdisciplinary items (16 multiple-choice and 9 constructed response). The pilot study involved 459 students and the revised version involved 792 students. Students were taking different levels of undergraduate courses in physics, biology, and physiology. Partial credit Rasch model was applied to the data sets. All but one item had good model-data fit. The Wright map also showed good targeting between item difficulty distribution and student ability distribution. The person reliability was 0.84. The correlation between disciplinary understanding and interdisciplinary understanding was 0.51.

What Do You Know About Alternative Energy? (Cheong, Johari, Said, & Treagust, 2015)

"What do you know about alternative energy?" is a diagnostic instrument of high school student alternative conceptions of alternative energy. The development and validation of the instrument involved repeated cycles of three broad steps: defining the content, obtaining information about students' conceptions, and developing the diagnostic instrument. An initial list of 123 propositional content knowledge statements about alternative energy was derived from textbooks and reliable internet sources. After rounds of reviews by university content experts, 47 propositional statements were used as the foundation for developing the instrument. Common alternative conceptions identified in the literature and Grades 10 and 11 students' open-ended responses for the reasoning part of the instrument were used to create the second-tier choices. The draft instruments were piloted with Grades 10, 13, and undergraduate students. The final version of 20-item, two-tier multiple-choice diagnostic instrument was administered to 491 Grades 10 and 11 students in Brunei. The content covered the following topics: (a) sources of energy, (b) conversion of energy in generation of electricity and efficiency of energy from power plants, (c) greenhouse emission, (d) costs of electricity generation, and (e) disadvantages and advantages of these alternative energy sources. Exploratory factor analysis resulted in a three-factor solution, explaining 37% of total variance. The three factors could be labeled as sources of alternative energy ($\alpha = 0.61$, 5 items), consequences of utilization of alternative energy ($\alpha = 0.241$, 3 items), and process to consider about alternative energy ($\alpha = 0.241$, 3 items). The alpha reliability for all the items together was 0.4.

Science Content Knowledge Test (Maerten-Rivera, Huggins-Manley, Adamson, Lee, & Llosa, 2015)

The Science Content Knowledge (SCK) test is a 30 item paper-and-pencil test to assess elementary science teachers' science content knowledge.

Items were selected from well-known, large-scale standardized tests such as NAEP, TIMSS. The initial 34 items were piloted with 144 K–6 teachers, 137 middle school students, and 30 college students majoring in elementary education. The psychometric properties of the test and items based on the pilot test were reviewed, and a panel of researchers, district personnel, and classroom teachers reviewed the pilot test results and selected 30 items to form the final version of the test. The 30 items included 24 multiple-choice and six constructed response items. The final version was given to 291 elementary teachers. Unidimensionality of the measure was established by confirmatory factor analysis. Rasch analysis was conducted to establish evidence for construct validity. Evidence of convergent validity was established by correlating elementary teachers' SCK ability estimates with scores based on classroom observations. Personal reliability at time points 0, 1, 2, and 3 for Project 1 were 0.72, 0.65, 0.54, and 0.58; Personal reliability at time points 0 and 1 for Project 2 were 0.77 and 0.76. There was a statistically significant correlation between SCK scores based on the paper-and-pencil test and SCK scores based on observation.

Diagnostic Test of the Water Cycle (Romine, Schaffer, & Barrow, 2015)

Diagnostic Test of the Water Cycle (DTWC) is a 15-item, three-tier diagnostic test measuring elementary education and middle/secondary science education preservice teachers' scientific understanding and confidence around the water cycle. The third tier is a four-point Likert scale consisting of "guessing," "uncertain," "confident," and "very confident." Selection of correct responses for both the first and second tiers and a confidence level of "confident" or "very confident" are considered indicative of proper scientific knowledge or understanding. Validation involved 130 university students from two colleges. Rasch modeling was used to validate the instrument. A scale of student misconception was also validated. Results suggest that DTWC scale had a good unidimensionality, local independence, and model-data fit. The DTWC had a person reliability of 0.55 when used as a single-tier instrument, 0.64 when used as a two-tiered instrument, and 0.73 when used as a three-tiered instrument with the third tier as confidence integrated into mastery.

Physics Teachers' Pedagogical Content Knowledge Test (Borowski, Fischer, Gess-Newsome, & von Aufschnaiter, 2016)

The physics teachers' Pedagogical Content Knowledge (PCK) test consists of 17 items, most of which are open-ended and some are close-ended items. Pedagogical Content Knowledge is defined as teachers' knowledge of and reasoning behind, and planning for teaching a particular topic of physics. About two-thirds of the PCK test items cover mechanics and the rest cover other areas of physics. The subtopics in mechanics were velocity/speed, forces as well as the relationship between forces, energy, and power. The items assess teachers' knowledge of student learning difficulties related to the topics and instructional approaches to deal with the difficulties. Five vignettes of description of realistic teaching sequences or materials are used to contextualize items. Each open-ended item was scored as correct, partly correct, or incorrect. Content validity of items was examined by six university physics educators and three experienced physics teachers. Multidimensional Rasch modeling was used to establish the evidence of internal structure. The correlation between the dimensions of content knowledge (CK) and PCK was 0.521 ($p < 0.01$), between CK and pedagogical knowledge (PK) was 0.202 ($p < 0.01$), and between PCK and PK was 0.232 ($p < 0.01$). The person reliability for the PKC scale was 0.98. Non-physics teachers achieved lower values on the PCK scale than physics teachers; teachers at lower level school tracks and with less university content preparation scored lower on the PCK test than teachers of higher level track and with more university content preparation; in-service physics teachers achieved better than preservice physics teachers; and physicists who neither taught at school nor at university scored lower than physics teachers teaching at higher level school tracks.

Understanding of Metallic Bonding (Cheng & Oon, 2016)

Understanding of Metallic Bonding (UMB) includes two-pairs (two parts) of items, one multiple choice and one explanation. The development of the instrument was based on ontological categories of scientific concepts and students' understanding of metallic bonding as reported in the literature. Part one measures students' understanding of metallic bonding and part two measures students' explanation of malleability of metals. Student understanding was scored by three levels from structure to process to interaction. Data were from 2006 high school students taking the chemistry course in Hong Kong. Partial credit Rasch modeling was used to validate the scale. Model-data fit, dimensionality, invariance property in terms of DIF, and item difficulty and student ability targeting in terms of the Wright map were examined.

Inter-Disciplinary Energy Concept Assessment (Park & Liu, 2016)

Inter-Disciplinary Energy Concept Assessment (IDEA) is based on the hypothesis that the development of understanding of the energy concept is discipline specific and forms a progression across science content both within a discipline and across disciplines. The IDEA consists of four linked forms, one each in environmental science, biology, chemistry, and physics. In addition, IDEA was linked to another energy concept assessment instrument called Energy Concept Assessment (ECA) for middle and high school students (Neumann, Viering, Boone, & Fischer, 2013). The initial instrument included 47 items; most of them came from various published items of large-scale assessments including AAAS, NAEP, and TIMSS; 27 items were developed by the authors. Open-ended questions were scored based on rubrics. Five experts reviewed the items for content validity. The initial instrument was then given to 164 students taking the first-year introductory courses in one of the four science disciplines. Unidimensional partial-credit Rasch model was applied to pilot data. The revised items were then given to 356 students in 12 first-year introductory science courses at six universities. Four-dimensional partial-credit Rasch analysis was conducted with the field test sample. There was a good model-data fit. The Expected A Posteriori/Plausible Value (EAP/PV) reliability coefficients for environmental science form was 0.772, biology 0.761, chemistry 0.758, and physics 0.756. Correlation between student scores on the four forms ranged from 0.772 to 0.872.

Assessment of Civic Scientific Literacy (Naganuma, 2017)

Assessment of Civic Scientific Literacy (ACSEL) measures adults' scientific literacy in three components: using scientific evidence, explaining scientific evidence, and making decisions based on objective information. Authentic problems faced by adult citizens and consumers were used to develop assessment items. Three contexts were used: environmental quality, health, and the frontiers of science and technology and greenhouse effect (GE), smoking and lung cancer (SLC); and genetically modified organisms (GMO) were chosen as representatives of the contexts. The constructed response question format was adopted. Responses from participants were scored using a scoring rubric. A total of 10 items were developed, with 3–4 items per context. Two of the items were from PISA and the rest were created by the author. Experts reviewed the items for content validity. A pilot study with 10 graduate students from various university majors was conducted. The validation study involved 401 adults in Japan. Cronbach's alphas for Using Scientific Evidence, Explaining Scientific Inquiry, and Making Decisions were 0.730, 0.631, and 0.388. Inter-rater reliability coefficients for all items ranged from 0.42 to 0.82.

Measurement of Student Understanding of Interdisciplinary Science (Yang, He, & Liu, 2017)

This instrument contains 20 items measuring elementary and middle school (Grades 4–8) students' understanding of six crosscutting concepts in the context of disciplinary ideas (e.g., cause and effect in the context of biodiversity) and science and engineering practices (e.g., scale, proportion, and quantity in the context of experiment design and data representation). The six crosscutting concepts are patterns, cause and effect, scale, proportion and quantity, matter and energy, structure and function, and systems and system models. Items forming the instrument were selected from various published instruments; they were in either multiple-choice or two-tier multiple choice format. Content validity was established by expert reviews. The instrument was administered in both the Fall and Spring semesters to 801 diverse Grades 4 through 8 students in a large urban school district in Northeastern United States. Rasch modeling was used to establish construct validity. Results showed that the measure had good unidimensionality, model-data fit, and item-ability range targeting. The person reliability was slightly over 0.60. There was a statistically significant positive correlation between student Rasch scale scores and their understanding of nature of science ($r = 0.216$, $p < 0.01$), and between student Rasch scale scores and self-efficacy ($r = 0.218$, $p < 0.01$). Student Rasch scale scores increased significantly from sixth grade to seventh grade and from seventh grade to eighth grade, while no statistically significant difference was found for any grade between Fall and Spring semesters.

The Pedagogical Content Knowledge in Biology Inventory (Großschedl, Welter, & Harms, 2019)

The Pedagogical Content Knowledge in Biology Inventory (PCK-IBI) is a 34-item paper-and-pencil based instrument targeting preservice and inservice biology teachers. Items are in a mixture of close-ended and open-ended formats. The development went through three stages. Pedagogical Content Knowledge is conceptualized to consist of four facets: knowledge of students' understanding, knowledge of instructional strategies, knowledge of curriculum, and knowledge of evaluation in five content areas of biology. The initial version of PCK-IBI was administered to 274 preservice biology teachers and Rasch modeling resulted in a revised instrument. Intraclass correlation coefficient for coding open-ended questions was 0.97. The revised version was then given to 432 preservice teachers and Rasch modeling resulted in further revisions. The instrument was found unidimensional with acceptable fit statistics. The DIF contrasts ranged from 0.01 to 0.70 with negligible DIF for 31 items. The EAP/PV and Weighted Likelihood Estimate (WLE) reliabilities were 0.66 and 0.72. Further confirmatory factor analysis supported the three-factor solution of PCK, PC, and Subject Matter Knowledge (SMK). The final version was given to 178 preservice and inservice teachers for further validation study. There was a statistically significant difference in PCK between preservice teachers in academic track and in nonacademic track; PCK scores were significantly positively correlated with number of semesters studied. There was a negative correlation between PCK and GPA. Cognitive ability scores were positively correlated with PCK scores; subscale scores of motivation for choosing teacher education, that is, utility and low difficulty of the study, were negatively correlated with PCK scores.

Commentary

The most important consideration when choosing a measurement instrument is evaluation of the measured construct of conceptual understanding. Different instruments may define understanding in quite different ways; thus it is important to ensure the appropriateness of the defined conceptual understanding for the intended uses. This evaluation is essentially an issue of construct validity. Once the measured construct of conceptual understanding is deemed appropriate, the next consideration is the intended population of the measurement instrument. An instrument validated for one population may not be valid for a different population. Only after the evaluation of the above two issues, can the focus of instrument evaluation then shift to reported technical properties of items (e.g., item difficulty and discrimination) and measures of the instrument (e.g., content validity, criterion-related validity, reliability, etc.).

Given that there can be a variety of different ways of establishing validity and reliability, it is important to examine the relevance of reported validity and reliability evidence to your intended use of the instrument. For example, an instrument validated through its ability to predict students' performances on a state test may not necessarily be appropriate for its use as a diagnostic test because the validity of a diagnostic test should more appropriately be established through the instrument's ability to assist in planning instruction. Similarly, if the reported reliability is solely based on the internal consistency (e.g., Cronbach's alpha), and your intended use is a repeated measurement of the targeted conceptual understanding for a same sample, then test–retest reliability should be expected.

Some of the above described instruments for assessing conceptual understanding are diagnostic tests. Diagnostic tests focus on analysis of students' selections of incorrect responses, and the analysis is typically done on the individual item level. If a diagnostic test is used for summative assessment for which analysis is typically based on total scores, then it is important to examine the evidence of unidimensionality—all items of the instrument measure the same construct. Evidence for unidimensionality is typically established based on factor analysis. If no evidence for unidimensionality is reported, then a summative use of the instrument may be questionable. For example, Huffman and Heller (1995) found that fewer than six FCI items converged on any single factor, and that different samples produced different factor patterns, indicating lack of unidimensionality of FCI. If this is the case, then using total scores of FCI for summative assessment may be questionable. Some authors of diagnostic instruments explicitly warn against uses of diagnostic instruments beyond the diagnosing purpose. For example, Hufnagel et al. (2000) stated about ADT 2.0, a diagnostic test for astronomy concepts, that

> the ADT should *not* be used as a graded test, or to assess the abilities of individual students. It cannot reliably assess any one concept, as that would require multiple questions on one concept. It also may not predict student course success for a number of reasons... The ADT is not intended to guide content selection, nor does it represent a fair sample of typical course content. (p. 155)

If an instrument does not entirely meet your selection criteria, you may decide to modify the instrument and go through a validation process to establish its validity and reliability. It is common that an instrument does not meet all the criteria or conditions of an intended use, and it is necessary for researchers to further validate the instrument to justify its new use. On the other hand, because statistics based on CTT are always sample dependent, and in many cases the samples used for validation are local or convenient samples, it is always necessary to continue validating an instrument even if it meets your selection criteria. Because of this limitation with CTT, it is clear that in recent years there is an increasing use of Rasch modeling in developing measurement instruments for conceptual understanding.

Developing Instruments for Measuring Conceptual Understanding

State the Purpose and Intended Population of Measurement

The very first step in developing a standardized measurement instrument for measuring conceptual understanding is to clearly identify the purpose for which the measurement instrument will be used. Typically, conceptual understanding measurement can be used for diagnostic, formative, or summative purposes; also measurement scores may be used for criterion-referenced or norm-referenced interpretations. Stating clearly intended uses of the standardized measurement instruments will help guide the subsequent instrument development and validation process; it will also help develop a user's guide for appropriate uses of the instrument. Related to the purpose is the intended population of the measurement instrument use; the two are dependent on each other. A measurement instrument is developed for specific purpose and for a specific population; the two go hand in hand.

Define the Construct to be Measured

In order to develop an instrument to measure conceptual understanding, it is necessary to explicitly identify the construct and define it. The construct for a conceptual understanding measurement instrument is the understanding of a specific science concept, such as understanding of forces, understanding of evolution, and so forth. Because there are various theories about understanding, it is necessary to adopt a theory about understanding before a definition of the construct is developed. For example, if the revised Bloom's taxonomy is adopted as a framework of understanding and the measurement target is the concept of mechanical forces, the next consideration is describing the internal hierarchy among specific elements of understanding. Because the purpose of developing a measurement instrument of conceptual understanding is to differentiate different levels of conceptual understanding among the target student population, there must be a theory suggesting which specific elements of understanding are more advanced and which specific elements of understanding are more primitive. For example, in terms of conceptual understanding of mechanical forces, a hierarchy of understanding may be conceptualized as follows (Figure 3.1).

Figure 3.1 suggests that students' understanding of mechanical forces can be differentiated along a linear dimension from remembering or recognizing different types of forces (gravitational, mechanical, magnetic, etc.) to explaining various motions (e.g., linear, projectile, circular, simple

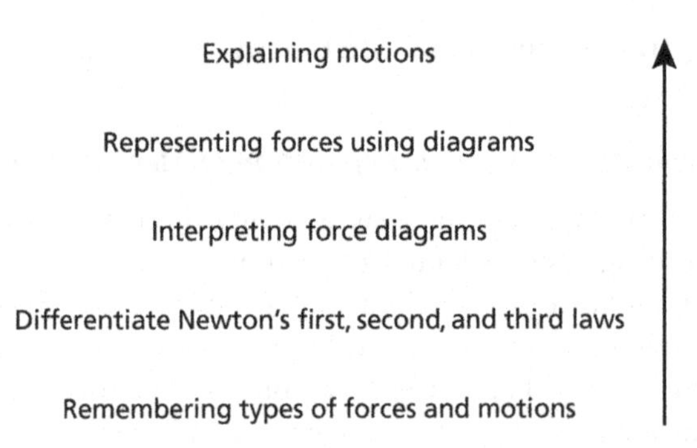

Figure 3.1 A hierarchy of elements in understanding of mechanical forces. The hierarchy is unidimensional from least competence at the bottom to most competent on the top.

harmonic, etc.). This defined construct explains difficulty levels of assessment items and students' understanding levels, which will guide subsequent stages of the measurement instrument development.

Identify the Performances of the Defined Construct

Based on the defined construct, the domain of the measurement target will then be defined. The domain of the measurement target is what the instrument will cover; it consists of specific observable performances subjects will demonstrate. For example, the domain of students' conceptual understanding of the concept of mechanical forces may be defined as students' abilities to answer questions related to Newton's first law, second law, and third law, the superposition principle, and kinds of forces.

Stating the observable performances of students in terms of the measurement target does not mean that the instrument will measure all the performances. One approach commonly taken in developing measurement instruments is domain sampling. That is, what a measurement instrument measures is a representative sample of the universe of performances that define the measurement target. Since it is impossible for any measurement instrument to measure all the performances of a target, if the measured performances are representative of all possible performances, then the instrument can still be used to make valid inferences about students' conceptual understanding of the target.

Take the concept of mechanical forces as an example again. If the measurement instrument is for diagnostic uses, then the universe of student performances should consist of all the known students' preconceptions or misconceptions that a given population of students typically demonstrate and are important to consider during curriculum and instructional planning. This universe of students' preconceptions may be identified through a comprehensive review of the literature on students' alternative conceptions of forces. This results in a list of students' alternative conceptions related to such aspects as Newton's first law, second law, and third law, the superposition principle, and kinds of forces.

However, if the intended use of the instrument is summative, then the domain of student performances for understanding of forces will be quite different. One possible way to define the domain is to analyze the expected curriculum standards. In the curriculum standards, the expected students' understanding of forces may be already specified. For example, the Next Generation Science Standards (Achieve, Inc., 2013), states performances of student understanding of mechanical forces (HS-PS-2) include:

a. Analyze data to support the claim that Newton's second law of motion describes the mathematical relationship among the net force on a macroscopic object, its mass, and its acceleration;

b. Use mathematical representations to support the claim that the total momentum of a system of objects is conserved when there is no net force on the system; and

c. Apply scientific and engineering ideas to design, evaluate, and refine a device that minimizes the force on a macroscopic object during a collision.

If the intended use of the measurement instrument is formative, then the above identified universe of student behaviors may not be enough, because they do not reflect the progression of student learning. One possible description of such a learning progression is as follows:

Facet Cluster on Forces

00 Students can identify forces on an object and compare their relative sizes.
- 01 Students can identify the sources of forces on an object.
- 02 Students can correctly identify the direction a force is acting.
- 03 Students can compare the relative sizes of forces on static objects.

40 The student reports that objects cannot exert forces along or parallel to its surface.

50 For an object at rest or moving horizontally, the student believes the downward force is greater than the upward force.

60 The student believes that force is a property of an object and its size is indicated by the magnitude of other properties of the object.
- 61 If an object has more mass than another object, it also has more force.
- 62 If an object is more active (moves faster) than another object, it also has more force.

70 The student reports an energy source as a force.
- 71 The engine or battery exerts a force on the object.

80 The student believes that passive objects cannot exert a force even though they touch another object.
- 81 For an object at rest on a surface (e.g., a book on a table), the surface (e.g., table) does not exert an upward force.
- 82 Passive objects (e.g., ropes) connecting two other objects do not exert forces, but instead transmit the active force. (For example, in the situation of a person pulling on a rope connected to a cart, the student identifies the person as exerting the force on the cart, not the rope.)

90 The student believes that motion determines the existence of a force or forces.
- 91 If an object is moving there is a "force of motion."
- 92 When the "force of motion" runs out, the object will stop.
- 93 If an object is not moving, no forces are involved in the situation.

Sources: http://www.diagnoser.com/ (accessed Jan. 30, 2019)

In the above facet cluster on force, explicit learning goals and various intermediate understandings, that is, different sorts of reasoning, conceptual, and procedural difficulties, form a progression. Each cluster contains the intuitive ideas students have as they move toward scientifically accurate learning targets. Each facet has a two-digit number. The 0X and 1X facets are the learning targets. The facets that begin with the numbers 2X through 9X indicate ideas that have more problematic aspects. In general, higher facet numbers (e.g., 9X, 8X, 7X) are the more problematic facets. The X0s indicate more general statements of student ideas. Often these are followed by more specific examples, which are coded X1 through X9. Therefore, nine categories of student performances are differentiated in this example. The nine categories may define the universe of student performances of conceptual understanding of force.

Similarly, for criterion-referenced and norm-referenced uses, the universe of student performances for a measurement instrument of conceptual understanding needs to be defined accordingly. No matter for what use the measurement instrument will be, the universe of student performances should be defined in operational terms. Operationally defined domain is better to inform the development and validation of the measurement instrument.

Once the measurement domain is defined, a sample of the domain, that is, what the measurement instrument will cover, will then be defined as a table of test specification. A table of test specification is also called a test blueprint. It consists of a topic dimension and a cognitive reasoning dimension. A table of test specification intends to provide guidance in the next step—constructing items. In order for the specification to be more informative, a detailed table of specification including type of items and total point of the instrument may be developed. Table 3.2 is a detailed table of test specification.

Table 3.2 states that there is a total of 50 points for the entire instrument. From Table 3.2 we see that the measurement instrument will include 35 items, among which 26 are multiple-choice, 7 short constructed-response,

TABLE 3.2 A Sample Table of Test Specification With Type of Items and Total Points

	Remembering	Interpreting	Explaining	
	Multiple-Choice	Short Constructed-Response	Extended Constructed-Response	Subtotal
Vectors and scalars	5 (5)	2 (3)	2 (10)	
Linear motion	8 (8)	2 (4)		
Projectile motion	4 (4)	1 (2)		
Circular motion	5 (5)	1 (3)		
Spring	4 (4)	1 (2)		
Subtotal	26 (26)	7 (14)	2 (10)	**35 (50)**

Note: # of Items (points)

and 2 extended constructed-response questions. Each multiple-choice question is worth 1 point, each short-constructed response question is worth 1 to 3 points, and each extended constructed-response question is worth 10 points.

One important decision when developing a table of test specification, like Table 3.2, is deciding the total number of items for the instrument or test length. An important consideration for the total number of items is the ability range of the intended population. Obviously, before a measurement instrument is developed, the exact ability range of the population's abilities is unknown, thus the best estimation is needed. Once the possible range of the population's abilities is estimated, then the total number of items should be decided to cover the entire range of the population's abilities. There is no need for too many items at each main ability level; typically 2–3 items for each main ability level is sufficient. Thus, the wider the population's ability range, the more items the measurement instrument will need.

Based on the table of test specification, a preliminary set of items may be constructed. Guidelines for writing different types of items, such as multiple-choice and true–false are reviewed in Chapter 1, and additional guidelines for other types of items such as constructed-response and performance assessment questions are available from other literature (e.g., Liu, 2009). When writing items, it is important to follow the table of test specification. If a group of item writers are involved in this step, it is helpful for them to spend time discussing the table of test specification so that everyone understands what the table specifies. A sample item may be written for each of the cells, and agreed to be used as a reference for writing other items. The initial items, together with their scoring keys/rubrics, define the space of subject performances on the measurement construct.

After the initial set of items are written, they should be reviewed for content accuracy, appropriateness or relevance to the test specifications, technical item-construction flaws, grammar, offensiveness or appearance of "bias," and level of readability (Crocker & Algina, 1986). Given the breadth of item review, a panel of experts is needed. The item review panel should possess a combination of expertise that includes content, pedagogy, and measurement. The content expert is best suited to review the content accuracy of items; the pedagogy expert is best suited to review the appropriateness of items for the target population; the measurement expert is best suited to review technical soundness of items. Additional experts may also be asked to review for other specific aspects of items such as readability and cultural bias. Item review may result in specific suggestions to revise items. Only after items have passed the item review will they then be put into pilot-testing.

Conduct Pilot-Test/Field-Test of Items

Once items are considered acceptable by a panel of experts, they may then be given to select target students. This step is called item tryout or pilot-testing. The number of subjects involved in this step does not need to be high. One important factor in considering the sample for tryout is representativeness—how the subjects represent the variation in the target population in terms of the measurement construct. In order to obtain maximal feedback from subjects on quality of items, some additional questions may be posed to subjects to respond to as part of answering the items. For example, DeBoer, Dubois, Hermann, and Lennon (2008) used the following set of questions to accompany each item in their pilot-testing of items:

1. Is there anything about this test question that was confusing? Explain.
2. Circle any words on the test question you don't understand or aren't familiar with.
3. Is answer choice A correct? Yes No Not Sure
4. Is answer choice B correct? Yes No Not Sure
5. Is answer choice C correct? Yes No Not Sure
6. Is answer choice D correct? Yes No Not Sure

(Questions 3–6 are each followed by the statement: Explain why or why not.)

7. Did you guess? Yes No
8. Should there be any other answer choice?
9. Was the picture or graph helpful? Why or why not? [If there was no picture or graph . . .] Would a picture or graph be helpful?
10. Have you studied this topic in school? Yes No Not Sure
11. Have you learned about it somewhere else? Yes No Not Sure Where? (TV, museum visit, etc.)?

In addition to probing questions like those above, interviewing a few representative students about the items, that is, orally administering the items and asking for justification, may also be helpful. The purpose of the above processes is to obtain as much information as possible, both qualitative and quantitative, about the processes of students answering the items in order to judge if the items operate as expected.

Rasch analysis (next section) will be conducted next. Rasch analysis results are very informative about the quality of items, such as suggesting potential deficiencies in certain items. Qualitative and quantitative analysis results of pilot-test data described above will result in specific suggestions for revising the items.

After items are revised, they are then assembled into an instrument for field-testing. The assembled instrument should include every element in the final form including instructions for subjects/examinees, instructions for test administrators, the entire set of items, and the scoring guide. The mechanical aspects of test assembly to consider may include item ordering, item grouping, and answering format. In general, items should be grouped by major item formats, such as selected response questions, short constructed-response questions, and extended constructed response questions, and should be placed in order of increasing difficulty within each item group. For a longer test and for a population of older age (e.g., high school students and older), a separate answering sheet may be provided for students to write on, but for a short test and for a younger population (e.g., elementary school students), answering may be done together with questions.

Field-testing of items involves a larger sample of subjects than that used in the pilot-testing. One major consideration in selecting the field-testing sample is the match between the range of subjects' abilities and the range of item difficulties. The required sample size also depends on the number of items. Typically, the sample size is over 200, or 5 to 10 times as many subjects as items. Items for field-testing should be assembled and presented in a format similar to that of the final instrument; thus items should not contain probing questions, as is the case in pilot-testing.

Conduct Rasch Analysis

After pilot-testing/field-testing, a data file containing subjects' responses to items is then prepared. The data file should include subjects' original response codes (e.g., A, B, C, D), not the scored points (e.g., 0, 1). The data file is then submitted to a Rasch analysis to produce a series of tables and graphs that describe the degree of model-data fit for items and the

```
File  Edit  Format  View  Help
&INST                          ; shows this is a control file (optional)
TITLE='FCI Pretest Rasch Analysis'  ; Report title
NAME1=1                        ; First column of person label in data file
ITEM1=3                         ; First column of student item responses in data file
NI=30                           ; Number of items
CODES=ABCDE                    ; Valid response codes in the data file
KEY1=CACEBBBBEADBDDAABBEDEBBACECEBC
DATA=FCIpretest.TXT              ; Specifies data source
&END                           ; Item labels for the 30 items follow
Q1
Q2
Q3
Q4
Q5
Q6
Q7
Q8
Q9
Q10
Q11
Q12
Q13
Q14
Q15
Q16
Q17
Q18
Q19
Q20
Q21
Q22
Q23
Q24
Q25
Q26
Q27
Q28
Q29
Q30
END NAMES                       ; END NAMES or END LABELS must come at end of list
```

Figure 3.2 Winsteps control file. It describes where is the data file, data structure (how many items, which columns are person IDs, and which columns are item responses), answer keys, and item labels.

measurement instrument as a whole. The following illustrative analysis was based on a sample of 50 first year undergraduate students responding to the 30-item Force Concept Inventory (Hestenes & Halloun, 1995). The data file and Winsteps control file are shown in Figures 3.2 and 3.3.

Review Fit Statistics of Items, the Wright Map, Dimensionality Plots, and Reliability

First, item fit statistics are reviewed, and potentially misfitting items are identified for possible revisions. Figure 3.4 is a portion of an output table.

Figure 3.4 shows that Infit MNSQs, Infit ZSTDs, and Outfit ZSTDs for all items are within the acceptable range, indicating that these items seem to have a good model-data fit. However, Outfit MNSQs for items Q5, Q25,

```
File  Edit  Format  View  Help
 1CBCACBBBEEDDCDADBCDDDDBADACDEE
 2CAEAEBBBDEBCABCCDEDAABDCBACDBE
 3ABAAEABBECCCBCCADDEADDDCDABDDE
 4CDCEDBBAECCCCAAADEABBACBDBCEBE
 5DDBADADBEDBCBDABDDCCABDCDBCEDE
 6CACADADBEDCBBDAADDBDEDCCBECEBE
 7CDDACBBBBACCCACCDCEEADAADACDDE
 8CDCAEBBBCACBCACCDEECEDBAEDCDDE
 9CACEDBBBEACBDDAADBEDEBCADECEBE
10AABACBADEECCABACDEDAADDCBDBADE
11DAAAABCBCADCCCCABCEEBDCACEBDBE
12CCCADBBEEACBDDCADDEDADDADACEDC
13CDCACBBEEABCBBCCDCADEBBABBCDBE
14CACEDBEBEADBDDAADBEDEBBADBCEBC
15CDBABBBBCADBCDAADDEDCABADBCEBC
16CDCAEBBBCACBCDCADEBCABDAADCDDC
17CACAEBBBEADBCDAADEEDCBBADABDBE
18CACEDBBBCADBDDAADAADDBDCDACEDC
19CACEABBBEAABDDAAAAEBCABADACEAC
20BBCAEBBABCDCCBAADEEDDBCDDAADDC
21CABAEBBDBACBBACCDBEDDDDCDAADDE
22CDCABBAABABCCDCCDEDCCDAADDCDDE
23CBCEBBBAEEBBCCCADDDACAAADDCDDE
24CAAAEBBAEABBBABADDACBACADBCDDE
25CACACBBBEACBDDCCDCEDCACADACDBC
26CEAADEEECACBABCCDEEDCDCDDDCDBE
27CBCAEBDBEABBCCAAAEEDEBBAEDCDBE
28CBCEEBBABACBCBBADDACEDCABACBDE
29CCCADBBBEACBCDCCDDCEDBBABBCEDE
30CBCECBBBCEBCBACADEDCCABEDABEEE
31CDCAEBBBBAABCDDCDEEDCBCEBBCBAE
32CACAEABEEABBCACCDECDCBBABACEBE
33CACAEBBBBABBCDCADEEDEBBADBCEDE
34CACEABBBEADBDDAABBEDCDDADBCEBE
35CADADBBEBEBBDDAAEAAEADBAEAADAC
36CABABAABDCBBCDCADCDCDCCACBADAE
37CBEACBBABABBCBCCDAEDCBBAEBCDDE
38CBAADBEABDBBBACCDCEDBBDCBBCEBE
39CACEDBBBCADBDCCADDEDEBBADECEDC
40CACADBBBEACBCDCCDEEDCABADBCDDE
41CDCBEBEBCACBCDCADEEDCBCADACDDE
42DDCABBBBEADBDDAADEEDEBBADBCEDC
43CCCADBBBDCCCCDAADBBDEBBABBCDBE
44CBCADBBBCAABCDAADDECCADAEBCDBE
45ABCEABBBCABBCBAADDCCCBBADACEAE
46CACEBBBBEADBDDAADBEDEBBACECEBC
47EDAACBEBCCDBDDCDDCEDCABADACDDE
48CADEDBEDEACBCDAABBDCCBBACECEBE
49CBCAEEEBCADBDDCADEEDEBBADACCDC
50DBEACABEEDBBCADDDCEEDADCEBCBDE
```

Figure 3.3 Data file for Winsteps Rasch analysis. The data file contains student IDs and their raw responses to items.

Q7, Q17 and Q9 are greater than 1.3, which are outside the acceptable range. Because Outfit MNSQs are unweighted mean residuals, these large fit statistics could be due to a few students' unusual response patterns, that is, outliers. In addition to examining the fit statistics, PTMEA CORR (point-measure correlation) is a measure of the relationship between subjects' scores on the item and their Rasch measures; any negative correlation may signal potential misfitting as well. In order to further find out if these items are misfitting, or just simply not have been responded to normally by a few individuals, item characteristic curves (ICCs) and students' response patterns may be reviewed. Figure 3.5 is the ICC for Q5.

ENTRY NUMBER	RAW SCORE	COUNT	MEASURE	MODEL S.E.	INFIT MNSQ	INFIT ZSTD	OUTFIT MNSQ	OUTFIT ZSTD	PTMEA CORR.		ITEM
5	6	50	2.23	.49	1.19	0.7	2.10	1.6	A	.22	Q5
25	4	50	2.81	.59	1.09	0.3	1.95	1.2	B	.26	Q25
7	36	50	-1.46	.34	1.20	1.2	1.91	1.8	C	.16	Q7
17	4	50	2.81	.59	1.47	1.1	1.39	0.7	D	.16	Q17
9	24	50	-0.21	.32	1.21	1.6	1.40	1.5	E	.26	Q9
1	41	50	-2.13	.39	0.92	-0.3	1.28	0.7	F	.32	Q1
2	21	50	0.10	.32	1.07	0.6	1.20	0.9	G	.38	Q2
19	29	50	-0.71	.32	1.19	1.6	1.15	0.6	H	.28	Q19
29	19	50	0.31	.33	1.13	0.9	1.12	0.6	I	.37	Q29
15	21	50	0.10	.32	0.98	-0.1	1.11	0.5	J	.45	Q15
22	24	50	-0.21	.32	1.08	0.7	1.07	0.3	K	.38	Q22

Figure 3.4 Sample item-fit statistics for selected FCI items. Fit statistics are within the highlighted columns. The "Entry Number" represents the item estimation order, "Raw Score" is the sum of total credits earned on the item by all examinees, "Count" is the total number of responses by examinees, "Measure" is the Rasch item difficulty, "Model S. E." is the standard error of the Rasch item difficulty, and "PTMEA Corr." is the correlation coefficient between examinees' item scores and their Rasch scale scores.

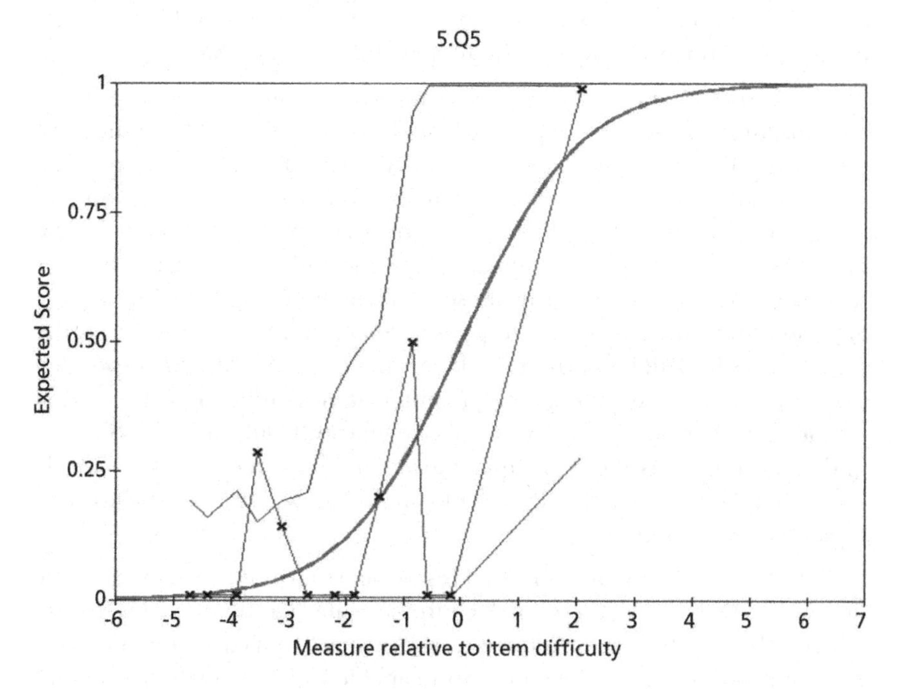

Figure 3.5 Sample item characteristic curve. Scatterplots are actual observations, and the trace line is the expected pattern.

In Figure 3.5, the *x*-axis is the difference between students' ability measures and the item difficulty measure, and the *y*-axis is the probability of students responding to the item. The observed pattern is in scatterplots, and the model expected pattern is in the smooth line. Students were grouped into 13 ability levels and are represented by "x" along the *x*-axis. As we can see, Q5 is a very difficult question, as most students' abilities were below the item difficult level (the difference is below 0). For students whose ability measures higher than the item difficulty measure, we should expect their probability to answer the question correctly to be greater than 0.5, and the bigger the difference is, the closer to 1 their probabilities should be. As can be seen from Figure 3.5, only for one group of students whose abilities are above the item difficulty, the probability to answer this question correctly is close to 1, indicating a good fit. For the 12 groups of students whose abilities were below the item difficulty, we should expect their probabilities to answer the item correctly to be smaller than 0.5, and the lower their abilities were the closer to 0 their probability should be. Overall, observed probabilities should fall within the 95% confidence interval band around the expected smooth line. As can be seen from Figure 3.5, one group of students' probability was outside the band, indicating a misfit for this group of students. Because the sample size was 50, the size for each group was about four students. Based on this, about 4 out of 50 subjects did not fit the model, indicating that, overall, the fit may still be acceptable.

In Figure 3.6, items are arranged vertically from the easiest (Q6) to the most difficult (Q25). Horizontally, subjects are arranged from the most capable (Person 46) to the least capable (Person 10). Thus, diagonally we should expect correct responses (1s) in the upper left corner region and incorrect responses (0s) in the lower right corner region. Those 0s in the upper left corner region are unexpected, indicating irregular responses by particular subjects. Similarly, those 1s in the lower right region are unexpected, indicating irregular responses. Specifically for Q5, which is the fourth most difficult item on the FCI, Persons 15, 23, 36, and 22 responded to the question in an unexpected manner. It is helpful to find out what reasons may have contributed to the irregularity. Could it be typo errors in data inputting? Is there any qualitative data from pilot-testing available to help explain the irregularity? Would an interview of those subjects shed light on their responses?

Another way to identify unusual response patterns is through examination of the Guttman scalogram. A Guttman scale (Guttman, 1944) is a set of items that produce an orderly response pattern from correct to incorrect when subjects are arranged from most capable to least capable and items

```
MOST UNEXPECTED RESPONSES
ITEM         MEASURE   |PERSON
                       |41 34341111424 43 1131 422 233322 2 5 1
                       |649429898759736521635187834086012265030
                   high---------------------------------------
   6 Q6      -2.29 o|...........0..0.0......................
   1 Q1      -2.13 F|....0..........0.....0.................
  27 Q27     -2.13 N|.........0..........00.................
  12 Q12     -1.58 h|.............0.........................
   7 Q7      -1.46 C|.0....0....00.0........................
  24 Q24     -1.46 b|........0.....0........................
   3 Q3      -1.13 c|......0...0............................
   8 Q8      -1.13 M|......0................................
  19 Q19     -0.71 H|......0.0...........................11.
  14 Q14     -0.61 f|.....0.................................
  20 Q20     -0.61 j|......00...............................
   9 Q9      -0.21 E|.....0.............................1111
  22 Q22     -0.21 K|...0...............................1...
  23 Q23     -0.11 L|..00...................................
   2 Q2       0.10 G|.................................1....1
  15 Q15      0.10 J|...................................1..1
  28 Q28      0.20 g|............................1.1....1...
  29 Q29      0.31 I|............................1....11....
   4 Q4       1.02 n|........................111...1........
  11 Q11      1.02 i|.....................1.1...1...........
  13 Q13      1.02 a|....................1..1...............
  21 Q21      1.02 m|...................1..1.1..............
  30 Q30      1.02 O|..................1.1......1...........
  18 Q18      2.01 d|.............1.................1.......
   5 Q5       2.23 A|..........1..............1...1..1......
  26 Q26      2.23 l|..............1......1.................
  17 Q17      2.81 D|0.....1..........1...1.................
  25 Q25      2.81 B|......1..............1.................
                    |-----------------------------------low-
                    |419343411114246431113184224233322225531
                    |64 42989875973 52 6351 783 086012 6 0 0
```

Figure 3.6 Sample most unexpected response patterns. The 0s and 1s in the circled ranges are unexpected.

are arranged from easiest to most difficult. Figure 3.7 is a portion of the Guttman scalogram of responses by the same group of university students on the FCI test.

In Figure3.7, subjects are arranged vertically from the most capable (Person 46) to the least capable (Person 40); items are arranged horizontally from the easiest (Q6) to the most difficult (Q25). It shows that overall the pattern is a gradual change diagonally from 1s in the upper left to 0s in the lower right. Most subjects' response patterns contain occasional irregularities such as Person 46's response to Q17, Person 14's response to Q7, and Person 9's responses to Q23, Q11, Q30, and Q5. Because the Rasch model is a stochastic model, it treats occasional irregularities as random effects. However, extensive irregularities such as those circled for Person 34

```
GUTTMAN SCALOGRAM OF RESPONSES:
PERSON |ITEM
       |  21 21  1112 22 122 11231 212
       |617274038694092325894131085675
       |------------------------------
    46 +111111111111111111111111111101
    14 +111101111111111111111111110000
     9 +111111111111111011111011010100
    34 +111111111111111001111111001001 0
    39 +111111111110101110101111100100
    42 +101111111111111101100111101000
    19 +111111111111010111101010100000
    48 +111101100101011111111000010111
    15 +111111101111100101110100101000
    17 +110111111111111111010100000000
    18 +111101111011010111011101000000
    33 +111111111111101110100001000000
    25 +111111110111110010010010100000
    27 +111101111110111101010001000000
    49 +011101111111101100000111100000
    12 +111111110111110000100010100000
    43 +111011011101101101010001010000
     6 +011100011101110011110001000100
    40 +111111110111101100000000000000
```

Figure 3.7 A sample guttman scalogram. Persons are arranged from *most able* (top) to *least able* (bottom), and items are arranged from *easiest* (left) to *most difficult* (right). The expected response patterns should be from 1s on the upper left to 0s on the lower right. The circled responses are unexpected.

and Person 6 are treated as errors; they will result in high Outfit MNSQs. Identifying those extensive irregularities can help pinpoint problematic items for revisions or extreme behavior subjects or outliers for deletion. For example, Person 6 seems to have responded randomly; this subject might not be taking the test seriously and his/her responses could be removed from analysis.

One final aspect of examination of model-data fit for items is the average ability measure order of item choices (for multiple-choice questions) or scoring categories (for scoring rubrics, please refer to Chapter 4). For multiple-choice questions, we should expect the correct choice, that is, the answer key, to have the highest average ability measure among all choices; similarly, average ability measures on a scoring rubric should gradually increase from the lowest score to the highest score. Figure 3.8 is portion of a sample Rasch analysis output table showing the average ability measures of item choices for four FCI questions.

Figure 3.8 shows that for Q7, the correct answer was B because its value was 1 while other choices had a value of 0. Thirty-six students selected the correct answer B; their average ability measure was –0.11. However, seven students selected the incorrect answer E; their average ability measure was

ENTRY NUMBER	DATA CODE	SCORE VALUE	DATA COUNT	%	AVERAGE MEASURE	S.E. MEAN	OUTF MNSQ	PTMEA CORR.	ITEM
5 A	C	0	9	18	-1.03	0.30	0.4	-0.30	Q5
	E	0	15	30	-0.54	0.21	0.6	-0.16	
	D	0	16	32	0.05	0.29	1.6	0.16	
	A	0	4	8	0.43	0.52	1.8	0.16	
	B	1	6	12	0.49	0.89	2.3	0.22	
25 B	B	0	10	20	-0.74	0.25	0.4	-0.20	Q25
	E	0	6	12	-0.69	0.34	0.4	-0.14	
	A	0	1	2	-0.44		0.4	-0.02	
	D	0	29	58	-0.11	0.22	1.2	0.12	
	C	1	4	8	0.85	1.24	2.1	0.26	
7 C	A	0	3	6	-1.71	0.39	0.3	-0.30	Q7
	C	0	1	2	-0.63		0.9	-0.04	
	D	0	3	6	-0.46	0.61	1.4	-0.05	
	E	0	7	14	-0.09	0.51	3.5	0.05	
	B	1	36	72	-0.11*	0.21	1.1	0.16	
17 D	E	0	1	2	-0.63		0.4	-0.04	Q17
	D	0	43	86	-0.33	0.19	2.0	-0.19	
	A	0	2	4	0.63	0.31	1.4	0.14	
	B	1	4	8	0.43*	0.52	1.4	0.16	

Figure 3.8 Sample Rasch output for item choice average ability measures. The correct answer should have the highest average ability measure, while distractors should have lower aberage ability measures.

–0.09, which is higher than –0.11, indicating that choices B and E need to be revisited to increase the item's discrimination. Similarly, choices B and A for Q17 also need revisited to increase the item's discrimination. Another aspect to look at in terms of item choices or scoring categories is the distribution of observations—how many subjects have selected each of the choices or scored each of the score categories. Ideally, there should be close to an even distribution of subjects over the choices or categories. Unusually low observations could indicate that the choices were flawed—either containing clues or being too obvious to be incorrect. Choices A for Q25, C for Q7, and E for Q17 had only one observation, indicating that those questions may not need these choices.

Following examination of model-data fit of items, an examination of model-data fit for the entire instrument is next. First, the Wright map should be examined for the match in distributions between person abilities and item difficulties, and any gaps should be identified. Figure 3.9 shows the Wright map for the FCI test based on a sample of 50 university introductory physics students' responses.

Figure 3.9 shows that although overall students' abilities spread about evenly over a range from –2.5 to 4.5 logits, there are a few gaps in items. First, there are no items for one very advanced student with an ability about

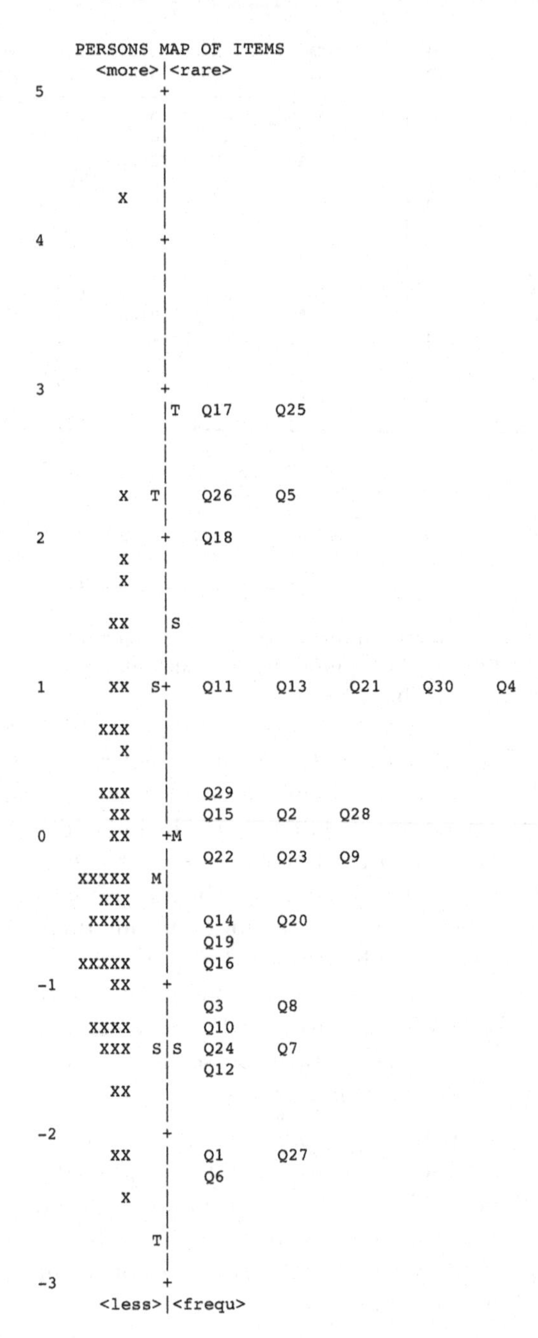

Figure 3.9 Wright map for the FCI test. Numbers on the left define a Rasch scale. Items and persons are distributed along the Rasch scale according to their relative measures.

4.5 logit. Second, there is a large item gap between 0.5 logit and 1.0 logit, and another large gap between 1.0 logit and 2.0 logit. Subjects whose abilities fall within those gaps will not be differentiated well, which will result in large measurement errors. Addition of items at the corresponding difficulty levels is necessary.

Third, the dimensionality of items should also be examined. This is usually through examination of factor loadings of items on a potential additional dimension within the item residuals. Figure 3.10 is the dimensionality map that shows how items are correlated with a potential additional construct within the item residuals.

Figure 3.10 shows that the unexplained variance was 30.0, and this unexplained variance explained by the first contrast was 3.3, which is greater than the expected 2.0, suggesting one possible additional dimension. In the

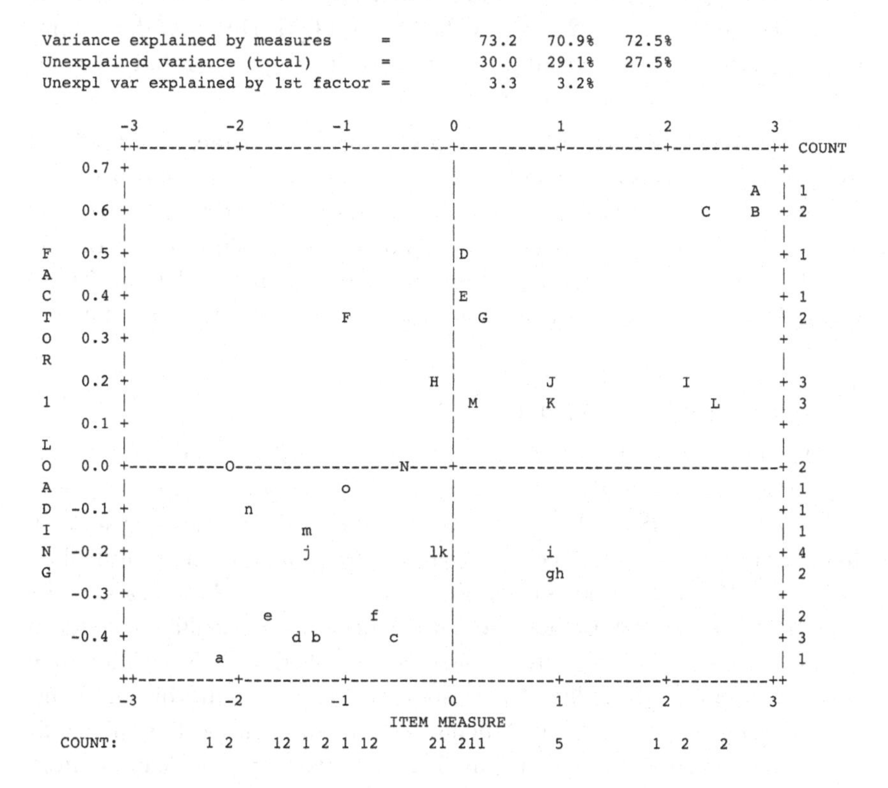

Figure 3.10 Dimensionality of item residuals. Unexplained variance explained by the first contrast should be smaller than 2. Loadings are correlation coefficients between item measures and their measures on an additional dimension. Items with correlation coefficients beyond the ±0.4 range are potentially measuring an additional dimension.

```
+-----------------------------------------------------------------------------+
| PERSONS       50 INPUT           50 MEASURED            INFIT          OUTFIT  |
|             SCORE    COUNT    MEAURE    ERROR     IMSQ    ZSTD    OMNSQ   ZSTD|
| MEAN         13.9     30.0     -0.24     0.49     0.99    -0.1     1.06     0.1|
| S.D.          5.5      0.0      1.23     0.10     0.28     1.3     0.68     1.2|
| REAL RMSE     .50   ADJ.SD   1.13 SEPARATION 2.23   PERSON RELIABILITY     .83|
|-----------------------------------------------------------------------------|
| ITEMS        30 INPUT            30 MEASURED            INFIT          OUTFIT  |
| MEAN         23.1     50.0      0.00     0.38      0.99   -0.1     1.06     0.1|
| S.D.         11.7      0.0      1.44     0.09      0.15    0.8     0.36     0.7|
| REAL RMSE     .40   ADJ.SD   1.39 SEPARATION 3.51   ITEM   RELIABILITY     .93|
+-----------------------------------------------------------------------------+
```

Figure 3.11 Summary of Rasch modeling statistics. Separation index is the ratio of adjusted standard deviation over root mean square error.

dimensionality graph, although most items fall within the range of small factor loadings (–0.4, +0.4), a few items, items A, B, C, and *a* are outside the range. Those items are Q25, Q17, Q26, Q2, and Q27. However, these items do not seem to have any common characteristics, suggesting that they may not measure an additional construct.

One last aspect to examine is the overall person separation and reliability. Figure 3.11 shows a summary of Rasch analysis for the FCI test based on a sample of 50 subjects. We can see that overall the test differentiated subjects well—with a separation index of 2.24 based on the observations, or 2.36 based on model expectations if there was a perfect model-data fit. Those separation indices are equivalent to the Cronbach's alpha of 0.83 and 0.85.

Examine Invariance Properties

The purpose of Rasch modeling is to develop an instrument that produces invariant measures. In order to test if measures are indeed invariant, both item measure invariance and person measure invariance may be examined. Item measure invariance may be examined by comparing item difficulties based on different groupings of subjects, such as gender, ethnicity, and socioeconomic background. Because we don't expect item difficulties to differ in terms of certain groupings, this analysis is also called analysis of differential item functioning (DIF). When DIF exists for an item, the item is biased against a particular group, because the difficulty measures are statistically significantly different between subsamples. Figure 3.12 presents sample DIF statistics from the Rasch modeling of FCI on two subsamples: males and females.

Figure 3.12 shows that item difficulty measures for males and females are different with varying sizes. For Q1, the difference between the difficulty measures is 1.31 logit in favor of females, that is, the item is more difficult

PERSON CLASS	DIF MEASURE	DIF S.E.	PERSON CLASS	DIF MEASURE	DIF S.E.	DIF CONTRAST	JOINT S.E.	t	d.f.	Prob.	MantelHanzl Prob.	MantelHanzl Size	ITEM Number	ITEM Name
F	-2.98	.77	M	-1.67	.47	-1.31	.91	-1.45	48	.1540	.0320	-.	1	Q1
M	-1.67	.47	F	-2.98	.77	1.31	.91	1.45	48	.1540	.0320	+.	1	Q1
F	0.23	.50	M	0.01	.43	0.22	.65	0.34	48	.7344	.8174	.00	2	Q2
M	0.01	.43	F	0.23	.50	-0.22	.65	-0.34	48	.7344	.8174	.00	2	Q2
F	-0.96	.49	M	-1.26	.44	0.30	.66	0.45	48	.6535	.5378	.00	3	Q3
M	-1.26	.44	F	-0.96	.49	-0.30	.66	-0.45	48	.6535	.5378	.00	3	Q3
F	0.75	.52	M	1.29	.53	-0.54	.74	-0.73	48	.4672	.7013	-.	4	Q4
M	1.29	.53	F	0.75	.52	0.54	.74	0.73	48	.4672	.7013	+.	4	Q4

Figure 3.12 Sample Rasch modeling statistics for differential item functioning. Statistics in the circled columns show the statistical signifcance of the difference between the measures obtained from different groups.

for males than for females. Although the t-test shows that the difference is not statistically significant, the probability for Mantel Haenszel test, a chi-square statistics, is statistically significant ($p = 0.032$, which is < 0.05), indicating that Q1 is potentially gender biased. For other items, although there are differences in difficulty measures between males and females, the differences are not statistically significant based on both t-tests and Mantel Haenszel tests, thus those items are not gender biased. A cautionary note is that the sample size in this example ($n = 50$) is small; there may not be enough statistical power to reject the null hypothesis.

Similarly, person measure invariance can be examined by conducting differential person functioning (DPF). In order to conduct this analysis, items will be grouped based on item characteristics, such as item format (e.g., multiple-choice vs. open-ended), item mode (e.g., paper-pencil vs. hands-on), and so on. Persons with significantly different abilities on different groups of items will be identified, and reasons for such differences should be examined. Any systematic differences should be considered as bias.

Establish Validity Claims

Through pilot-testing, field-testing, and, if necessary, further testing of the instrument, researchers will become satisfied with Rasch analysis results at one time point. The measurement instrument development proceeds with formally establishing validity claims for publication. Validity claims include two components: establishing interpretative arguments and establishing validity arguments. Interpretative arguments include collecting evidence related to test content, response processes, internal structure, external structure, and consequence.

Test Content

Evidence for the validity related to test content may be established by examining the agreement between the coverage of the test items and the assessment domain as defined by the construct and the table of test specification. The alignment between the domain defined by the test specification and the domain of the test items may be examined qualitatively and/or quantitatively. Qualitative analysis involves experts' judgment on the match between the two. Quantitative analysis may involve calculating an alignment index (e.g., Porter, 2002). If the alignment is significantly high, then the test scores can be generalized into the target assessment domain. In Rasch analysis, the person and item invariance properties provide direct evidence of test content representing the content domain of the defined construct.

Response Processes

Rasch analysis provides various statistics and graphical representations to help evaluate examinees' response patterns. In addition, qualitative studies may also help establish evidence related to the validity of response processes. Qualitative studies seek to understand how the measurement instrument works in specific context and with individual subjects. A case study of a few subjects taking the measurement instrument may be conducted to help understand how individuals perform on the test in specific context, which may have already been part of the item tryout. The data of a qualitative validity study are often in the form of narrative observations, or interpretative descriptions of subject behaviors on the measurement instrument. The qualitative data may then be categorized in order to identify common themes. The themes may further be analyzed to identify relationships among them. The categories, themes, and relationships identified from the qualitative interpretation can greatly enhance the validity arguments.

Internal Structure

Rasch modeling for developing a measurement instrument specifically adopts a theory-based interpretation approach. When applying Rasch models to develop measurement instruments, the first step is to identify a valid theory. Based on the theory, a number of inferences are then derived. The inferences include linearity of a latent construct, the hierarchy of behaviors of items, and the overall coherence of the items, that is, unidimensionality. Items and the measurement instrument are specifically constructed according to the above inferences that are consistent with characteristics of Rasch models, so that the probability of a subject answering an item correctly is solely determined by the difference between an item's latent difficulty and a subject's latent ability. Fit statistics, the Wright map, dimensionality map, DIF, and other statistics and graphs can help establish evidence related to instrument's internal structure.

External Structure

External structure is about whether or not the subjects' Rasch measures are correlated with other relevant variables. For conceptual understanding measurement, relevant variables may be subjects' course performance, achievement, and future tests of conceptual understanding, and so forth. Establishing validity related to external structure typically involves collection of additional data related to the relevant variables, and statistical analysis usually involves correlation or regression. For example, if the person ordering based on the Rasch measures is consistent with the teacher's ordering of the students, then there is evidence for the criterion-related validity of Rasch measures.

One powerful design for validity study related to extrapolation is the Multitrait-Multimethod (MTMM) design (Campbell & Fiske, 1959). Table 3.3 is a hypothetical design and its results.

In the validity study represented in Table 3.3, conceptual understanding of two concepts, Concept 1 and Concept 2, are involved, and two methods, Method A (e.g., multiple-choice test) and Method B (e.g., constructed response test), are used. This design implies that collection of three additional data sets is necessary for the same sample of subjects, one for conceptual understanding of another concept using two different test methods, and another for conceptual understanding of the same concept using a different test method. Each subject will now have four test scores, corresponding to the four scenarios of the design: same test method but with two different concepts and same concept but with two different methods (A1, A2, B1, and B2). Correlation between two sets of subjects' scores can be computed and the results may be presented in the format of Table 3.3. In Table 3.3, values on the diagonal in parentheses refer to the reliability of each test. Values in bold are correlation coefficients between scores on the same concept based on two different methods. A validity claim may be made if this correlation is high. This type validity is also called convergent validity. Values in italics are correlation coefficients between two sets of scores on two different concepts based on a same assessment format. This type correlation should be moderate (may or may not be statistically significant depending on the nature of the two concepts), and should definitely be lower than the previous correlation for the same concept with different test formats. Finally, values in normal font are correlation coefficients between two sets of scores for two different concepts based on two different test methods, and should not be significant. This type validity is also called discriminant validity or divergent validity. As can be seen, a MTMM design can help make a variety of validity claims pertaining to external structure.

TABLE 3.3 A Hypothetical MTMM Validation Design

	Method A		Method B	
	Concept 1 (A1)	Concept 2 (A2)	Concept 1 (B1)	Concept 2 (B2)
A1	(.80)			
A2	*.60*	(.85)		
B1	**.76**	.22	(.90)	
B2	.20	**.85**	*.50*	(.83)

Consequences

Consequences are related to uses of measurement results. For example, a conceptual understanding test may be used to place subjects into different levels of courses, or different instructional approaches. Two types of evidence are needed: one pertaining to the decision making process and the other to the intended and unintended consequences. A decision-making process based on test scores must be based on a clearly stated rationale and procedure. Evidence may come from the literature or from some commonly accepted theories. New empirical evidence to justify the decision-making process may also be established. Similarly, evidence for the intended and unintended consequence should be established empirically when possible, but may also be done through literature.

Although establishing evidence related to consequences requires additional data collection, enhancing interpretation of Rasch measures, that is, translating quantitative measures into qualitative categories, can help produce positive evidence related to consequences. Interpretation of Rasch measures may be enhanced using both internal and external criteria. Internal criteria include review of the clarity of the qualitative categories, justification of cutoff scores, and distribution of subjects among different categories. External criteria include correlation studies between categorization of subjects and their performances on other relevant measures.

Reliability Claims

Reliability is a necessary condition for validity, thus a reliability study is an important component of applying Rasch models to develop measurement instruments. Rasch modeling provides various statistics pertaining to reliability of items, persons, and the entire measurement instrument. Specifically, the standard errors of measurement (SEs) for individual items and subjects are calculated, and large SEs are indications of item misfitting. Reliability of the entire measurement instrument is examined based on the person separation index and its equivalent Cronbach's alpha. A high separation index (e.g., > 2.0) and alpha (e.g., > 0.80) are indications of high reliability.

In addition to the reliability study as part of Rasch modeling, traditional reliability studies may also be conducted. For example, inter-rater reliability may be established during the item writing process involving scoring rubrics. If more than one form of an instrument is available, then the equivalence between two sets of subjects' scores from two forms can be established by calculating the correlation coefficient. Further, inconsistency to scores may also be caused by the timing of a test. This reliability study involves administering the same instrument twice within a certain time interval, and the correlation coefficient between subjects' scores on the two test administrations is calculated as a measure of reliability. The final potential source for subjects' score inconsistency is the media used to present the test. For example, a paper-and-pencil test may be converted into a computer-administered test, or an instrument in English for one culture may be translated into another language for another culture. In these cases, a same sample of subjects must take both forms of the test and the correlation coefficient between the two sets of scores is computed as a reliability measure.

In fact, all the above traditional reliability studies may be incorporated into Rasch modeling by examining person measures invariance. This is because Rasch modeling can produce two sets of person ability measures that are invariant from time to time, or from test form to test form if model-data fit is good and the instrument is unidimensional. This property of Rasch measures is called person measure invariance property. The same method as described in Chapter 2 on item measure invariance property can be used to examine the person measure invariance property.

Establish Validity Arguments

While it is important that a variety of validity evidence is collected about a measurement instrument as described above, it is also important to examine the consistency among this variety of validity evidence. Thus, in addition to establishing interpretative arguments, a validity claim for a measurement instrument must also involve evaluation of the interpretative arguments as a whole. Important aspects to consider when evaluating interpretative arguments are: (a) clarity, (b) coherence, and (c) plausibility of both inferences and assumptions. A claim about the validity of a measurement is sound when all the interpretative arguments are clearly stated, coherent, and plausible and the assumptions underlying the arguments are sound. Because of the potential complexity of the interpretative arguments, it is unlikely that a validity claim can be made in a binary fashion, that is, either having validity or having no validity. It is possible that a validity claim is categorical with specific qualifiers or conditions.

Develop Guidelines for Use of the Instrument

The final step in developing a measurement instrument is the creation of a user's manual or guide. In addition to the actual measurement instrument that consists of items, a user's guide should also state the purpose or use of the instrument and the test domain including the table of test specification, and document item properties, and validity and reliability evidence. The user guide may also include item scoring rubrics, and rules for determining performance categories based on total test scores. If relevant, the user's guide should also suggest ways for making valid decisions based on total scores, and when possible point out the potential invalid decisions made based on total scores. The overall purpose of a user's guide is to help users decide if the instrument is appropriate for the intended use and, if so, how to use it.

When a measurement instrument has been developed by applying Rasch models, another important consideration is score reporting. Because Rasch modeling requires application of computer programs, we should not expect users to conduct Rasch modeling in order to obtain subjects' ability measures. A raw-score to Rasch measure conversion table should be provided as part of the user's guide. With this conversion table, users will still score subjects using scoring keys/rubrics and obtain raw total scores for the subjects. Users will then consult the score conversion table to find out equivalent Rasch ability measures, which is also called Rasch scale scores. Rasch scale scores will then be used in subsequent statistics analysis. Table 3.4 is a sample score conversion table from raw total scores to Rasch ability measures or scale scores for the FCI test.

In Table 3.4, raw total scores range from 0 to 30, because FCI is a 30-multiple-choice-item test with each item worth one point. Each raw total score is correspondent with one Rasch scale score, that is, measure, together with its standard error of measurement. By consulting this table, users of FCI can find out their subjects' Rasch scale scores based on their raw total scores, and use Rasch scale scores for subsequent statistical analysis. Once again, although measures in Table 3.4 contain negative values, ranging from –5.39 to 5.6, Rasch measures can be scaled to any range with any mean through a simple linear transformation.

TABLE 3.4 Raw Total Score to Rasch Scale Score Conversion Table

Score	Measure	SE	Score	Measure	SE	Score	Measure	SE
0	–5.39E	1.85	11	–0.82	0.44	22	1.40	0.49
1	–4.13	1.04	12	–0.63	0.44	23	1.65	0.51
2	–3.35	0.76	13	–0.44	0.44	24	1.92	0.54
3	–2.87	0.65	14	–0.25	0.44	25	2.23	0.57
4	–2.49	0.58	15	–0.06	0.44	26	2.57	0.61
5	–2.18	0.54	16	0.13	0.44	27	2.98	0.67
6	–1.91	0.51	17	0.33	0.44	28	3.51	0.79
7	–1.67	0.48	18	0.52	0.45	29	4.32	1.06
8	–1.44	0.47	19	0.73	0.46	30	5.60E	1.86
9	–1.23	0.46	20	0.94	0.47			
10	–1.02	0.45	21	1.16	0.48			

Chapter Summary

Conceptual understanding can be defined in many ways. One way is to conceptualize it as a process. An example of this conceptualization is the six facets of understanding that include explanation, interpretation, application, perspective, empathy, and self-knowledge. Understanding may also be defined as a state, for example, as meanings that are (a) resonant with or shared with others, (b) without internal contradictions, (c) without extraneous or unnecessary propositions, and (d) justified by the conceptual and methodological standards of the prevailing scientific paradigm. Understanding may further be conceptualized as both a process and a state. A representative definition of this conceptualization is the six elements of understanding; they are proposition, string, image, episode, intellectual skill, and motor skill. Besides the above conceptualizations, understanding is also domain specific, that is, specific to content domains. There are various domains of science, such as physical science, life science, and earth and space science. Each domain consists of a list of fundamental concepts with specific conceptual understanding. The revised Bloom's taxonomy can represent the three characteristics of conceptual understanding, that is, process-oriented, state-oriented, and domain specific. The revised Bloom's taxonomy of learning outcomes in the cognitive domain consists of two dimensions, one dimension for types of knowledge and another for cognitive process skills. The types of knowledge include factual knowledge, conceptual knowledge, procedure knowledge, and meta-cognitive knowledge. The cognitive process skills include: (a) remember, (b) understand, (c) apply, (d) analyze, (e) evaluate, and (f) create.

Science educators have developed various instruments for measuring conceptual understanding. When using an instrument to measure a conceptual understanding, it is important to review the intended uses of the measurement instrument and the evidence for validity and reliability. Many current measurement instruments for conceptual understanding may not possess desirable validity and reliability, because our notions of validity and reliability have been evolving; continuously validating a current measurement instrument is always necessary.

Using Rasch models to develop a measurement instrument for conceptual understanding starts with a theory about conceptual understanding of a concept. The theory must be valid, and should imply a linear construct that differentiates different levels of conceptual understanding among subjects. Based on the defined construct and different levels of performances of conceptual understanding, a test specification table is then created to guide the development of an item pool. Items will be reviewed by a panel of experts for content validity. Items may then be tried out with a small number of subjects to see if the items function as expected. The final set of items is then administered to a representative sample of subjects whose ability range matches the difficulty range of items; this process is called pilot-testing and field-testing. Pilot-testing and field-testing data form the basis for Rasch analysis. During Rasch analysis, model-data fit of items, Wright maps, item dimensionality, person reliability, as well as differential item functioning are examined. Rasch analysis may result in revisions of items and possibly further testing. The final set of items demonstrating a good fit with a Rasch model and being unidimensional forms a measurement instrument. Although much of the evidence for the validity and reliability of the measurement instrument is obtained during the Rasch modeling, additional validity and reliability studies may also be conducted. The final step in developing a measurement instrument for conceptual understanding is development of documentation for measurement instrument uses. Score reporting should be an important component of documentation.

Exercises

1. Force Concept Inventory (FCI; Hestenes, Wells, & Swackhamer, 1992/1995) is one of the most popular conceptual tests in science education. Based on principal component factor analysis, Huffman and Heller (1995) found that fewer than six FCI items converged on any single factor, and that the two samples suggested different factors. Huffman and Heller interpreted their results to suggest that FCI is not unidimensional. However, Hestenes and Halloun (1995) counter-argued that a single factor solution of factor analysis consistent with the Newtonian force concept could be expected if non novices (i.e., those who scored 60%–80% on the inventory) or a physicist data sample were used. What are your views on the nature of unidimensionality of FCI? What are the implications of this debate on the application of FCI for educational research?
2. Figure 3.10 was based on a group of college students' responses to a FCI pretest at the beginning of the introductory college physics course. What does Figure 3.10 suggest in terms of dimensionality of FCI? Is the suggestion consistent with Huffman and Heller's (1995) conclusion? If there is a discrepancy between the two, explain why?
3. Identify an instrument described in this chapter that was developed using Rasch modeling. Examine its development process and compare it to the steps for developing conceptual understanding tests using Rasch modeling described in this chapter, what are similarities and differences? How can you justify the differences if any?

4. Locate one instrument introduced in this chapter and any technical documentation or reports containing information on the validity and reliability of the instrument. Apply Standard 1 Validity (Joint Committee of the AERA, NCME and APA, 2014) to review the instrument; is the instrument valid and reliable? Administer the instrument to a group of intended subjects to collect data, and then conduct Rasch modeling based on the collected data. Are the conclusions about the validity and reliability of the instrument based on Rasch modeling of your data different from what were reported by the author(s)? If yes, explain why?

References

Achieve, Inc. (2013). *Next generation science standards.* Washington, DC: Author.

Adadan, E., & Savasci, F. (2012). An analysis of 16–17 year-old students' understanding of solution chemistry concepts using a two-tier diagnostic instrument. *International Journal of Science Education, 34*(4), 513–544.

Anderson, D. L., Fisher, K. M., & Norman, G. J. (2002). Development and evaluation of the conceptual inventory of natural selection. *Journal of Research in Science Teaching, 39*(10), 952–978.

Anderson, L. W., & Krathwohl, D. R. (Eds). (2001). *A taxonomy for learning, teaching, and assessing: A revision of Bloom's taxonomy of educational objectives.* New York, NY: Longman.

Arslan, H. O., Cigdemoglu, C., & Moseley, C. (2012). A Three-tier diagnostic test to assess pre-service teachers' misconceptions about global warming, greenhouse effect, ozone layer depletion, and acid rain. *International Journal of Science Education, 34*(11), 1667–1686.

Bailey, J. M., Johnson, B., Prather, E. E., & Slater, T. F. (2012). Development and validation of the star properties concept inventory. *International Journal of Science Education, 34*(14), 2257–2286.

Beichner, R. J. (1994). Testing student interpretation of kinematics graphs. *American Journal of Physics, 62*(8), 75–762.

Borowski, A., Fischer, H. E., Gess-Newsome, J., & von Aufschnaiter, C. (2016). Developing and evaluating a paper-and-pencil test to assess components of physics teachers' pedagogical content knowledge. *International Journal of Science Education, 38*(8), 1343–1372.

Bunce, D. M., & Gabel, D. (2002). Differential effects on the achievement of males and females of teaching the particulate nature of chemistry. *Journal of Research in Science Teaching, 39*(10), 911–927.

Caleon, I., & Subramaniam, R. (2010). Development and application of a three-tier diagnostic test to assess secondary students' understanding of waves. *International Journal of Science Education, 32*(7), 939–961.

Campbell, D. T., & Fiske, D. W. (1959). Convergent and discriminant validation by the multitrait-multimethod matrix. *Psychological Bulletin, 56,* 81–105.

Cheng, M. M. W., & Oon, P.-T. (2016). Understanding metallic bonding: Structure, process and interaction by Rasch analysis. *International Journal of Science Education, 38*(12), 1923–1944.

Cheong, I. P., Johari, M., Said, H., & Treagust, D. (2015). What do you know about alternative energy? Development and use of a diagnostic instrument for upper secondary school science. *International Journal of Science Education, 37*(2), 210–236.

Cheong, I. P., Treagust, D., Kyeleve, I., & Oh, P.-Y. (2010). Evaluation of students' conceptual understanding of Malaria. *International Journal of Science Education, 32*(18), 2497–2519.

Chu, H.-E., Lee, E. A., Ko, H. R., Hee, S. D., Moon, N., Min, B. M., & Hee, K. K. (2007). Korean year 3 children's environmental literacy: A prerequisite for a Korean environmental education. *International Journal of Science Education, 29*(6), 731–746.

Crocker, L., & Algina, J. (1986). *Introduction to classical & modern test theory.* Fort Worth, FL: Holt, Rinehart, and Winston.

DeBoer, G. E., Dubois, N., Hermann Abell, C., & Lennon, K. (2008, March–April). *Assessment linked to middle school science learning goals: Using pilot testing in item development.* Paper presented at the National Association for Research in Science Teaching Annual Conference, Baltimore, MD.

Ding, L., Chabay, R., Sherwood, B., & Beichner, R. (2006). Evaluating an electricity and magnetism tool: Brief electricity and magnetism assessment. *Physical Review Special Topics—Physics Education Research, 2*(1), 010105.

Doran, R. L. (1972). Misconceptions of selected science concepts held by elementary school students. *Journal of Research in Science Teaching, 9*(2), 127–137.

Engelhardt, P. V., & Beichner, R. J. (2004). Students' understanding of direct current resistive electric circuits. *American Journal of Physics, 72*(1), 98–115.

Fives, H., Huebner, W., Birnbaum, A. S., & Nicolich, M. (2014). Developing a measure of scientific literacy for middle school students. *Science Education, 98*(4), 549–580.

Großschedl, J., Welter, V., & Harms, U. (2019). A new instrument for measuring pre-service biology teachers' pedagogical content knowledge: The PCK-IBI. *Journal of Research in Science Teaching, 56*(4), 402–439.

Guttman, L. (1944). A basis for scaling qualitative data. *American Sociological Review, 9*(2), 139–150.

Haidar, A. H., & Abraham, M. R. (1991). A comparison of applied and theoretical knowledge of concepts based on the particulate nature of matter. *Journal of Research in Science Teaching, 28*(10), 919–938.

Haslam, F., & Treagust, D. F. (1987). Diagnosing secondary students' misconceptions of photosynthesis and respiration in plants using a two-tier multiple choice instrument. *Journal of Biological Education, 21*(3), 203–211.

Hestenes, D., & Halloun, I. (1995). Interpreting the force concept inventory: A response to Huffman and Heller. *The Physics Teacher, 33*(8), 502–506.

Hestenes, D., Wells, M., & Swackmaher, G. (1992). Force concept inventory. *The Physics Teacher, 30*(3), 141–158.

Huffman, D., & Heller, P. (1995). What does the Force Concept Inventory actually measure? *The Physics Teacher, 33*, 138–143.

Hufnagel, B., Slater, T., Deming, G., Adams, J., Adrian, R. L., Brick, C., & Zeilik, M. (2000). Pre-course results from the astronomy diagnostic test. *Publication of the Astronomical Society of Australia, 17*, 152–155.

Joint Committee of American Educational Research Association, American Psychological Association, & National Council on Measurement in Education. (2014). *Standards for educational and psychological testing.* Washington, DC: American Psychological Association.

Lasry, N., Rosenfield, S., Dedic, H., Dahan, A., & Reshef, O. (2011) The puzzling reliability of the Force Concept Inventory. *American Journal of Physics, 79*(9), 909–912.

Libarkin, J. C., & Anderson, S. W. (2006). The geoscience concept inventory: Application of Rasch analysis to concept inventory development in higher education. In X. Liu & W. J. Boone (Eds.), *Applications of Rasch measurement in science education* (pp. 45–73). Maple Grove, MN: JAM Press.

Liu, X. (2009). *Essentials of science classroom assessment.* Thousands, CA: SAGE.

Maerten-Rivera, J. L., Huggins-Manley, A. C., Adamson, K., Lee, O., & Llosa, L. (2015). Development and validation of a measure of elementary teachers' science content knowledge in two multiyear teacher professional development intervention projects. *Journal of Research in Science Teaching, 52*(3), 371–396.

Maloney, D. P., O'Kuma, T. L., Hieggelke, C. J., & van Heuvelen, A. (2001). Surveying students' conceptual knowledge of electricity and magnetism. *Physics Education Research, American Journal of Physics Supplement, 69*(7), S12–23.

McClary, L. M., & Bretz, L. (2012). Development and assessment of a diagnostic tool to identify organic chemistry students' alternative conceptions. *International Journal of Science Education, 34*(15), 2317–2341.

Mintzes, J. J., & Wandersee, J. H. (1998). Reform and innovation in science teaching: A human constructivist view. In J. J. Mintzes, J. H. Wandersee, & J. D. Novak (Eds.), *Teaching science for understanding: A human constructivist view* (pp. 30–59). San Diego, CA: Academic Press.

Mulford, D. R., & Robinson, W. R. (2002). An inventory for alternate conceptions among first-semester general chemistry students. *Journal of Chemical Education, 79*(6), 739–744.

Nadelson, L. S., & Southerland, S. A. (2009). Development and preliminary evaluation of the measure of understanding of macroevolution: Introducing the MUM. *The Journal of Experimental Education, 78*(2), 151–190.

Nadelson, L. S., & Southerland, S. (2012). A more fine-grained measure of students' acceptance of evolution: Development of the inventory of student evolution acceptance—I-SEA. *International Journal of Science Education, 34*(11), 1637–1666.

Naganuma, S. (2017). An assessment of civic scientific literacy in Japan: Development of a more authentic assessment task and scoring rubric. *International Journal of Science Education, Part B, 7*(4), 301–322.

National Research Council. (1996). *National science education standards.* Washington, DC: National Academy Press.

Nehm, R. H., & Schonfeld, I. S. (2008). Measuring knowledge of natural selection: A comparison of the CINS, an open-response instrument, and an oral interview. *Journal of Research in Science Teaching, 45*(10), 1131–1160.

Neumann, K., Viering, T., Boone, W., & Fischer, H. (2013). Towards a learning progression of energy. *Journal of Research in Science Teaching, 50*(2), 162–188.

Odom, A. L., & Barrow, L. H. (1995). Development and application of a two-tier diagnostic test measuring college biology students' understanding of diffusion and osmosis after a course of instruction. *Journal of Research in Science Teaching, 32*(1), 45–61.

Oliver-Hoyo, M., & Sloan, C. (2014). The development of a visual-perceptual chemistry specific (VPCS). *Journal of Research in Science Teaching, 51*(8), 963–981.

Opfer, J. E., Nehm, R. H., & Ha, M. (2012). Cognitive foundations for science assessment design: Knowing what students know about evolution. *Journal of Research in Science Teaching, 49*(6), 744–777.

Park. M., & Liu, X. (2016). Assessing understanding of the energy concept in different science disciplines. *Science Education, 100*(3), 483–516.

Porter, A. C. (2002). Measuring the content of instruction: Uses in research and practice. *Educational Researcher, 31*(7), 3–14.

Romine, W., Barrow, L. H., & Folk, W. R. (2013). Exploring secondary students' knowledge and misconceptions about influenza: Development, validation, and implementation of a multiple-choice influenza knowledge scale. *International Journal of Science Education, 35*(11), 1874–1901.

Romine, W. L., Schaffer, D. L., & Barrow, L. (2015). Development and application of a novel Rasch-based methodology for evaluating multi-tiered assessment instruments: Validation and utilization of an undergraduate diagnostic test of the water cycle. *International Journal of Science Education, 37*(16), 2740–2768.

Romine, W. L., & Walter, E. M. (2014). Assessing the efficacy of the measure of understanding of macroevolution as a valid tool for undergraduate non-science majors. *International Journal of Science Education, 36*(17), 2872–2891.

Sadler, P. M. (1998). Psychometric models of student conceptions in science: Reconciling qualitative studies and distractor-driven assessment instrument. *Journal of Research in Science Teaching, 35*(3), 265–296.

Sbeglia, G. C., & Nehm, R. H. (2019). Do you see what I-SEA? A Rasch analysis of the psychometric properties of the Inventory of Student Evolution Acceptance. *Science Education, 103*(2), 287–316.

Shen, J., Liu, O. L., & Sung, S. (2014). Designing interdisciplinary assessments in sciences for college students: An example on osmosis. *International Journal of Science Education, 36*(11), 1773–1793.

Thornton, R. K., & Sokoloff, D. R. (1998). Assessing student learning of Newton's laws: The force and motion conceptual evaluation and the evaluation of active learning laboratory and learning curricula. *American Journal of Physics, 66*(4), 338–352.

Tongchai, A., Sharma, M. D., Johnston, I. D., Arayathanitkul, K., & Soankwan, C. (2009). Developing, evaluating and demonstrating the use of a conceptual survey in mechanical waves. *International Journal of Science Education, 31*(18), 2437–2457.

Wheeler, A. E., & Kass, H. (1978). Student misconceptions in chemical equilibrium. *Science Education, 62*(2), 223–232.

Wiggins, G., & McTighe, J. (2005). *Understanding by design.* Alexandria, VA: Association for Supervision and Curriculum Development.

White, R., & Gunstone, R. (1992). *Probing understanding.* London, England: The Falmer Press.

Williamson, V., Huffman, J., & Peck, L. (2004). Testing students' use of the particulate theory. *Journal of Chemical Education, 81*(6), 891–896.

Wuttiprom, S., Sharma, M. D., Johnston, I. D., Critaree, R., & Soankwan, C. (2009). Development and use of a conceptual survey in introductory quantum physics. *International Journal of Science Education, 31*(5), 631–654.

Wittmann, C. M. (1998). *Making sense of how students come to an understanding of physics: An example from mechanical waves* (Doctoral thesis). University of Maryland. Retrieved from http://www.physics.umd.edu/perg/dissertations/Wittmann/

Yang, Y., He, P., & Liu, X. (2017). Validation of an instrument for measuring students' understanding of interdisciplinary science in grades 4–8 over multiple semesters: A Rasch measurement study. *International Journal of Mathematics and Science Education, 16*(4), 639–654.

Yeo, S., & Zadnik, M. (2001). Introductory thermal concept evaluation: Assessing students' understanding. *The Physics Teacher, 39*, 496–504.

4

Using and Developing Instruments for Measuring Affective Variables

Affective variables are both outcomes and processes in science education; they form an important domain of measurement. This chapter will first review various theoretical frameworks related to affective variables. It will then introduce published standardized instruments for measuring affective variables. Finally, this chapter will describe the process for developing new instruments for measuring affective variables using the Rasch modeling approach.

What Are Affective Variables?

Affective variables refer to constructs related to emotions, attitudes, feelings, and beliefs that are parts of both learning processes and learning outcomes. One most influential conceptualization of affective variables is the *taxonomy of affective education objectives* (Krathwohl, Bloom, & Masia, 1964). This taxonomy perceives affective variables as a continuum from external

Using and Developing Measurement Instruments in Science Education, pages 121–192

influences to internal driving forces. This continuum is also hierarchical. A lower degree of feeling and emotion is characterized by passively attending to external stimuli, and a higher degree of feeling and emotion is characterized by proactively acting and creating external stimuli. Figure 4.1 presents the Krathwohl et al.'s taxonomy of affective educational objectives.

Figure 4.1 shows that the entire affective domain is conceptualized as a continuum of internalization from receiving to responding, valuing, organization, and characterization. The continuum is hierarchical, that is, higher categories subsume all lower categories. Each category also contains a few hierarchical subcategories. Specifically, interest relates to subcategories 1.1 to 3.2; appreciation relates to subcategories 1.3 to 3.2; attitude and valuing relate to subcategories 2.2 to 4.1, and adjustment relates to subcategories 2.2 to 5.2.

Definitions of the categories and subcategories shown in Figure 4.1.

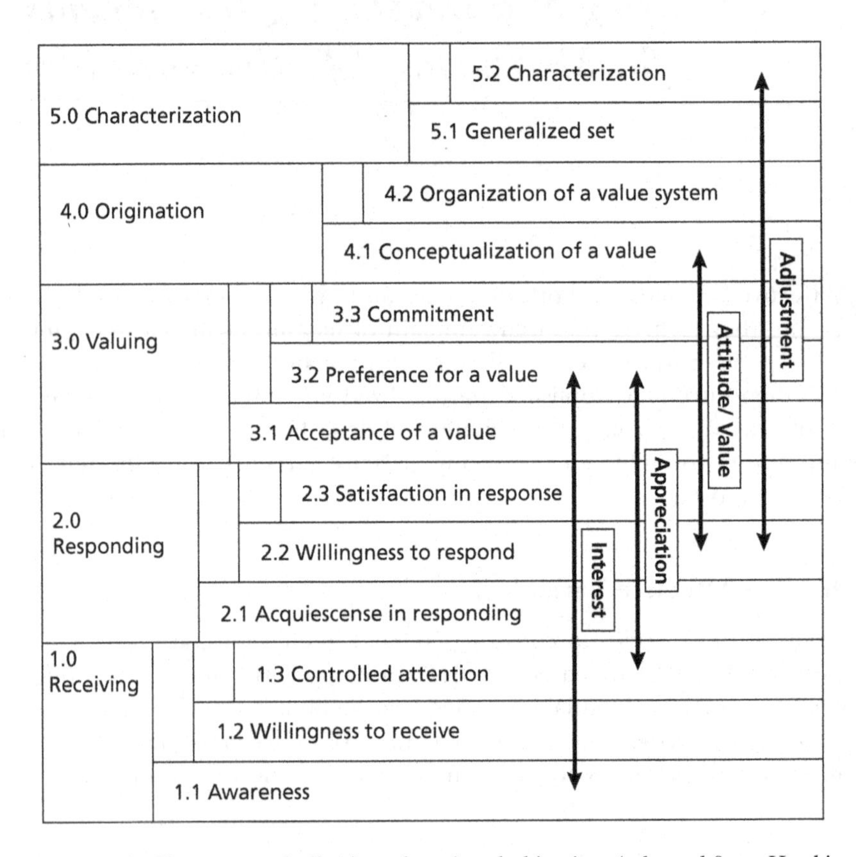

Figure 4.1 Taxonomy of affective educational objectives (adapted from Hopkins, 1998, p. 274). The hierarchical progression from receiving (1.0) to characterization (5) is internalization.

As can be seen from Figure 4.1, the Krathwohl et al.'s taxonomy of affective educational objectives is based on a continuous process of internalizing values to form a coherent value system, which in turn governs a person's behaviors. The whole taxonomy can be considered as a spiral consisting of multiple cycles of the behavior → internalized value → behavior process. Therefore, although the taxonomy is for the affective domain, it also involves the other two domains—the cognitive and psychomotor domains. This is because without reasoning, it is impossible to internalize values in order to form a value system; similarly, without action (which involves psychomotor skills), it is impossible for a person to act according to his/her value systems.

Affective outcomes are typical expectations in science curriculums for students. For example, the *Benchmarks for Science Literacy* (American Association for the Advancement of Science, 1993) state that by the end of 12th grade, students should "know why curiosity, honesty, openness, and skepticism are so highly regarded in science and how they are incorporated into the way science is carried out; exhibit those traits in their own lives and value them in others" (p. 287); students should also "view science and technology thoughtfully, being neither categorically antagonistic nor uncritically positive" (p. 287). In science education, two of the most researched affective variables are attitude and motivation.

Attitude

Various definitions of attitude are available. One definition is by Eagly (1992) who defines attitude as a tendency or state internal to a person that biases or predisposes the person toward favorable or unfavorable responses. Simpson, Koballa, Oliver, and Crawley (1994) define *attitude* as a predisposition to respond positively or negatively to things, people, places, events, or ideas. Science education researchers have studied various constructs related to attitude. In general, these attitudinal constructs can be grouped into two categories: those that are related to students, and those that are related to teachers. Examples of constructs of attitude related to students are student attitude toward school science, student attitude toward science and technology, student attitude toward science labs, student attitude toward science teaching, and student attitude toward science careers. Examples of constructs of attitude related to teachers are teachers' attitude toward inquiry, elementary teachers' attitude toward teaching science, teachers' attitude toward science curriculum, and teachers' attitude toward computer integration. Regardless of the attitude being measured, it is important to note that "attitude cannot be separated from its context and the underlying body of influences that determine its real significance" (Osborne, Simon, & Collins, 2003, p. 1055).

Research in science education has differentiated between "attitude towards science" and "scientific attitude" (Gardner, 1975). *Scientific attitude, or science attitude,* refers to evidence-based reasoning and argumentation, search for clarity and internal consistency, open-mindedness and skepticism, and willingness to change when data contradict their own views. On the other hand, *attitude toward science* refers to feelings, beliefs, and values held by a person about science. Attitude toward science is not a unitary construct; it consists of a number of interrelated sub-constructs such as perception of the science teacher, anxiety toward science, the value of science, self-esteem related to science, motivation towards science, enjoyment of science, attitude of peers and friends toward science, attitude of parents toward science, the nature of the classroom environment, achievement in science, and fear of failure in courses (Osborne et al., 2003). Similarly, Simpson and Troost (1982) identified 15 categories of attitudes toward science, including science affect, science self-concept, general self-esteem, locus of control, achievement motivation, science anxiety, emotional climate of the science class, physical environment of the science class, other students in the science class, science teacher, science curriculum, family—general, family—science, friends and science, and school. Further, attitudes toward science are students' expressed preferences and feelings; they are not necessarily student behaviors related to science (Osborne et al., 2003).

As for teachers, research has shown that teachers' beliefs affect their attitudes. For example, based on a socio-cultural model of beliefs, science teachers' attitudes toward science are strongly influenced by science teachers' epistemological beliefs, while science teachers' beliefs are closely linked to teachers' content knowledge, confidence, self-efficacy, experience, and social context (Jones & Carter, 2007).

A related concept to attitude is values. While attitude primarily deals with things, people, and places, values deal with abstract ideas such as science and technology. Values emphasize more on affect and cognition and less on immediate behavior (Simpson et al., 1994).

Motivation

Although there is no universally agreed upon definition of motivation (Jones & Carter, 2007), *motivation* may be considered as "an internal state that arouses, directs, and sustains students' behavior" (Koballa & Glynn, 2007, p. 85). Motivation focuses on the desire to act or not to act; it leads to behaviors (Simpson et al., 1994). According to Koballa and Glynn (2007), there are four *approaches to defining motivation*: the behavioral, humanistic, cognitive, and social. The behavioral approach to motivation focuses on extrinsic and material incentives and reinforcement; the humanistic approach focuses on human's intrinsic needs such as self-actualization and self-determination; the cognitive approach emphasizes goals, plans, expectations, and attributions; and finally the socio-cultural approach focuses on individuals' identities and interrelations with others within a community. For example, taking primarily a cognitive approach, Zusho, Pintrich, and Coppola (2003) conceptualize motivation processes in relation to student science achievement to include four components: the self-efficacy beliefs, task value beliefs, goal orientation, and affects. The self-efficacy beliefs are students' judgments of their capabilities to perform a task, as well as their beliefs about their agency in the course. Task value beliefs refer to students' beliefs on the utility and importance of a course. Goal orientations are individuals' purposes when approaching, engaging in, and responding to achievement situations. Specifically, two achievement goals, that is, mastery and performance goals, are found to be important determinants of students' motivation and performance. Finally, affects refer to students' interests and anxiety. Science educators study motivation in order to explain why students strive for particular goals when learning science, how intensively they strive, how long they strive, and what feelings and emotions may characterize them in these processes. Like attitude, there are various constructs related to motivation, such as motivation to learn, motivation to teach, motivation to implement a new curriculum, to name a few. Motivation is influenced by attitude.

Instruments for Measuring Student Affective Variables

A variety of standardized measurement instruments related to affective variables are available in science education. The following summative descriptions introduce instruments published over the past 50 years in major science education research journals, mostly in *Journal of Research in Science Teaching*, *Science Education*, and the *International Journal of Science education*. The majority of the instruments have been developed based on CTT. They are for various intended uses and based on various theoretical frameworks of affective variables. Each instrument description is information only, not intended to be a critical review of its strength and weakness. Also, descriptions do not follow a same format or even in similar length; how and how much an instrument is described depend on what is reported in the publication. In general, when available each description contains information about the instrument's intended population, purpose, composition (e.g., number and type of items), validation process, and key indices of validity and reliability. Instruments are described in their publication years and in alphabetical order of authors' last names.

Instruments for Measuring Student Attitudes

Science Support Scale (Schwirian, 1968)

The Science Support Scale (Tri-S) is a 40-item Likert-scale survey of university nonscience major students' attitudes toward science. Attitude toward science is conceptualized as particular cultural values which are more conducive to the growth and development of the scientific institution in a society than are their opposites. Specifically, these values are (a) rationality, (b) utilitarianism, (c) universalism, (d) individualism, and (e) a belief in progress and meliorism. An initial list of 60 questions were created corresponding to the above five values. This original scale was tried with 196 university nonscience majors. Item analysis based on item discrimination and subscale internal consistency reduced the original number of items to 40 (8 items per subscale). Split-half internal consistency reliability corrected by the Spearman-Brown formula for the five subscales ranged from 0.59 to 0.87.

The Scientific Attitude Inventory (Moore & Sutman, 1970; Moore & Foy, 1997)

The Scientific Attitude Inventory (SAI) measures high school students' attitude toward science. It consists of six positive and six corresponding negative intellectual and emotional attitudes toward science. These six pairs of attitudes are: (a) nature of the laws and/or theories of science—changeable vs. unchangeable truths, (b) the basis of scientific explanation—observation vs. authority, (c) characteristics needed to operate in a scientific manner—independence of thought vs. group thinking, (d) the type of activity engendered in science—idea generating vs. technology developing, (e) dependent on public understanding vs. no need for public understanding, and (f) science as a career—interesting and rewarding vs. dull and uninteresting. Validation involved expert panel reviews and pilot-testing by 10th grade students in three sequences of biology courses. Analysis of variance controlling for IQ and science GPA showed that the groups who received instruction relevant to the attitudes scored significantly higher on the post-test than on the pretest. The final form of SAI consists of 60 items with five items for each of the six pairs of attitude subscales. The test–retest reliability based on pre- and post-test scores was 0.934.

The SAI-II is a revision of SAI; it consists of 40 items. The SAI-II contains improvement in readability and elimination of gender-biased language. The SAI-II has also replaced the four-category Likert responses in SAI with five-category Likert responses. The intended attitudes to be measured in SAI-II remain to be six pairs of opposing attitudes. The alpha reliability was 0.78. However, factor analysis did not result in the hypothesized 12-attitude structure.

Although SAI as well as SAI-II has been widely used in science education research, there are questions about its construct validity both conceptually (Munby, 1983a, 1997) and empirically (Lichtenstein et al., 2008; Moore & Foy, 1997). Munby (1983a) conducted a conceptual analysis of items in SAI and concluded that many SAI items may not tap into attitude related to science, raising doubt about its construct validity. Revisions to SAI to produce SAI-II failed to address the issue of construct validity (Munby, 1997). In a recent validation study through both exploratory and confirmatory factor analysis (CFA), Lichtenstein et al. (2008) failed to produce the claimed constructs corresponding to the 12 attitudes or 6 pairs of attitudes. At most, Lichtenstein could only identify 2 meaningful constructs that may be called "Science is Rigid" (6 items, $\alpha = 0.59$) and "I Want to be a Scientist" (8 items, $\alpha = 0.85$).

Specific Interest in Biological, Physical, and Earth Sciences (Skinner & Barcikowski, 1973)

Specific Interest in Biological, Physical, and Earth Sciences (SIBPE) measures intermediate level (Grades 7 and 8) students' attitudes toward specific disciplines. The items ask students to indicate how many times they had done each of the 70 activities during a specific time period. The instrument contains 12 general science questions with 12 items from each of the science subject areas (i.e., biology, chemistry, physics, and earth science). Validation of the instrument involved panel reviews and pilot-testing with 2,137 seventh and eighth grade students participating in an NSF funded science program. Factor analysis was conducted in subsamples of seventh grade males, females, and eighth grade males and females. Eight similar main factors were present in all the subsamples.

The Biology Attitude Scale (Russell & Hollander, 1975)

The Biology Attitude Scale (BAS) is a 30-item instrument for measuring university students' attitudes toward biology. The scale was given consecutively once a week for 7 weeks to a group of students taking a biology course, and the stability of students' scores was very high (averaged 0.80 over 7 weeks). The reliability of the Likert scale was consistently over 0.90.

Biology majors scored significantly lower after the biology course; the life science majors scored statistically the same from pre- to post-course; and education majors scored statistically significantly higher on the post-course survey than on the pre-course survey. Correlation between BAS and another scale based on semantic differential was high ($r = 0.80$).

The Environmental Concern Scale (Weigel & Weigel, 1978)

The Environmental Concern Scale (ECS) is composed of 16 statements focusing on a wide range of conservation and pollution issues. The general public is the target audience. The statements are presented in a five-point Likert scale ranging from "strongly agree" to "strongly disagree." Each item had a high correlation with the scale score (0.12 to 0.42). The correlation between the scale scores and scores on a criterion measure was 0.58. Prediction validity was established based on correlation between scale scores and behaviors on petitioning ($r = 0.5$, $p < 0.01$), the litter pickup ($r = 0.36$, $p < 0.05$), recycling ($r = 0.39$, $p < 0.01$), and the comprehensive behavioral index ($r = 0.62$, $p < 0.01$). Six week test–retest reliability was 0.83. Alpha internal consistency was 0.85.

Test of Science-Related Attitudes (Fraser, 1978)

The Test of Science-Related Attitudes (TOSRA) is a 70-item Likert scale instrument measuring junior high school students' science related attitudes. It has seven scales: social implications of science, normality of scientists, adoption of scientific attitudes, enjoyment of science lessons, leisure interest in science, and career interest in science. The TOSRA is an extension of a previous instrument by adding two new scales, and by providing uniform instructions, number of items across scales and response formats. Initial items were reviewed by a group of science teachers for content validity, and field-tested for item quality. The main validation study involved 1,337 Grades 7–10 students in Australia. Cronbach's alpha coefficients for the above seven scales ranged from 0.66 to 0.93 for the Grade 7 sample, from 0.64 to 0.93 for the Grade 8 sample, from 0.69 to 0.92 for the Grade 9 sample, and from 0.67 to 0.93 for the Grade 10 sample. Inter-correlations among TOSRA scales were calculated as indices of discriminant validity. The correlations were generally low ranging from 0.10 to 0.58 (mean = 0.33).

The Simpson-Troost Attitude Questionnaire (Simpson & Troost, 1982)

The Simpson-Troost Attitude Questionnaire (STAQ) is one of a series of measurement instruments developed as part of a large-scale longitudinal study on relationships between affective variables and student achievements. It is a 60-item Likert-scale survey of secondary school (Grades 6–10) students' attitude toward science. It contains 15 subscales related to: student attitude toward science, motivation to achieve in science, science anxiety, attitude toward science teachers, and attitude toward science curriculums. The validation of the instrument went through multistages and multi-years with thousands of students. A total of 12 editions of the instrument were piloted. The content validity was established through expert panel review; construct validity was established through item analysis and exploratory factor analysis. Internal consistency reliability for the subscales was claimed to be above 0.90. Studies using STAQ reported Cronbach's alpha coefficients for different subscales ranged from 0.33–0.95 (Owen et al., 2008).

Owen et al. (2008) conducted a validation study of the instrument with a sample of 1,754 secondary (Grades 6–8) students. The sample was randomly divided into two subsamples, one for exploratory and another for confirmatory factor analysis. The exploratory factor analysis found 10 interpretable factors, but only five of the factors had a Cronbach's reliability coefficient greater than 0.70. The five factors were called *motivating science class* (6 items), *self-directed effort* (4 items), *family models* (4 items), *science is fun for me* (4 items), and *peer models* (4 items). A confirmatory factor analysis with the above five factors had a very good model-data fit. Additional data collected on stability of scores found that stability was higher across high school grades than across middle school grades.

Women in Science Scale (Erb & Smith, 1984)

The WiSS is a Likert-scale survey of early adolescent girls' and boys' attitudes toward women in science. The WiSS assumes three dimensions of attitude toward women in science: (a) women possess characteristics that enable them to be successful in science careers, (b) women's roles as mothers and wives are compatible with successful science careers, and (c) women and men should have equal opportunities to pursue science careers. Pilot testing with middle school and high school students resulted in elimination of items with poor item qualities. Construct validity was established based on convergent and divergent validities. The final 27-item instrument had an overall alpha reliability of 0.92 and a test–retest (8 weeks interval) reliability of 0.82.

Owen et al. (2007) conducted a revalidation study of WiSS using a sample of 1,439 middle school students in Texas. A confirmatory factor analysis based on the claimed three dimensions showed strong inter-correlation (correlation coefficients ranged from 0.89 to 0.90) among the three dimensions, suggesting a significant overlap among dimensions within the instrument. After a series of exploratory and confirmatory factor analysis, the authors found that the 27-item original instrument could be shortened into a 14-item instrument with two meaningful dimensions: gender equality (8 items) and sexism (6 items). The correlation between the two dimensions was 0.78. The Cronbach's alpha coefficients for the two dimensions were 0.78 and 0.75.

Attitude Toward Science in School Assessment (Germann, 1988)

The Attitude Toward Science in School Assessment (ATSSA) measures secondary school students' general attitudes toward science, specifically, how students feel toward science as a subject in school. An initial list of 34 items was created. A panel review resulted in reduction of 10 items and revision of the remaining items. The remaining 24 items were pilot tested with a group of 125 science students in Grades 7 and 8. Principal component factor analysis found that 14 items loaded highly on the first factor, and the remaining items loaded on four other factors. The 14 items were selected to form the ATSSA instrument. This final instrument was subjected to four subsequent validation studies. Cronbach's alpha estimates from the four studies were all greater than 0.94; all 14 items loaded on only one factor with consistent factor loadings in all four studies. Percentages of variance accounted for by this factor were 64.9, 69.8, 67.4, and 59.2. Item-total correlations ranged from 0.61 to 0.89. Student attitude scores on ATSSA from two classes with known different attitudes toward science were found statistically significantly different. There was also a statistically significant correlation between scores of ATSSA and measures of formal logical reasoning and biology achievement tests.

Early Childhood Women in Science Scale (Mulkey, 1989)

The Early Childhood Women in Science Scale (ECWiSS) is a 27-item Likert-type instrument measuring early childhood children's (K–Grade 4) attitudes toward women scientists. Items are given orally, and students are asked to color the line-drawn faces corresponding to *strongly agree, agree, disagree,* and *strongly disagree.* The ECWiSS is modeled after WiSS for middle school students, and covers the same three dimensions: (a) women possess characteristics which enable them to be successful in science careers, (b) women's roles as mothers and wives are compatible with successful science career pursuits, and (c) women and men ought to have equal opportunities to prepare for and pursue science careers. The Cronbach's alpha for the entire instrument was 0.90 and the Guttman split-half reliability coefficient was 0.87. Construct validity of the instrument was established through a causal comparative design and a principal component factor analysis. The scale was able to detect statistically significant differences between groups (e.g., gender, grade level, academic ability, and social economic background). Scores on ECWiSS were also statistically significantly correlated with measures obtained from an occupational inventory and teachers' observational ratings. Principal component factor analysis supported the original hypothesized three dimensions.

Attitudes Toward Science Inventory (Gogolin & Swartz, 1992)

The Attitudes Toward Science Inventory (ATSI) is a 48-item Likert-type instrument measuring both college science majors' and nonmajors' attitudes toward science. It comprises six scales with eight items per scale. The six scales are: perception of the science teacher, anxiety toward science, value of science in society, self-concept in science, enjoyment of science, and motivation in science. Items were based on modification of a previously published instrument. Alpha coefficients for the six scales based on pretest ranged from 0.73 to 0.90; and from 0.77 to 0.88 based on post-test. Item-scale correlation was all above 0.3 with the exception of one item.

Attitude Toward Science Questionnaire (Parkinson, Hendley, Tanner, & Stable, 1998)

The Attitude Toward Science Questionnaire (ASQ) is a 33-item 5-point Likert scale instrument for measuring 13–14-year-olds' attitudes to science. The development of items started with asking approximately 100 students to write down statements about science and their science lessons. Those statements that matched the intended measurement objectives were then used to create item statements. Half of the statements were phrased positively and the other half negatively. Principal component factor analysis revealed that there were six factors; they are: enjoyment, level of difficulty, importance, reading and writing, practical work in science, and time. The number of items per factor ranged from 3 to 9. The overall Cronbach's alpha coefficient for the final questionnaire was 0.92. A sample of 72 pupils representing those with high, medium, and low attitude toward science was interviewed to provide additional evidence of validity for the instrument.

Secondary School Students' Attitude Toward Science (Francis & Greer, 1999)

This is a 20-item unidimensional instrument arranged for scoring on a 3-point Likert scale. The 20 items were selected from the original 60 items based on principal component factor analysis. The content validity of the scale was supported by the observation that items recording the largest item-scale correlation were clearly central to the domain of affective science-related attitudes. A statistically significant correlation was found between students' attitude scores on the instrument and their numbers of science-related subjects studied. The alpha coefficients for different grades of students ranged from 0.88 to 0.91.

Attitude Toward Science (Pell & Jarvis, 2001)

This instrument assesses attitudes to science of 5- to 11-year-old children. Items are stated in the five-point "smiley" face Likert-scale format with only positively worded statements. The instrument includes three main scales: (a) being in school, (b) science experiments, and (c) what science is. Pilot-testing studies helped revise items and scales. Five subscales were derived: (a) liking school, (b) independent investigator, (c) science enthusiast, (d) social context, and (e) difficult science. The alpha reliability coefficients were 0.65, 0.63, 0.74, 0.68, and 0.63. There was a statistically significant correlation among subscales *a* to *d*. Items of *a* to *d* together (25 items) formed the Science Interest scale with an alpha coefficient of 0.82.

The Attitude Scale (Kesamang & Taiwo, 2002)

The Attitude Scale (AS) measures Botswana junior high school students' attitudes toward science. An initial set of 26 items covering a wide range of areas related to likes and dislikes of science and the importance of science to mankind were created, and an expert-panel review of the items resulted in deletion of 6 items. The 20-item instrument was then pilot tested with 60 Grade 7 students. Principal component factor analysis revealed two factors related to "likes of science" and "dislikes of science." The split-half reliability was found to be 0.75 after the Spearman-Brown adjustment.

Shark Attitude Inventory (Thompson & Mintzes, 2002)

The Shark Attitude Inventory (SAI) contains 39 Likert-type questions asking students to state their agreement by selecting one of *strongly agree, slightly agree, neutral, slightly disagree,* and *strongly disagree.* The initial set of 48 questions was created based on modification of propositions in a published instrument on attitudes toward animals. Responses by a sample of nonscience majors taking a college introductory biology course were used to refine the instrument. Principal component factor analysis revealed four factors with an eigenvalue equal to or greater than 2. The four factors were labeled as utilitarian/negative, naturalistic, scientific, and moralistic. The Cronbach's alpha for the above four subscales were 0.43, 0.73, 0.65, and 0.36.

Chemistry Attitudes and Experiences Questionnaire (Dalgety, Coll, & Jones, 2003)

The Chemistry Attitudes and Experiences Questionnaire (CAEQ) is a survey instrument for first-year chemistry students. It was based on a modified theory of planned behavior that involves the following constructs: attitude toward chemistry, chemistry self-efficacy, and chemistry learning experiences. The attitude construct includes four subscales: attitude toward chemists, attitude toward chemistry knowledge, attitude toward chemistry methods, and attitude toward chemistry values. Initial items (131) for the subscales came from previously published attitude toward science instruments and were reduced to 20 items (5 per subscale) after interviews with chemistry students and faculty. The chemistry self-efficacy construct includes four subscales: learning chemistry theory self-efficacy, applying chemistry theory self-efficacy, learning chemistry skills self-efficacy, and applying chemistry skills efficacy. Initial items (61) for the self-efficacy subscales came from a similar instrument in biology, and 20 items, or 5 per subscale, resulted after interviewing chemistry faculty and students. The learning experiences construct includes seven subscales based on different classes and instructors: lecture learning experiences, tutorial learning experiences, laboratory learning experiences, laboratory books learning experiences, lecturer learning experiences, tutor learning experiences, and demonstrator learning experiences. Initial items (70) were selected from other related instruments, and interviews of faculty and students resulted in 35 items (5 per subscale). The attitude and self-efficacy items were presented in a semantic differential format with two bipolar adjectives and 7 points in-between. The learning experience items were stated in a Likert-scale format with five categories from *strongly disagree* (1) to *strongly agree* (5).

A pilot study involving 129 first-year chemistry students from one institution in New Zealand was conducted, and principal component factor analysis revealed the same number of factors as the defined subscales. Nineteen students who completed the survey were also interviewed about their perceptions of the survey. Based on the pilot study results, the instrument was finalized for validation with a larger sample from two institutions. The final instrument contains 21 items on attitude toward chemistry, 17 items on self-efficacy, and 31 items on learning experiences. Principal component factor analysis was conducted again to confirm the construct validity of the subscales. The averaged Cronbach's alpha reliability coefficient over all the subscales was 0.74 at the start of the year ($n = 332$) and 0.84 at the end of the semester ($n = 337$). Chemistry majors had a statistically significantly more positive attitude toward chemistry, higher chemistry self-efficacy, and more positive learning experiences than nonmajors at both administrations of the survey. Moreover student learning experiences had a statistically significant correlation with students' attitude toward chemistry and self-efficacy in chemistry.

Changes in Attitudes About the Relevance of Science (Siegel & Ranney, 2003)

Changes in Attitudes About the Relevance of Science (CARS) is a 59-item Likert-scale survey of middle and high school students' attitudes toward the relevance of science to everyday life. The 59 items are divided into three equivalent forms with 17 unique items plus 8 common items for all three forms. The formation of three equivalent forms was based on Rasch difficulty estimates of items. Most items fit the Rasch model well. Alpha reliability coefficient for each of the forms was above 0.8. The results also showed that CARS was useful to measure the effects of innovative science curriculums on students.

Attitude Toward Chemistry (Salta & Tzougraki, 2004)

Attitude Toward Chemistry (ATC) is a 23-item Likert-scale survey of high school chemistry students' attitudes toward chemistry. It contains four subscales: the difficulty of chemistry courses (6 items), the interest in chemistry courses (9 items), the usefulness of chemistry courses for future careers (3 items), and the importance of chemistry for student life (5 items). Development of the instrument went through three stages taking place in Greece. Stage 1 involved an expert panel review of items to ensure content validity. Pilot testing was used to revise items and study preliminary properties of the instrument. The main study was used to establish construct validity through exploratory factor analysis and reliability through computing Cronbach's alpha coefficients. The four factors explained 47% of total variance. The alpha coefficients for the four subscales were 0.87, 0.89, 0.71, and 0.67 respectively.

Attitude Toward Critical School Science Activity and Attitude Toward Progressive School Science Activity (Zacharia & Calabrese Barton, 2004)

Attitude Toward Critical School Science Activity (ATCSSA) and Attitude Toward Progressive School Science Activity (ATPSSA) are two parallel forms of an instrument for measuring urban middle school students' attitudes toward progressive school science (PSS) and critical school science (CSS). The PSS is represented by current major reform documents such as National Science Education Standards (NRC, 1996), while the CSS is represented by such science curriculum emphases as critical science, feminist science, and multicultural science. Each form contains 13 items in both close-ended (i.e., four point Likert-scale format) and open-ended formats. The 13 items in each form respond to a scenario reflective of either PSS or CSS. Principal component factor analysis showed that items 9–13 highly loaded on one dominant factor, that is, attitude toward either PSS or CSS.

Cronbach's alpha was above 0.90 for both forms. Inter-rater reliability for open-ended questions was all above 0.90. Results showed that some students had a negative attitude toward PSS but a positive attitude toward CSS.

The Colorado Learning Attitude About Science Survey (Adams, Perkins, Podolefsky, Dubson, Finkelstein, & Wieman, 2006)

The Colorado Learning Attitude About Science Survey (CLASS) is a 42-item Likert scale measuring high school and college physics students' attitude toward science. Validation involved interviewing both students (34) and physics experts (16), and factor analysis of pilot-testing data. Reliability was established by comparing surveys of two different physics classes in different semesters. Results were very similar for both calculus-based and algebra-based physics classes.

Environment Literacy Instrument for Korean Children (Chu, Lee, Ko, Hee Moon, Min, & Hee, 2007)

Environment Literacy Instrument for Korean Children (ELIKC) measures third grade (8–9 years old) Korean students' environmental literacy. It includes the following scales: Environmental Knowledge (24 items), Environmental Attitude (22 items), Behaviors to Solve Environmental Problems (16 items), and Skills to Solve Environmental Problems (7 items). The knowledge and skill scales use multiple-choice questions, while attitude and behaviors scales use the five category Likert-scale questions. The draft instrument including both multiple-choice and open-ended questions was pilot-tested with 250 second grade students at the end of the school year. Items were reviewed by four science education professors and three environmental education researchers for content validity. The final version was administered to 969 third-grade students. The Cronbach's alphas for the four scales were 0.65 (knowledge), 0.81 (attitude), 0.74 (behaviors), and 0.46 (skills).

Affective Characteristics Scale (Gungor, Eryrilmaz & Fakroglu, 2007)

Affective Characteristics Scale (ACS) is a set of 53 items scored on a 5-point Likert scale with each item labeled from 1 (*strongly disagree*) to 5 (*strongly agree*). It measures secondary school physics students' attitude toward science. It includes 10 subscales: importance of physics, personal interest, situational interest, extracurriculum activities, student motivation, achievement motivation, self-efficacy, self-concept, course anxiety, and test anxiety. Initial items came from various published relevant instruments in science and mathematics. Translation from English into Turkish was conducted with the assistance of a language expert. An expert panel review and exploratory factor analysis during the pilot study resulted in the revision and deletion of some initial items. Cronbach's alpha coefficients for the 10 subscales based on the final 53 items ranged from 0.84 to 0.92.

Attitude to Science Measures (Kind, Jones, & Barmby, 2007)

Attitude to Science Measures (ASM) assesses secondary school students' attitudes toward science. It has the following subscales: learning science in school, self-concept in science, practical work in science, science outside of school, future participation in science, importance of science, general attitude towards school, and combined interest in science. The instrument uses a Likert-scale format, with each measure being made up of a series of statements asking students to state their level of agreement to the statements by choosing one response from *strongly agree, agree, neither agree nor disagree, disagree,* and *strongly disagree.* Statements were asked to capture various attributes of the attitude object and express different evaluative dimensions.

Reliability analysis and factor analysis were conducted on pilot study data; items that reduced the internal reliability of attitude measures or did not group together with other items were either removed from the measures, or their wording was modified. In the final validation study, the revised questionnaire was given twice to students: 2 weeks before and 2 weeks after a university mobile lab outreach program.

Principal component factor analysis using oblique rotation on seven factors found that the extracted factors did indeed correspond to the seven expected areas of attitudes toward science. However, one item "scientists have exciting jobs" did not load on any of the factors, and this item was removed from the final instrument. Further principal component factor analysis on items of individual subscales found that each subscale was unidimensional. For all the attitude subscales, the Cronbach's internal reliability coefficient was above 0.7. Correlation between the seven subscale measures ranged from 0.3 to 0.7.

Science Achievement Influences Survey (Odom, Stoddard, & LaNasa, 2007)

Science Achievement Influences Survey (SAIS) is a 31-item survey to measure middle school students' self-reported attitudes toward science (in a Likert-scale format from *strongly agree* to *strongly disagree*) and frequency of exposure to classroom teacher practices and peer and home support (in a five option rating scale from *less than once a month* to *more than once a week*). Factor analysis using principal component extraction with varimax rotation identified five factors: attitudes about science (7 items), peer participation

(4 items), student-centered teaching practices (7 items), home support (4 items), and teacher-centered teaching practices (4 items). Cronbach alpha coefficients ranged from 0.607 to 0.883 for the above subscales.

Attitude to Science Instrument (shortened version; Caleon & Subramaniam, 2008)

ASI, shortened version, is a 17-item Likert scale survey of upper elementary school students' attitudes toward science. The questions include three of five subscales from its original instrument with minor modifications in presentation formats (e.g., using all positive statements and smiley faces). The three subscales correspond to enjoyment of science—interest in out-of-school science activities, social implications of science—the value students give to science in improving lives, and career preference for science—preference, interest, and enjoyment in science careers. There are six items for the first two subscales and five items for the last subscale. The Flesch-Kincaid readability level of the instrument is 3.9. Cronbach's alpha internal consistency coefficients were 0.77 for enjoyment of science, 0.72 for social implications, 0.87 for career preference, and 0.89 overall. A statistically significant difference was found concerning enjoyment of science between gifted students and non-gifted students, and between above-average students and average-achievement students. Similarly, a statistically significant difference was found in both career preference and social implications between gifted students and average students, and between above-average students and average students.

Attitude Toward Chemistry Lessons Scale (Cheung, 2009)

The construction of the Attitude Toward Chemistry Lessons Scale (ATCLS) was based on a theoretical model with four dimensions: liking for chemistry theory lessons, liking for chemistry laboratory work, evaluative beliefs about school chemistry, and behavioral tendencies to learn chemistry; it was modified from the Test of Science-Related Attitudes (TOSRA) developed by Fraser (1978). The 20-item trial version of ATCLS was pilot tested on a convenience sample of 777 Secondary 4–7 chemistry students in Hong Kong, and the 12-item final version of ATCLS was administered to 954 Secondary 4–7 chemistry students in Hong Kong on a seven-point Likert-scale with labels of *strongly disagree, moderately disagree, slightly disagree, not sure, slightly agree, moderately agree,* and *strongly agree.* A semi-structured interview with 10 Secondary 4–7 students randomly selected from 10 chemistry classes in two secondary schools in Hong Kong was conducted first to clarify students' ideas and to pose follow-up questions. Two science educators were invited to determine the content validity by classifying the 20 items to the four dimensions. Based on the data of the final version, all of the 12 item-total correlations were moderately positive ranging from 0.55 to 0.76; and alpha reliability coefficients were 0.86, 0.84, 0.76, and 0.76 for the four dimensions. Results of the confirmatory factor analysis of data based on the final version showed that the standardized factor loadings were reasonably high and statistically significant on the expected four dimensions.

Asian Student Attitudes Toward Science Class Survey (Wang & Berlin, 2010)

The Asian Student Attitudes Toward Science Class Survey (ASATSCS) is an instrument designed to measure the attitudes toward science class of fourth- and fifth-grade students in an Asian school culture. The development focused on three science attitude components: science enjoyment, science confidence, and importance of science as related to science class experiences.

The first version of the ASATSCS contained 32 items, including 16 positive and 16 negative statements. The draft of the ASATSCS was written in English and then translated into Chinese. The researcher translated the English version of the survey into Chinese and then it was back-translated by a bilingual professional translator. An expert panel of two science educators, two science teachers, and one doctoral student in science education evaluated the instrument for content or face validity. The initial version of ASATSCS was administered to 265 fourth- and fifth-grade students from three elementary schools in Taiwan. A principal components analysis was first performed to identify clusters of items that had variance in common, and the examination of the scree plot suggested that an extraction of one component accounted for 32.91% of the variance. The loadings of the 30 items ranged from 0.34 to 0.74 and therefore met the criteria for adequate loadings. Two items did not reach the criterion of 0.33 and were consequently deleted. The final version of the ASATSCS includes 30 items, and all items in the ASATSCS use a 5-point Likert-type response format from *strongly disagree* to *strongly agree.* A Cronbach's alpha coefficient for internal consistency revealed an overall instrument reliability estimate of 0.93.

Genetically Modified Organisms Attitudes Scale (Herodotou et al., 2012)

Genetically Modified Organisms Attitudes Scale (GMOAS) is a 16-item Likert scale instrument for measuring secondary school students' attitudes towards GMOs. Development of the instrument was initiated with a thorough review of the literature on GMOs and biotechnology; this process resulted in 78 items. The majority of these items were adapted from existing instruments, with a smaller number of new items drafted by the authors. The initial pool of items was reviewed by an expert panel consisting of university researchers, secondary biology teachers, and primary school teachers; this resulted in 31 items to form the initial version of the instrument. The initial version was then administered to 1,111 secondary school students (Grades 10 through 12) in Cyprus; the valid sample size was 975. Principal component analysis with varimax rotation resulted in a five-factor solution explaining 45.08% variance. After a few rounds of eliminating non-fitting items, the final principal component analysis of 16-item sample data resulted in a three-factor solution explaining 45.84% variance. The three factors were labeled as: GMO implications on health, interest in the topic of GMOs, and GMO implications on the environment. The Cronbach's alpha for the three subscales ranged from 0.58 to 0.78; the alpha for the entire scale was 0.78.

Animal Welfare Attitude Scale (Mazas, Manzanal, Zarza, & María, 2013)

Animal Welfare Attitude Scale (AWA) measures attitudes of secondary-school and university students towards animal welfare. Items are in Likert-scale. Four components are identified to define the construct of the attitude towards animal welfare; they are: animal abuse for pleasure or due to ignorance, leisure with animals, farm animals, and animal abandonment. The initial version of the scale of 60 items was judged by an expert panel for content validity. Then the scale was given to 61 students who were asked to examine it critically and interviews were held. The final version of the scale contains 29 items. A sample of 329 students, aged between 11 and 25 from secondary schools, and a university in Spain were used to validate the scale.

The items in the final version of the scale had item-total correlations between 0.27 and 0.60. The Cronbach's alpha was 0.89. In exploratory factor analysis, five major factors with eigenvalues greater than 1 were found. The proportion of variance explained by the first factor had a value of 26.58%. The second factor extracted had an explained variance of 8.8% and the third factor, 5.4%. When applied to 1,007 students in a following study, the scale had a Cronbach's alpha ranging from 0.69 to 0.84 for the four subscales.

Arabic-Speaking Students' Attitudes Toward Science Survey (Abd-El-Khalick, Summers, Said, Wang, & Culbertson, 2015)

Arabic-Speaking Students' Attitudes toward Science Survey (ASSAS) is an instrument to measure Qatari student attitudes toward science in Grades 3–12. Based on the theory of reasoned action and planned behavior

(TRAPB; Ajzen & Fishbein, 2005), a total of 74 five-point Likert-scale items were adopted or revised from existent instruments. The initial pool of items was reviewed by a panel of 10 international experts to establish face and content validity, resulting in a 60-item initial version of ASSAS. The initial version was piloted with 369 Qatari students. Exploratory factor analysis was conducted on the pilot data, suggesting a five-factor solution. The results supported a 5-factor model for the scale; the five factors are: attitudes toward science and school science, unfavorable outlook on science, control beliefs about ability in science, behavioral beliefs about the consequences of engaging with science, and intentions to pursue science. Item difficulties were also reviewed for different grades. Thirty-five of the initial 60 items were retained and used for a validation study. The revised version was administered to a sample of 3,027 students in Grades 3–12 in various types of schools. Confirmatory factor analysis was conducted on the validation sample data. After removing 3 items loading on multiple factors, the analysis showed that the 32-item instrument had a good model-data-fit. The Cronbach's alpha reliability coefficients for the five factors ranged from 0.61 to 0.81.

Attitude Profiles Toward Evolution–Creation Controversy (Konnemann, Asshoff, & Hammann, 2016)

A set of four scales were developed to measure student attitude profiles toward evolution-creation controversy. The profile consists of attitudes toward the evolutionary theory, attitudes toward the biblical accounts of creation, attitudes toward creationist beliefs, and attitude toward scientific beliefs. A small exploratory study with open-ended and closed-ended questions was conducted to generate items of the scales ($n = 31$; Grades 10 and 12), followed by a series of studies that focused on the development, pilot testing, and validation of different scales ($n = 842$; Grades 9–13). The main validation study was conducted with 1,672 German students (Grades 10–13). All scales contain both positive and negative scoring items; all Likert-scale items used four points of *strongly agree* (4), *agree* (3), *disagree* (2), and *strongly disagree* (1). Classical analysis of items and scales (i.e., item means, standard deviation, item discrimination, and internal consistency) was conducted. Multidimensional Rasch modeling was conducted to test dimensionality. There was a positive intercorrelation between attitudes toward evolutionary theory and scientific beliefs, between attitudes toward biblical accounts of creation and creationist beliefs, negative correlation between attitudes toward evolutionary theory and the biblical accounts of creation, between attitudes toward the biblical accounts of creation and scientific beliefs, and between scientific and creationist beliefs. The final version of the attitudes toward evolutionary theory (AE) had 23 items; attitudes toward biblical accounts of creation (AC) had 17 items, creationist beliefs (CR) had 14 items, and scientific beliefs (SC) had 10 items. The Cronbach's alphas for the above scales were 0.92, 0.94, 0.94, and 0.83.

Behaviors, Related Attitudes, and Intentions Toward Science (Summers & Abd-El-Khalick, 2018)

The scale of Behaviors, Related Attitudes, and Intentions toward Science (BRAINS) was based on the theory of reasoned action and planned behaviors (TRAPB). Initial development of items for the constructs underlying TRAPB began with a review of previous widely used attitude instruments and appropriate items were selected for the constructs to be measured. A total of 62 items were identified; 16 of them were modified and 12 additional items were created. The 74 items with a 5-point response scale from *strongly disagree* to *strongly agree* formed the initial version of the instrument. Expert panel review of the initial item pool was conducted to establish content validity. An online administration with audio playing of questions was implemented. The online version was piloted with 151 students from third to seventh grade students, and a subsample of the surveyed students were selected for responding to a few open-ended questions about the survey. The revised 59-item instrument was administered to a representative, random sample of 1,291 Illinois students in Grades 5 through 10. Confirmatory factor analysis and subsequent refinement resulted in a 30-item instrument with five factors consistent with the TRAPB framework: attitudes toward science, behavioral beliefs about science, intentions to engage in science, normative beliefs, and control beliefs. Reliabilities based on CFA were 0.91 for attitude toward science, 0.88 for intention, 0.82 for behavioral beliefs, 0.87 for control beliefs, and 0.70 for normative beliefs.

Attitude Toward Science (Syed Hassan, 2018)

Attitude toward Science was developed in Malaysian Secondary school context. It contains 20 items in a 5-point Likert scale format. Attitude toward science is conceptualized as consisting of the following components: well-being—favorable attitude towards life and creator, student engagement in subject, knowledge about the importance of career in science, scientific inquiry through ICT integration, and supportive learning environment. The initial instrument included 30 items. The validation sample was 350 lower secondary students (age 13 to 15 years) from three schools. Confirmatory factor analysis was used to validate the instrument. The data fit the model adequately. All items had good dispersion with standard deviation ranging from 0.81 to 1.12. Reliability for all 30 items was 0.98. However, only the 20 best items conforming to the five factor model of science attitude were selected to form the instrument; the other 10 items not conforming to the model were eliminated. Correlation between factors ranged from 0.447 to 0.808.

Instruments for Measuring Students' Motivation, Interest, and Self-Efficacy

Scientific Curiosity Inventory (Campbell, 1971)

Scientific Curiosity Inventory (SCI) measures junior high school students' curiosity about science. Items correspond to the first three levels of the Krathwohl's affective domain taxonomy, that is, receiving, responding, and valuing. The questions ask students to indicate how far they would go to satisfy their curiosity about a set of content questions related to physics, chemistry, biology, and earth science. Validation of the instrument involved an expert panel review and pilot-testing with a heterogeneous junior high school sample. Items included in the instrument had biserial correlation coefficients from 0.39 to 0.69 with a mean value of 0.54. The adjusted split-half reliability coefficient was 0.89, and standard error of measurement was 1.66.

Self-Efficacy Instrument (Baldwin, Ebert-May, & Burns, 1999)

The Self-Efficacy Instrument (SI) is a 23-item rating-scale survey for measuring university non-biology majors' self-efficacy related to biology. Each item asks students to think about how confident he/she feels in carrying out a given task by selecting from *totally confident, very confident, fairly confident, only a little confident,* and *not at all confident.* The development of the instrument underwent three phases: developing items, identifying dimensions of the biology self-efficacy, and establishing criterion-related validity. The item development phases involved review of literature, interview of select students, and pilot-testing of the draft instrument. The revised instrument was then pilot tested again with a bigger university non-biology major sample. Principal component factor analysis suggested that there were three dominant factors as expected. Factor 1 related to students' sense of perceived confidence in writing and critiquing his/her biology ideas through laboratory reports as well as using analytical skills to conduct experiments in biology; Factor 2 related to perceived confidence in generalizing skills learned from one biology course to other biology/science courses; Factor 3 related to students' perceived confidence in his/her ability to apply biological concepts and skills to everyday events. The alpha internal consistency coefficients for the above three scales were 0.88, 0.88 and 0.89. Correlation between scores on the SI and scores on the biology achievement tests was small (0.18 to 0.27), demonstrating good discriminant validity.

Social-Cultural Scale (Kesamang & Taiwo, 2002)

The Social-Cultural Scale (SCS) measures Botswana junior high school students' Setswana customs and beliefs as they relate to identifiable practices at home, at the lands and at the cattle posts, as well as beliefs about death, rainfall, drought, and so forth. A total of 30 Likert-scale items were initially constructed. A five-member expert panel reviewed the items, resulting in the removal of 10 items. A pilot study involving 60 Grade 7 students revealed that the instrument had a split-half reliability of 0.86 after the Spearman-Brown adjustment. Principal component factor analysis resulted in six factors.

Cognitive Conflict Levels Test (Lee et al., 2003)

The Cognitive Conflict Levels Test (CCLT) is a 12-item classroom survey of high school students' levels of cognitive conflict. Cognitive conflict is defined by four components; they are recognition of contradiction, interest, anxiety, and cognitive reappraisal of situation. There are 3 items for each of the above four components; each item consists of a statement (e.g., the result of the demonstration confuses me) and a five-choice rating scale from 0 (*not at all true*) to 4 (*very true*). A six-expert panel reviewed the items. Development of the instrument went through three pilot-tests, all taking place in South Korea. Pilot-Test 1 involved 152 10th grade students plus an interview of selected students; Pilot-Test 2 involved 88 10th grade students as well as interview of a few selected students. Items were revised based on content review, item analysis, construct validation, and reliability review. The final pilot-test involved 279 10th grade students. All pilot-tests followed the following procedures: (a) assessment of student preconceptions and strength in their beliefs (about 7 minutes), (b) demonstration of an anomalous situation (about 2 minutes), and (c) administration of CCLT (about 4 minutes). Factor analysis showed that there were four dominant factors in student responses to CCLT; the four factors explained over 72% total variance. Correlation between the four subscales ranged from –0.04 to 0.48. Cronbach's alpha internal consistent coefficients for the four subscales ranged from 0.69 to 0.87.

Students' Motivation Toward Science Learning (Tuan, Chin, & Shieh, 2005)

Students' Motivation Toward Science Learning (SMTSL) is a 35-item questionnaire measuring secondary school students' motivation to learn science. It contains six subscales: (a) self-efficacy (7 items), (b) active learning strategies (8 items), (c) science learning value (5 items), (d) performance goal (4 items), (e) achievement goal (5 items), and (f) learning environment stimulation (6 items). All items are in a Likert-scale format, with 1 for *strongly disagree*, 2 = *disagree*, 3 = *no opinion*, 4 = *agree*, and 5 = *strongly agree*. Validation took place in Taiwan with a sample size of 1,407. Principal component factor analysis found six factors as expected. Cronbach's internal consistency reliability coefficients for the six subscales ranged from 0.7 to 0.91; the alpha for the entire instrument was 0.91. Correlation coefficient among the subscales ranged from 0.09 to 0.51. In addition, students' SMTSL scores were found to be statistically significantly correlated with their attitude toward science scores and with their science achievement scores.

Intrinsic Motivation for Learning Science (Juriševi, Glažar, Puko, & Devetak, 2007)

The 125-item Intrinsic Motivation for Learning Science (IMLS) questionnaire assesses intrinsic motivation for learning biology (IMLS biology), physics (IMLS physics), and chemistry (IMLS chemistry), as well as general intrinsic motivation for studying (IMLS general learning), and motivations for learning mathematics (IMLS mathematics) and foreign languages (IMLS foreign language). In the part of the IMLS for chemistry, special attention is directed to the assessment of students' intrinsic motivation for learning chemical concepts on the three levels of chemical representation (i.e., macro, submicro, and symbolic). Each item is stated in a Likert-scale format, with 1 = *strongly disagree*, 2 = *disagree*, 3 = *sometimes disagree/sometimes agree*, 4 = *agree*, and 5 = *strongly agree*. The internal consistency (Cronbach α) of the IMLS was 0.78. The subscales measuring intrinsic motivation for learning different levels of chemical concepts correlated statistically significantly ranging from 0.21 to 0.60 ($n = 140$). Student intrinsic motivation for learning chemistry was statistically significantly correlated with students' understanding of chemistry concepts ($r = 0.30$). Statistically significant differences were found between intrinsic motivation for learning chemistry and intrinsic motivation for learning biology and physics; a statistically significant difference was also found between intrinsic motivation for learning chemistry and general intrinsic motivation for learning.

Affective Assessment in Cell Biology (Kitchen ey al., 2007)

Affective Assessment in Cell Biology (AACB) is a set of three instruments for assessing three affective variables related to a university cell biology course. The three instruments are: (a) attitude toward college courses that focus on the development of thinking skills versus courses that focus on factual recall, (b) self-efficacy toward understanding basic concepts in cell biology and solving problems that require the interpretation of experimental data, and (c) interest in learning various concepts and skills in cell biology. The attitude instrument contains 10 adjective pairs forming a semantic differential scale. The scale is presented in two sets, one referring to the development of thinking skills, and another to the focus on factual recall. Item-total correlation ranged from 0.63 to 0.79 for items in the analytic set, and from 0. 25 to 0.70 for items in the recall set. The overall internal consistency reliability was 0.92 for the analytic skill set and 0.87 for the recall set. Principal component factor analysis revealed that the majority of items loaded on one factor.

The self-efficacy (confidence) instrument assesses students' self-efficacy regarding cell biology based on seven tasks: (a) a graphical figure on conventional coordinates, (b) an electropherogram, (c) a textbook figure, (d) a table with several columns of experimentally derived numbers, (e) a page of text, and (f) and (g) the title pages from two published papers on cellular biology topics. The tasks represent what practicing biologists engage daily in research. Students are asked to rate their confidence on a 5-point scale from 0 (*not confident at all*) to 4 (*extremely confident*) in either reading or comprehending the item (textual excerpt) or drawing conclusions from the data presented in graphs and tables. The item-total correlation was above 0.6 for six items and 0.48 for one item. Internal consistency reliability was 0.87.

The interest instrument contains two sets of 11 items, each measuring interest in cell biology topics and interest in cognitive reasoning skills in cell biology. The item-total correlation coefficients for all items in the interest in topics scale were above 0.45, and the internal consistency reliability was 0.93. Similarly, the item-total correlation coefficients for all items in the interest in cognitive skills scale were above 0.50, and the internal consistency reliability was 0.91.

Sources of Science Self-Efficacy Scale (Britner, 2008)

Sources of Science Self-Efficacy Scale was adapted from a scale to measure the same construct in mathematics. It consists of four subscales measuring effects of mastery experiences (8 items), vicarious experiences (5 items), social persuasions (8 items), and physiological states (8 items). The physiological state subscale is also called the science anxiety scale. Exploratory factor analysis was conducted to validate the above subscale structure. Items for each of the subscales loaded on one factor. Loadings for mastery experience items ranged from 0.56 to 0.85, vicarious experience items from 0.49 to 0.85, social persuasion items from 0.63 to 0.86, and physiological state items from 0.61 to 0.87. Cronbach's alpha reliability indices for the above subscales were 0.87 for mastery experiences, 0.81 for vicarious experiences, 0.92 for social persuasions, and 0.92 for physiological states.

Conceptions of Learning Science Questionnaire (Lee, Johanson, & Tsai, 2008)

The Conceptions of Learning Science Questionnaire (COLS) is a 31-item survey on high school student conceptions of learning science. Conceptions of learning science were identified from previous phenomenographic studies, and include views of learning science as: (a) memorizing, (b) preparing for test, (c) calculating and practicing tutorial problems, (d) increasing knowledge, (e) applying, (f) understanding, and (g) seeing in a new way. Initially, six to eight items were constructed for each of the above seven categories, and all items followed a Likert scale format with five categories from *strongly agree, agree, no opinion, disagree,* and *strongly disagree.* An expert reviewed the items for content validity. Construct validity was established through both exploratory and confirmatory factor analyses. Categories of *understanding* and *seeing in a new way* were combined after factor analysis. Cronbach's alpha coefficients for the final six subscales were 0.85, 0.91, 0.90, 0.84, and 0.91, and the alpha for the entire instrument was 0.91.

Intellectual Risk Taking, Interest in Science, Creative Self-Efficacy, and Perceptions of Teacher Support Survey (Beghetto, 2009)

This 18-item survey measures elementary school (Grades 3–6) students on four constructs: intellectual risk taking, interest in science, creative self-efficacy, and perceptions of teacher support. Each item is rated by students from 1 (*not true*) to 5 (*very true*). Intellectual risk taking is measured based on students' reports of engaging in intellectually risky learning behaviors (e.g., sharing tentative ideas, asking questions, willingness to try and learn new things) when learning science. Interest in science is measured based on student content-specific feeling-related (e.g., I like science) and value-related (e.g., Science is important to me) responses. Creative self-efficacy is measured based on student beliefs about their ability to generate novel and useful ideas in science and whether they view themselves as having a good imagination in science. Finally, perceptions of teacher support are related to key teacher support aspects identified in the literature. Principal component analysis found that the above four factors explained 49.5% of the total variance and items loaded highly on their expected factors. Cronbach's alpha coefficients for the above four scales were 0.80, 0.77, 0.83, and 0.77. Student intellectual risk taking scores were statistically significantly correlated with measures of student science abilities, science interest, and creative self-efficacy in science.

Test for Ethical Sensitivity in Science Plus (Fowler, Zeidler, & Sadler, 2009)

Test for Ethical Sensitivity in Science (TESSplus) is an adapted version of the Test for Ethical Sensitivity in Science (TESS; Clarkeburn, 2002) with the addition of one more scenario. It assesses high school students' ability to recognize the moral aspects associated with scientific issues. TESSplus consists of two socio-scientific scenarios which students read and are asked to list up to five possible questions or issues they would raise before reaching a decision about the scenario. One scenario, taken from the original TESS, describes a situation involving the development of pharmaceutical milk using genetically modified cows (genetic modification scenario). The other scenario describes a situation involving reproductive cloning for infertile parents (reproductive cloning scenario). Scoring of the TESSplus is done by rating each response on a scale of 0 to 3 points based on the degree of moral considerations present (0 being *none* to 3 being *strong* moral consideration). Scores from both scenarios are added to provide a measure of moral sensitivity. Two researchers were involved in scoring the TESSplus. Inter-rater consistency was 97%.

Science Motivation Questionnaire (Glynn & Koballa, 2006; Glynn, Taasoobshirazi, & Brickman, 2009)

The Science Motivation Questionnaire (SMQ) is a 30-item Likert-scale instrument for measuring students' motivation to learn science in college science courses. It has the following subscales: (a) intrinsically motivated science learning (5 items), (b) extrinsically motivated science learning (5 items), (c) relevance of learning science to personal goals (5 items), (d) responsibility (self-determination) for learning science (5 items), (e) confidence (self-efficacy) in learning science (5 items), and (f) anxiety about science assessment (5 items). The overall Cronbach's alpha was 0.93.

An initial validity study (Glynn & Koballa, 2006) found that the Cronbach's coefficient alpha was 0.93. The SMQ scores were statistically significantly correlated with student college science GPAs and the student belief that science was relevant to one's career. Each item is rated from 1 (*never*) to 5 (*always*). Out of a total possible maximal score of 150, a score between 30 and 59 is interpreted as *never to rarely* motivated; a score between 60–89 is interpreted as *rarely to sometimes* motivated, a score between 90–119 is interpreted as *sometimes to often* motivated, and a score between 120–150 is interpreted as *often to always* motivated.

A further validation study with non-science majors (Glynn et al., 2009) was also conducted. Principal component factor analysis revealed five factors; they are: intrinsic motivation and personal relevance (10 items), self-efficacy and assessment anxiety (9 items), self-determination (4 items), career motivation (2 items), and grade motivation (5 items). The five factors explained 60% of total variance. Cronbach's coefficient alpha was found to be 0.91. Correlation between SMQ scores and students' college science GPAs, and between SMQ scores and student beliefs on the relevance of science to their careers was statistically significant. Student essays about their motivation to learn science and interviews with 48 students provided additional support to the construct and criterion-related to the instrument.

Science Motivation Questionnaire II (Glynn, Brickman, Armstrong, & Taasoobshirazi, 2011)

Science Motivation Questionnaire (SMQ II) is a 25-item multiple-choice instrument designed to measure the motivation of students to learn science in college courses. It was revised from Science Motivation Questionnaire (SMQ; Glynn & Koballa, 2006; Glynn et al., 2009). Specifically, 16 were from the original questionnaire and 9 were new. SMQ II assesses five motivation components: intrinsic motivation, self-determination, self-efficacy, career motivation, and grade motivation; it targets positive, mutually supporting motivators and includes items about self-efficacy but not anxiety, a negative motivator. In addition, one of the original scales, extrinsic motivation or "learning science as a means to a tangible end," has been transformed into two scales, grade motivation and career motivation. The validation sample included 680 undergraduate students with 367 in science majors and 313 in non-science majors.

Principal component analysis showed that all of the items met the criterion of loading at least 0.35 on their respective factor. The eigenvalue and percent of variance explained by each factor were: intrinsic motivation (35.33%), career motivation (11.31%), self-determination (8.84%), self-efficacy (7.13%), and grade motivation (5.03%). The cumulative percent of variance explained by the factors was 67.64%. The reliabilities (internal consistencies) of the scales, assessed by Cronbach's alphas, were: 0.92 for career motivation, 0.89 for intrinsic motivation, 0.88 for self-determination, 0.83 for self-efficacy, and 0.81 for grade motivation. The overall Cronbach's alpha of all 25 items was 0.92.

Students' Adaptive Learning Engagement in Science Questionnaire (Velayutham, Aldridge, & Fraser, 2011)

The Students' Adaptive Learning Engagement in Science (SALES) questionnaire is a 32-item instrument to measure salient factors related to the motivation and self-regulation of students in lower secondary science classrooms. The development of the SALES questionnaire followed a three-stage approach. Stage 1 involved identifying and defining salient adaptive learning engagement constructs; four constructs were identified: Learning goal orientation, task value, self-efficacy, and self-regulation of effort. Stage 2 involved writing individual items for each of the scales of the four constructs. Ten experienced science teachers were asked to assess the comprehensibility, clarity, and accuracy of items for each scale to establish the face validity. Stage 3 commenced with a pilot study conducted with 52 Grade 8 students, and 12 students were selected for semi-structured interviews. The final version of the survey was administered to 1,360 students from 78 classes in five public schools in the Perth metropolitan area of Australia.

Principal component analysis indicated that there were four significant factors with cumulative variance for all four factors to be 63.2%. All items loaded on one factor with a value above 0.50 and did not load on any other factors. Internal consistency reliability for each factor was above 0.90. The highest between-factor correlation was 0.57, suggesting appropriate discriminant validity. The concurrent validity was evidenced by significant differences among classes based on ANOVA, suggesting that each scale in the SALES survey differentiated significantly between classes. Predictive validity was determined by correlation between each of the SALES scale and students' science achievement using a one-tailed Pearson coefficient, in which all the scales had a statistically significant correlation with students' achievement scores.

Science Motivation Scales (Martin, Mullis, Foy & Stanco, 2012)

The 2011 TIMSS student survey includes subscales measuring three motivational constructs: intrinsic or interest value (student like learning science), extrinsic or utility value (student value science), and ability beliefs (student confidence in science). A total of 20 items were available for the above three constructs; each item was in a 4-point rating scale. Tee and Subramaniam (2018) conducted a Rasch analysis to validate the instrument. They used a sample of 10,447 students from the United States; 3,842 students from England; and 5,927 students from Singapore. Results suggest that most items fit the unidimensional rating scale Rasch model well; a few items were outside the acceptable fit statistics ranges, possibly measuring additional dimensions. The Wright map also showed a few large gaps in item difficulties. No item showed significant DIF between boys and girls. The person separation index was 2.12.

STEM Career Interest Survey (Kier, Blanchard, Osborne, & Albert, 2013)

The STEM Career Interest Survey (STEM-CIS) consists of 44 items, 11 in each of science, technology, engineering, and mathematics disciplines, in 5-point Likert scale format. The initial 30 items related to student self-efficacy, personal goals, outcome expectations, interest, personal inputs, and contextual supports and barriers. The initial instrument was pilot-tested with 61 students. The revised and expanded instrument was administered to 1,061 middle school students in Grades 6–8. Confirmatory factor analysis was conducted to subscales of science, technology, engineering, and mathematics with good model-data fit. The alpha for science subscale was 0.77, mathematics 0.85, technology 0.89, and engineering 0.86. Correlation between the above four subscales ranged from 0.72 to 0.82. Confirmatory factor analysis with one single factor for all items also showed good model-data fit.

Continuing Motivation for Science Activities (Fortus & Weiss-Wedder, 2014)

This questionnaire was based on a series of extracurricular science-related activities reflecting engagement with scientific content or practices, manifested across varying contexts when other alternatives exist and not required by school or faculty. Only activities adolescents may be engaged in independently were included. Items were responded by selecting among the following categories: *not true at all, not so true, somewhat true, true,* and *very true.* Initial items were tested with three science education researchers. The items were then validated using the cognitive pretesting procedure with Grades 5–7 students. The final version of the instrument includes 19 items covering the continuum from rejecting extracurricular science activities to embracing extracurricular science activities. The final version was given to 2,958 students from Grades 5 through 8 in both traditional and democratic schools in Israel. Rasch modeling was used to establish validity and reliability of the scale; both fit statistics and category thresholds were reviewed.

McGill Self-Efficacy of Learners for Inquiry Engagement (Ibrahim, Aulls, & Shore, 2016)

McGill Self-Efficacy of Learners for Inquiry Engagement (McSELFIE) is a 60-item survey to measure inquiry engagement self-efficacy of undergraduate students in natural science disciplines. Inquiry engagement is defined as carrying out the practices of science. Content validity was established with a sample of experts, and construct validity was established by confirmatory factor analysis with a sample of 110 undergraduate students. Three types of self-efficacy for inquiry engagement were conceptualized and confirmed by confirmatory factor analysis: the self-efficacy for the practices of science (POS), self-efficacy for achieving inquiry-learning outcomes (ILOs), and self-efficacy for demonstrating student personality characteristics (SPCs). The Cronbach's alphas for the three scales of the constructs were all above 0.95.

Self-determination, Purpose, Identity, and Engagement in Science (Skinner, Saxton, Currie, & Shusterman, 2017)

Self-determination, Purpose, Identity, and Engagement in Science (SPIRES) contains a set of measures related to undergraduate students' self-systems processes, engagement/disaffection, and identity as a scientist. It includes 11 short-form scales containing 4–5 items each. All items used a 5-point Likert scale ranging from *not true at all* (1) to *totally true* (5). An iterative confirmatory process was used to reduce the item pool for each measure from 6–12 to 4–5 items. The development of the instrument was informed by the self-determination theory (SDT). The SDT assumes that students come to university classes with fundamental psychological needs, intrinsic to all humans, whose fulfillment provides the motivation to engage in learning. The SDT focuses on three needs: (a) competence—the need to feel efficacious and capable, (b) autonomy—the need to experience one's true self, and (c) relatedness—the need to connect deeply with others and to belong. Participants were undergraduates enrolled in eight science courses in biology, chemistry, and physics. There were 856 students at Time 1 early in the semester and 574 students at Time 2 toward the end of the semester. Motivational scales were adapted from previously published measures; new items were also created as needed. Confirmatory factor analysis and measurement invariance analyses were performed in R software environment. Results showed that of 11 scales, all but two (i.e., competence and relatedness) demonstrated adequate properties of unidimensionality on at least one time point. Cronbach's alphas ranged from 0.67 to 0.91 at Time 1 and from 0.73 to 0.94 at Time 2. Six scales met criteria for strong evidence of invariance across disciplines. All three self-systems measures were positively correlated with both behavioral and emotional engagement measures and negatively correlated with both behavioral and emotional disaffection. The components of identity as a scientist were positively correlated with each other. Finally, the motivational measures were correlated significantly with students' final course grades at T2.

Young Children's Science Motivation (Oppermann, Brunner, & Eccles, 2018)

Young Children's Science Motivation (YCSM) is a 28-item one-to-one interview instrument. Young children's motivation is differentiated into outcome expectancy and subject task value beliefs. Outcome expectancy beliefs focus on children's self-confidence in science related activities;

subject task value beliefs focus on children's intrinsic values of science learning. Four rounds of pilot testing were conducted involving 18, 6, 9, and 55 children in preschool centers in Berlin, Germany. The final version of the instrument includes 28 items, 8 items on children's self-confidence in life science, 7 items on their self-confidence in physical science, 7 items on children's enjoyment in life science, and 6 items on their enjoyment in physical science. Items for self-confidence include four options from *very well* to *not at all*; items for enjoyment include four options from *very much* to *very little*. The options are presented together with a diagram of increasing size. In addition to the visual response format, small hand puppets are used to guide children through the interview. Administration of YCSM was conducted as one-on-one interview. The final validation study involved 283 preschool children from 22 centers with and 24 centers without an explicit science focus. The structure of children's motivational beliefs was investigated using confirmatory factor analysis, confirming the two factor structure. Cronbach's alpha for the self-confidence scale was 0.87, and for the enjoyment scale was 0.86. Based on CFA results, a shorter version of the instrument, with 5 items for each scale, was created. Confirmatory factor analysis also confirmed the two factor structure of the short version. Cronbach's alpha for the self-confidence scale was 0.68 and for the enjoyment scale was 0.74 for the shorter version.

Instruments for Measuring Teacher Affective Variables

Attitude Toward Science (Button & Stephens, 1963)

Attitude Toward Science (ATS) measures preservice elementary teachers' attitudes toward elementary school science. The development of the instrument followed the approach to developing an equal interval Thurston scale. One hundred subjects sorted the 50 statements into 11 groupings from 1 (*dislike*) to 11 (*like*). Categories 1–5 were considered unfavorable, 6 neutral, and 7–11 favorable. Q values (distance between the first quartile and the third quartile) were used to select the final 20 items to form the instrument. The final 20 items met the following requirements: (a) low Q values, (b) normal distribution of scale values from 1 through 11, and (c) an equal number of favorable and unfavorable statements. The test–retest reliability coefficient was 0.93.

An Attitude Inventory (Hoover & Schutz, 1963)

This attitude inventory is a 54-item survey for measuring science teachers' attitudes toward environmental conservation. Each item presents a brief hypothetical situation to which the subject might react on a 5-point

Likert scale. Pilot-testing involved 104 science teachers. Cluster analysis found 16 clusters. However, only 3 of 16 clusters had a KR-20 reliability greater than 0.70. Cluster 1 (18 items) was interpreted as assistance for the common good (KR-20 = 0.93); Cluster 2 (20 items) was interpreted as regulation for the common good (KR-20 = 0.93); and Cluster 3 (16 items) was interpreted as private rights versus conservation (KR-20 = 0.87).

Beliefs About Nature of Science, Nature of Children, and Role of a Teacher (Good, 1971)

This instrument measures elementary teachers' beliefs about nature of science, nature of children, and the role of a teacher. The initial version was tried with prospective elementary teachers in three quarters as both pre- and post-tests; items that showed significant increase from pre- to post-course were retained. The final version contains 30 items. Each item has five possible responses ranging from (1) *strongly agree* to (5) *strongly disagree.*

Checklist for Assessment of Science Teachers (Brown, 1973)

Checklist for Assessment of Science Teachers (CAST) has two forms: the supervisor's perceptions form (CAST: SP), and the pupil's perceptions form (CAST: PP). The CAST: SP consists of three subscales: (a) student–teacher relations, (b) classroom activities used by the teacher, and (c) teacher's personal adjustment. The CAST: PP consists of only subscales A and B above. Each subscale consists of five questions, and each question has five responses describing various characteristics of the teacher's instructions. Items were selected from the available related instruments at the time. A group of professors and doctoral students in science education rated teachers. Intraclass correlations ranged from 0.53 to 0.98 with a mode of 0.86 for the 15 items. KR-20 was 0.74 for CAST: PP.

The Science Attitude Scale (Shrigley, 1974; Thompson & Shrigley, 1986)

The Science Attitude Scale (SAS) is a five-category Likert-scale instrument for measuring elementary preservice science teachers' attitudes toward science. There are 23 items, 14 positive ones and 9 negative ones. It measures two related attitude domains: (a) attitude toward teaching science, and (b) attitude toward taking science courses. After the publication of the original instrument in 1974, the instrument was revised and revalidated (Thompson & Shrigley, 1986). First, the original 23 items were reviewed by an expert panel of three science educators who recommended that 10 of 23 items be eliminated. The remaining items were also revised and 36 new items were written. The items together related to four subcomponents of the attitude: (a) the comfort–discomfort of teaching science, (b) the handling of science equipment, (c) time required to prepare and teach science, and (d) the basic need American students have for science. The revised instrument was then pilot-tested with 83 preservice elementary teachers. Twenty-two items were retained based on the criteria that items had high item-total correlation, low percent of neutral responses, a skewed distribution among the five choices, and were related to one of four subcomponents. The retained 22 items were given to 226 preservice science teachers. The coefficient alphas for the four subcomponents/subscales ranged from 0.63 to 0.79; inter-correlation between the four subscales ranged from 0.46 to 0.70. Principal component factor analysis found that there were four dominant factors, but the four factors did not correspond to the defined four subcomponents. The first factor contained 8 items, which was interpreted as teacher anticipation or the preparation for the teaching of science. The second factor contained five items and the third factor contained 3 items; Factors 2 and 3 were interpreted as comfort–discomfort of teaching science. The fourth factor only had one item about children's curiosity about science. Coefficient alpha based only on the 16 items related to Factors 1–3 was 0.88. Correlation between scores on SAS and scores on a metrication attitude scale was 0.34, and the correlation between scores on SAS and scores on a reading attitude scale was 0.08, indicating convergent and divergent validity.

Inquiry Science Teaching Strategies (Lazarowitz & Lee, 1976)

Inquiry Science Teaching Strategies (ISTS) is a 40-item Likert scale measuring secondary science teachers' attitudes toward inquiry science teaching. Inquiry science teaching involves three areas: classroom teacher–student interaction, laboratory investigations, and textbook uses. Validation of the instrument involved expert panel reviews of items, and pilot-testing with preservice science teachers. Construct validity was established using five groups with various degrees of attitudes toward inquiry teaching and the five groups scored as expected with science experts scoring the highest and ordinary science teachers the lowest. Cronbach's alpha of the instrument based on the above groups ranged from 0.54 to 0.85.

Teacher Orientation to Science Instruction (Connelly, Finegold, Wahlstrom, & Ben-Peretz, 1977)

Teacher Orientation to Science Instruction (TOTSI) assesses science teachers' assumptions on the following four dimensions: nature of the learner, nature of scientific knowledge, cultural setting for the curriculum, and nature of the teacher. In addition, the instrument assesses interactions among these factors. TOTSI has 8 subtests, and each subtest has 3 to 8 items. The teacher indicates how well the statements match his or her own positions. Because there may be a difference between what a teacher actually does in class and what a teacher would like to do in class, every response has two columns: Column A relates to actual classroom situation and Column B relates to desired classroom situation. The instrument was field-tested with science teachers. Reliability coefficients ranged from 0.21 to 0.79 for various sets of items.

Moore Assessment Profile (Moore, 1977)

Moore Assessment Profile (MAP) is a 117-item science teacher needs assessment. Each item is followed by a continuum from 1 to 4, representing *no need, little need, moderate need,* and *much need.* Space is available for teachers to list additional needs. Validation involved panel review of items, and pilot-testing with elementary, middle, and high school science teachers. The reliability based on Hoyt's analysis of variance was 0.986. Factor analysis indicated that there were 13 main factors accounting for 73% of the total variance.

Science Teaching Efficacy Beliefs Instrument (Riggs & Enochs, 1990)

The Science Teaching Efficacy Beliefs Instrument (STEBI) measures elementary science teachers' self-efficacy related to teaching science. It contains two scales: the personal science teaching efficacy belief scale, and the science teaching outcome expectancy scale. Items are presented in a Likert-scale format with choices from *strongly agree* to *strongly disagree.* Initial items were reviewed by an expert panel. A preliminary study with 71 practicing elementary teachers resulted in further revisions of items and refinement of the instrument. The main study involved 331 subjects. Cronbach's alpha for the personal science teaching efficacy belief scale was 0.91, and 0.77 for the science teaching outcome expectancy scale. Principal component factor analysis showed that there were only two factors with an eigenvalue greater than one, supporting the hypothesized two scales of the instrument. Teachers' scores on the two scales were also found to be statistically significantly correlated with a list of criterion variables such as years of teaching, use of activity-based science instruction, and so forth.

Context Beliefs About Teaching Science (Lumpe, Haney, & Czerniak, 2000)

Context Beliefs About Teaching Science (CBATS) is a 26-item survey of science teachers' context beliefs. Each item asks teachers to state their degree of agreement in a Likert-scale format with the stated factor to enable them to be an effective teacher—the enabling subscale, and to indicate how likely each stated factor will occur in their schools by choosing one from *very likely, somewhat likely, neither, somewhat unlikely,* and *very unlikely*—the likelihood subscale. Three groups of teachers were involved in the development. Interviewing 130 science teachers generated a pool of items for creating the initial instrument. The initial instrument was pilot-tested with a sample of 71 teachers. Based on the item and test analysis of pilot-testing data, the instrument was revised and then given to a larger sample of science teachers. Factor analysis suggested that there were two dominant factors accounting for 23.9% variance. Science teachers' CBATS scores were found to be modestly correlated with their outcome expectancy scores ($r = 0.34$, $p = 0.000$) and with science teaching capability beliefs ($r = 0.19$, $p = 0.002$). In addition, CBATS scores were statistically significantly correlated with years of teaching, number of science methods courses taken, number of science teaching strategies used, and minutes spent on teaching science. Science teachers demonstrated statistically significant higher enabling beliefs than likelihood beliefs. The alpha for the enabling subscale was 0.86, for the likelihood subscale 0.85, and for the entire instrument 0.85.

Self-Efficacy Beliefs About Equitable Science Teaching and Learning (Ritter, Boone, & Rubba, 2001)

Self-Efficacy Beliefs About Equitable Science Teaching and Learning (SEBEST) assesses preservice elementary science teachers' personal self-efficacy and outcome expectancy beliefs with regard to teaching and learning science in an equitable manner when working with diverse learners. A seven-step plan was followed to develop the instrument. Step 1 involved the definition of the construct and content to be measured; a content matrix consisting of three dimensions, that is, positive/negative *x* personal self-efficacy/outcome expectancy *x* ethnicity/language minorities/gender/socioeconomic, was developed. Step 2 involved item preparation; 195 Likert-type items were developed. Step 3 involved review of draft items by 10 graduate students in science education. Step 4 involved five experts who reviewed the content validity of the revised items; 48 items were retained to form the draft instrument. Step 5 involved the first try-out of the draft instrument with 226 preservice elementary science teachers. Step 6 involved formulation of the instrument based on item and factor analysis. Principal component factor analysis identified four significant factors. Eleven items loaded on Factor 1 that was identified as personal self-efficacy associated with socioeconomic status, gender, and ethnicity. Ten items loaded on Factor 2 that was identified as outcome efficacy with language minorities, socioeconomic status, gender, and ethnicity. Six items loaded on factor 3 that was identified as personal self-efficacy associated with language minorities; and eight items loaded on Factor 4 that was identified as outcome expectancy across all content contexts. Internal consistency reliability coefficients for the four subscales ranged from 0.72 to 0.82. The reliability coefficient for the entire instrument was 0.87. Step 7 was a further study of reliability with different samples. Alpha coefficients were consistently high. Finally, Rasch rating scale modeling was applied to the data to provide additional evidence for the construct validity of the instrument.

Attitudes Toward Teaching of Environmental Risk (Zint, 2002)

Attitudes Toward Teaching of Environmental Risk (ATER) is a 14-item instrument for measuring science teachers' attitudes toward teaching environmental risks. The items are related to three theories about attitude: theory of reasoned action, theory of planned behavior, and theory of trying. Each item has seven response options described as *extremely, quite, slightly, neither, slightly, quite*, and *extremely*. Validation involved a mail-survey of over two thousand teachers with an adjusted return rate of 80%. Construct validity was established by confirmatory factor analysis. Composite reliability based on weighted factor loadings was reported to be high (over 0.90 for three major subscales).

The Attitudes and Beliefs About the Nature and the Teaching of Mathematics and Science (McGinnis et al., 2002)

This instrument measures preservice science teachers' attitudes and beliefs about the nature and the teaching of math and science. Most items are from the available related instruments at the time. All items are in the Likert-scale format. Phase 1 development involved expert reviews of draft items. Two additional phases of pilot-testing resulted in the final 37-item version of the instrument. Factor analysis revealed five main factors. Alpha reliability coefficients ranged from 0.60 to 0.81 for the above five factors. Results showed that preservice teachers who participated in reformed courses improved significantly toward the intended direction on their attitudes and beliefs about the nature and the teaching of mathematics and science.

Teacher Perceptions and Practices Regarding the Use of the History of Science in Their Classrooms (Wang & Marsh, 2002)

This survey measures elementary and secondary science teachers' values and practices of integrating history of science into science teaching. The survey covers three aspects regarding history of science: (a) conceptual understanding, (b) procedural understanding, and (c) contextual understanding. Thirteen items are included, with four on conceptual understanding, three on procedural understanding, and six on contextual understanding. The response to each item is from two 5-point scales, one is in the Likert scale (i.e., from *strongly disagree* to *strongly agree*) on values of integrating history of science, and another is a rating scale (i.e., from *rarely occurred* to *occurred frequently*) on practices of integrating history of science. Cronbach's alpha for the perceptions scale was 0.904, and alpha for the practice scale was 0.949.

Attitude to Science Education (Pell & Jarvis, 2003)

Attitude to Science Education (ASE) is a 49-item Likert scale survey of primary teachers' attitudes toward science teaching. It measures two main constructs: attitude toward science teaching and attitude toward professional preparation for science teaching. Initial questions (58) came from previously published instruments to make them appropriate for primary teachers in England and Wales. Eighteen teachers were involved in a pilot study to finalize the instrument containing 52 items. The main study involved 58 teachers. The pilot study and main study samples were pooled into one sample for the final validation study. Thirty-three items were related to effective science teaching in the classroom. Principal component factor analysis revealed three major factors forming three subscales: investigative, pupil-centered science pertaining to the value teachers put on encouraging pupil initiative, interest, and wonder; classroom management pertaining to value given to systematic, structured approaches to learning within the classroom; and general scientific method pertaining to views about the natural philosophy underlying an empirical, pupil-participative science. Cronbach's alpha coefficients for the above three subscales were 0.89, 0.83, and 0.83.

Nineteen items were related to preparation and professional aims. Principal component factor analysis revealed three major factors forming three subscales: in-service improvement pertaining to teachers' attitude toward the worth of in-service education with an emphasis on appropriate use of human and physical resources; theoretically-grounded science teaching pertaining to the extent teachers feel they should operate from a child-centered constructivist process; and testing pertaining to attitude toward a formative approach to assessment. Cronbach's alpha coefficients for the above three subscales were 0.86, 0.86, and 0.65. Combining items for the above two constructs by excluding three items formed the composite scale called "Attitude to Science Teaching." The Cronbach's alpha for this composite scale was 0.94.

Biology Teachers' Attitude Toward and Use of Indiana's Evolution Standards (Donnelly & Boone, 2007)

This instrument measures biology teachers' attitudes toward standard and evolution in particular. The instrument includes 24 items related to four subscales: US (teacher use of standards, 8 items), AS (teacher attitude toward standards, 7 items), UES (teacher use of evolution standards, 7 items), and ETP (teachers' evolution teaching practices, 5 items). Its development followed an iterative process guided by Rasch modeling. Initial items were developed based on hypotheses, and Rasch modeling was used to test the hypotheses. The initial draft form (Form A) was revised by modifying/deleting/adding items based on reviews of selected science teachers and university science educators, and on fit statistics and person-item maps produced by Rasch modeling. Data for Rasch modeling came from a stratified random sample of biology teachers in Indiana. Item reliability coefficients based on item separation indices for the revised form (Form B) were 0.98 (US), 0.97 (AS), 0.90 (UES), and 0.98 (ETP). The construct validity was claimed based on person-item map and its agreement with researchers' hypotheses.

Biotechnology Attitude Questionnaire (Prokop, Leskova, Kubiatlo, & Diran, 2007)

The Biotechnology Attitude Questionnaire (BAQ) is a 17-item Likert-scale survey to measure university preservice science teachers' attitudes toward biotechnologies. Questions for the instrument came from various previously published instruments. The content validity of the instrument was established through review by three experts in the fields of genetics and biology education who were asked whether the items were relevant to the goal of the questionnaire. The BAQ had a reliability (Cronbach's alpha) of 0.76. A statistically significant correlation was found between students' biotechnology knowledge and their attitude toward biology technology.

Nanotechnology Attitude Scale for K–12 Teachers (Lan, 2012)

The Nanotechnology Attitude Scale for K–12 teachers (NAS-T) was developed to assess K–12 teachers' attitudes toward nanotechnology. The 23 Likert-scale items on a five-point scale ranging from *strongly disagree* or *never* to *strongly agree* or *always* were grouped into three components: importance of nanotechnology, affective tendencies in science teaching, and behavioral tendencies to teach nanotechnology. The content validity and face validity of the NAS-T were established via two panels of experts, in which the first panel included two professors in the field of nanotechnology and one professor from the college of education, and the second panel included five K–12 teachers who have participated in the K–12 nanotechnology program. Validation sample involved 233 K–12 teachers who participated in a nanotechnology teacher training program.

The Cronbach's alpha values of three NAS-T subscales ranged from 0.89 to 0.95, and the Cronbach's alpha value of the NAS-T total scale was

0.94. The item-total scale correlation coefficients in the range of 0.41 to 0.72 supported the homogeneity of the 23 NAS-T items. Both exploratory and confirmatory factor analyses were conducted to establish construct validity. In exploratory factor analysis, three conceptually meaningful factors were extracted and explained 64.11% of the total variance. Each of the 23 items had factor loading higher than 0.3. The CFA results showed that three NAS-T factors were moderately correlated to each other and ranged from 0.27 to 0.61. Convergent validity was established based on positive Pearson zero-order correlations among three NAS-T subscale scores, between NAS-T scores and teacher science teacher efficacy in science, self-perception of own nanoscience knowledge, and their test scores on nanoscience knowledge.

Dimensions of Attitude Toward Science (van Aalderen-Smeets & van der Molen, 2013)

The Dimensions of Attitude Toward Science (DAS) questionnaire measures the attitude of in-service and preservice primary teachers toward teaching science. An initial version of the instrument was piloted with a sample of 64 in-service and preservice teachers, which resulted in deleting or rewriting a number of items. The revised DAS included a total of 28 items. For each item, respondents were asked to indicate to what extent they agreed or disagreed on a five-point Likert scale ranging from *totally disagree* (score 1) to *totally agree* (score 5). A total of 556 respondents (158 in-service and 398 preservice) were included in the final validation study.

Confirmatory factor analysis by setting extraction of seven factors showed that the rotated factor structure corresponded exactly to the hypothesized scale structure, and none of the items showed a cross loading on any of the other factors. The factor loadings varied from 0.37 to 0.90. The internal consistency by Cronbach's alpha for the seven scales ranged between 0.74 and 0.93, and item-total correlations computed within the factors ranged between 0.44 and 0.85.

Instruments for Measuring Other Affective Variables

The Measure of Acceptance of the Theory of Evolution (Rutledge & Warden, 1999)

The measure of Acceptance of the Theory of Evolution (MATE) contains 20 multiple-choice items. The items relate to fundamental concepts of evolutionary theory and the nature of science. The item format is 5-point Likert scale, with a balance of both positive and negative statements. The initial items were critically analyzed by a jury of five university professors who had expertise in the fields of evolutionary biology, science education, and philosophy of science. The revised instrument was then given to public high school biology teachers in Indiana; 552 teachers (53% return rate) completed the survey. Exploratory factor analysis showed that there was only one significant factor accounting for 71% of total variance. All items loaded on the factor with a value of 0.65 or greater. The alpha reliability was 0.98. The corrected item-total correlation was greater than 0.65 for all items.

The MATE was also given to university non-biology major students ($n = 61$; Rutledge & Sadler, 2007). The test–retest reliability was 0.92; Cronbach's alpha was 0.94. Romine, Water, Bosse, and Todd (2017) administered MATE to another sample of undergraduate students ($n = 194$) and used the Rasch model to analyze data. They found that MATE was best used as a two-dimensional instrument: (a) acceptance of evolution facts and data and (b) acceptance of the credibility of evolution and rejection of nonscientific ideas. The person reliability for the first scale was 0.87 and second scale 0.88.

Significant Person Influence on Attitudes Towards STEM (Sjaastad, 2013)

Significant Person Influence on Attitudes towards STEM (SPIAS) is an instrument designed to measure different modes of significant persons' influence on secondary school students' attitudes towards STEM. Based on Woelfel and Haller's (1971)'s model of interpersonal influence, SPIAS measures four sub-constructs, "defining the self," "defining the object," "modelling the self," and "modelling the object." Focus group interviews with Norwegian adolescents in a STEM mentoring program were used to generate statements. The initial version of SPIAS had 31 items, and a Likert scale was used. One hundred (100) students were included in the pilot study and the final version was reduced to 17 items.

A principal components analysis for the final version indicated that the items' loadings to the first component accounted for 20% of the variance, and the second component accounted for 14% of the variance. Rasch analysis was conducted to the data for each subscale; items retained in the final version demonstrated good model-data fit. Principal component analysis of Rasch residuals suggested that the unidimensionality for each of the four scales was good. Category thresholds were also appropriate. No DIF was found for any item. The person separation reliabilities for the four scales were 0.72, 0.76, 0.66, and 0.58.

Scientific Academic Emotion Scale (Chiang & Liu, 2014)

Scientific Academic Emotion Scale (SAES) is a questionnaire to reveal the academic emotions of university students studying science in three situations: attending science class, learning scientific subjects, and problem solving in science. The SAES includes 15 emotional subscales categorized as: (a) positive-activating emotions: pride, enjoyment, hope, outlook, and attention; (b) positive-deactivating emotions: relief, resilience, and context; (c) neutral emotions: self-awareness and social intuition; (d) negative-activating emotions: anger and anxiety; and (e) negative-deactivating emotions: shame, hopelessness, and boredom. The pilot study was administered at a university in South Taiwan, and a total of 150 participants were recruited. In the formal validation study, a total of 396 participants took part in the study. The final version included 123 items within 15 subtests.

Item analysis showed acceptable discrimination. Both exploratory and confirmatory factor analyses were conducted. Confirmatory factor analysis showed that the factor structure conformed to the hypothesis. The Cronbach's alphas within the scales were from 0.737 to 0.936.

STEM Awareness Community Survey (Sondergeld & Johnson, 2014)

STEM Awareness Community Survey (SACS) is an instrument to measure community stakeholders' STEM awareness. Based on open-ended question survey of business and informal education agencies, the STEM awareness construct is defined as consisting of four main themes: (a) industry engagement in STEM education (IE), (b) STEM awareness and resources (AR), (c) regional STEM careers and workforce (CW), and (d) preparation of students for success in colleges and careers (PR). Three parallel versions of the SACS were created to assess K–12 teachers, higher education faculty, members from the business community on their attitudes and beliefs about regional STEM awareness and support. The surveys were each composed of 63 items with the majority of them being on a traditional 1–5 point Likert scale from *strongly disagree* to *strongly agree*; a few items were selecting all appropriate options and open-ended responses. Rasch analysis of the pilot study data resulted in revisions of some items and the scales. The revised version included 39 totally Likert-scale items: 8 for IE, 13 for AR, 6 for PR and 12 for CW and all items were in a four-point format. Validation sample included 600 participants. Rasch analysis of the validation sample suggested that there was sufficient evidence for the unidimensionality and internal construct validity of the measures; the person reliability was 0.92.

Generalized Acceptance of EvolutioN Evaluation (Smith, Scott, & Devereaux, 2016)

Generalized Acceptance of EvolutioN Evaluation (GAENE) is an instrument to measure only evolution acceptance—not related to knowledge or religious beliefs. Acceptance is considered distinct from belief; it is the mental act or policy of deeming, positing, or postulating that the current theory of evolution is the best current available scientific explanation of the origin of new species from preexisting species. Item development was based on extensive student interviews, expert reviews, and multiple rounds of item revisions. The initial version included 41 items; the final version (GAENE 2.0) contained 35 items. Validation study of GAENE 2.0 with over 600 high school and post secondary students showed a reliability to be 0.945. Principal components analyses supported a one-factor solution. Rasch analysis also suggested strong evidence of construct validity.

Science Curiosity in Learning Environments (Weible & Zimmerman, 2016)

Science curiosity in Learning Environments (SCILE) is a 12-item scale to measure scientific curiosity in youth. Curiosity in science is defined as a desire for content-specific knowledge about natural phenomena. The initial items for the scale were adapted from a previously published general curiosity instrument by crafting the items appropriate for K–12 students and by incorporating scientific practices included in the U.S. science standards documents for K–12. Items were created to address three aspects of scientific curiosity: (a) stretching or exploring—seeking new information and experiences, (b) embracing—the acceptance of novel, uncertain, and unpredictable nature of everyday life, and (c) science and engineering practices. Items were stated in a five-point rating scale: *always, often, sometimes, not often,* and *never.* The initial set of 30 items was then administered to 663 youth aged 8–18 in the United States. Exploratory factor analysis was conducted to reduce the total items to 12. The 12 items loaded strongly onto one of the three factors. A follow-up confirmatory factor analyses confirmed the three-factor solution: stretching, embracing, and science practices. The scale provides a valid and reliable measure of youth's scientific curiosity for elementary, middle school, and high school students.

Self-Efficacy for Public Engagement With Science (Evia, Peterman, Cloyd, & Besley, 2018)

Self-efficacy for Public Engagement with Science (SEPES) is a 13-item instrument measuring scientists' self-efficacy in public engagement with science. An initial pool of 30 items was created based on Bandura's (2006) self-efficacy theory. A six-point Likert scale was used for each item: *strongly disagree, moderately disagree, mildly disagree, mildly agree, moderately agree,* and *strongly agree.* Response process validity evidence was gathered through think aloud interviews to understand how scientists interpreted and responded to the items; a total of 25 scientists were interviewed. Based on interviews,

19 items were retained for pilot test. The pilot test sample was recruited through e-mails and listservs. A total of 297 scientists responded. The internal consistency reliability of the 19 items was 0.90. A one-factor exploratory factor analysis accounted 33.6% variance by the dominant factor; all 19 items had positive factor loadings ranging from 0.39–0.70. The graded response model was used to reduce the 19-item instrument to a 13-item instrument.

The Organic Chemistry-Specific Achievement Emotions Questionnaire (Raker, Gibbons, & Cruz-Ramirez de Arellano, 2019)

The Organic Chemistry-Specific Achievement Emotions Questionnaire (AEQ-OCHEM) is a 216-item instrument measuring student achievement emotions when taking the undergraduate organic chemistry course. Items are grouped based on context (i.e., classroom, study, exam) and time (i.e., before, after, and during each context). Items are responded in 5-point Likert scale: *strongly agree* to *strongly disagree*. The development of AEQ-OCHEM was informed by the control-value theory (CVT). CVT assumes that affective, behavioral, and achievement relations are disciplinarily-, culturally-, and demographically-specific. Emotions have valence (i.e., positive or negative) and activation (i.e., activating or deactivating). The study sample included 533 students in three sections of a first-semester of a yearlong organic chemistry course at a research university in the United States. The Achievement Emotions Questionnaire (Pekrun, Goetz, Frenzel, Barchfeld, & Perry, 2011) was revised into the AEQ-OCHEM through an iterative process of revision and expert feedback. Confirmatory factor analysis revealed good model-data-fit for the factor structure of 8 subscales (anger, anxiety, boredom, enjoyment, hope, hopelessness, pride, and shame) in each of three contexts, that is, a total of 24 subscales. External validity was established by correlations between achievement emotions and other study measures including academic performance, Motivated Strategies for Learning Questionnaire, Academic Motivation Scale—Chemistry, and Regulation of Learning Questionnaire. Cronbach's alpha values for the 24 subscales were good and excellent.

Commentary

A large number of measurement instruments are available; they measure a variety of affective variables related to teachers and students. One major consideration when choosing a measurement instrument to use is the appropriateness of the defined construct. Because affective variables are diverse, and different theories may define a same affective variable quite differently, the applicability of the theory that was used to define the measured construct to the intended uses must be critically evaluated. Take attitude as an example. We can see from the above instruments that the instruments measure a wide variety of attitudes, reflecting diverse theoretical frameworks related to attitude. The diversity in theoretical frameworks requires that an attitude instrument has a clearly defined construct of attitude. This is an issue of construct validity. Unfortunately, many of the instruments have an unclear definition of attitude, thus the construct validity is questionable. This limitation in attitude instruments has been pointed out in many previous reviews of attitudinal scales. For example, based on a review of attitude instruments, Germann (1988) concluded that, in many of the instruments, "First, the construct of attitude has been vague, inconsistent, and ambiguous. Second, research has been conducted without a theoretical model of the relationship of attitude with other variables" (p. 689). One consensus emerging from the literature is that there is no such thing as a general attitude toward science; instead, attitude is quite domain and context specific. For example, attitude toward science and attitude toward chemistry may be quite different (Salta & Tzougraki, 2004); attitude toward the importance of science, attitude toward science careers, and attitude toward science curriculums may also be quite different from each other (Menis, 1989).

Examination of the literature has shown that the validity of the above instruments varies greatly. Researchers have raised questions about validity of many instruments. For example, based on a review of 56 instruments on attitudes toward science, Munby (1983b) concluded that most of the instruments lack construct validity. Laforgia (1988) claimed that many attitude instruments have inadequate criterion-related validity. Doran, Lawrenz, and Helgeson (1994) concluded that "ambiguity of terms and quality of instruments are two serious problems facing those interested in assessing attitudes to science" (p. 428). Blalock et al. (2008) conducted a comprehensive review of 66 measurement instruments published in peer-reviewed journals from 1935 to 2005; only 20 of the instruments were directly related to attitude toward science. They used a scoring rubric to rate every measurement instrument in terms of the theoretical background, reliability, validity, dimensionality, and development/usage with higher weights given for reliability, validity, and dimensionality. Scores for the 20 attitudes toward science instruments ranged from 3 to 22 out of a total of possible 28 points, with the highest score (22) for Attitude Toward Science in School Assessment (Germann, 1988). They expressed a number of methodological concerns with the published attitude instruments including absence of reliability and validity evidence, overall poor quality of validation studies, a nearly universal disregard for missing data, and a dominance of instrument validation in a single study. Many of the instruments rely heavily on content and face validity established by expert panel reviews while ignoring other types of evidence. Given the above, any instrument should be chosen with a critical eye. Also, continuing validating the chosen instrument based on new data should always be conducted.

It is common for attitude measurement instruments to adopt the Likert scale. Besides the potential issue with fakability (i.e., responses may not be truthful; Laforgia, 1988), one serious issue associated with the Likert scale is obtaining a total scale score by adding individual item scores. Values such as 1–5 assigned to five choices of a statement do not have the same origin and interval unit because they are not on a ratio or interval scale. Also, some items are easier to endorse, while others more difficult to endorse. The consequence of the above is that individual item scores should not be meaningfully added into a total score. This should make sense if we consider that individual items represent different levels of attitude, and a same value such as 2 can mean quite different things on different items. Consider "I am happy to attend the science class" and "I want to major in science at university." Choosing "*agree*" to the above two statements, thus receiving a same score of 4, would mean very different levels of attitude toward science. Adding two 4s together does not make logical sense. In order to address this issue, different ways of analyzing Likert scale data than using total scores should be adopted. For example, responses to different items in an attitude scale may be represented by a profile, so that the difference in profiles between different groups or between two time points can be meaningfully compared. Chi-square or non-parametric statistics, instead of parametric inferential statistics (e.g., *t*-test) may be applied for statistical testing. One best way currently available is to use Rasch modeling to convert raw scores into latent scores so that respondents' attitudes will be measured on a latent attitude scale. More and more recent instruments have been developed using Rasch modeling. Alternatives to the Likert scale should also be considered. Examples of such alternatives are the Thurston scale (1925), Guttman scale (1944), semantic differential (Osgood, Suci, & Tannenbaum, 1971), and checklist.

As for measurement instruments for teachers, it is important to make a differentiation between teacher beliefs and teacher practices (Connelly et al., 1977; Wang & Marsh, 2002). This distinction is very important because the two are not necessarily always the same. Understanding the discrepancy between teachers' beliefs and practices can inform ongoing science education reforms so that best practices promoted in university classrooms are actually implemented in K–12 classrooms. This issue also points to the critical importance of assessing actual teaching practices and their direct impact on student learning. Instruments for evaluation of teaching practices are introduced in Chapter 7.

Developing Instruments for Measuring Affective Variables

State the Purpose and Intended Population

The need for measuring affective variables comes from various demands. Many affective outcomes are part of national and state science curriculum standards, and measurement of these affective learning outcomes is an important part of program evaluation. For example, in the National Assessment of Educational Progress (NAEP) science assessment, the student background questionnaire typically includes a scale for measuring students' attitudes toward science. Information from this measurement can help science education policy making at the national level. Similarly, affective learning outcomes are an important part of a teacher education program, thus measurement of teacher affective learning outcomes may be a part of the teacher education program evaluation. Because the measurement of affective variables primarily relies on self-reports, measurement of affective learning outcomes for accountability purposes may not be appropriate.

In addition, because cognitive, affective, and psychomotor domains are closely related, measurement of affective variables related to both students and teachers is important for monitoring science teaching and learning processes. Research on science teaching and learning may require examination of the relationship between cognitive, affective, and psychomotor variables, and measurement of affective variables can help study the quantitative relationships among various cognitive, affective, and psychomotor variables.

Define the Construct To Be Measured

Affective variables are vast and diverse, it is critically important to clearly define the construct to be measured. Although it is possible to measure more than one construct in one measurement administration, essentially each measurement scale should measure one and only one construct. In this way, each measurement instrument is one scale and each measure is based on only one scale.

Identifying and defining the construct to measure must be consistent with theories. Different theoretical orientations suggest different variables related to the construct. For example, motivation from a behavioral orientation would focus on such variables as rewards, benefits, peer pressures, and so on; while a social-cultural orientation toward motivation would focus on such variables as peer interaction, cultural values, societal norms, historical evolution, and environmental settings. Because of the contextualized nature of affective variables, when defining the affective construct, it is important to consider the construct's relevance to specific science content domains, nature of the target students (e.g., elementary, secondary, and university science students) and teachers (e.g., preservice and in-service, or elementary and secondary science teachers), and its relations with other cognitive, affective, and psychomotor variables. For example, attitudes toward science may be differentiated as attitude toward school science as taught in the school, attitude toward science as it is involved in everyday life, attitude toward science as it is considered as a future career, and so forth. Whatever theory is used to define the construct to measure, keep in mind that developing a measurement instrument is not to validate the theory; on the contrary, the theory needs to be assumed in order to guide the development process of the measurement instrument.

One basic assumption of measurement is the difference in quantity of the defined construct among subjects. Any defined construct must imply a hierarchy of subjects' performances on the construct. For example, if the defined construct is secondary students' attitude toward science, then it is assumed by this construct that there are different levels of attitude toward science among secondary students, and this difference can be quantified on a unidimensional linear scale.

Identify Performances of the Defined Construct

Once the construct is defined, specific subjects' performances or behaviors related to the defined construct should then be identified. Because the defined construct implies a linear scale, the specific behaviors should form a clear hierarchy or progression. For example, interest in school science may be defined by a collection of student behaviors from attending science class on time, active participation in science activities, and making efforts to complete science assignments, and so on. These student behaviors can be hypothesized to form a progression according to their interest in school science. Figure 4.2 shows a sample hierarchy of behaviors that represent the interest in school science.

Once behaviors that define the affective construct are identified, the next step in development of the measurement instrument is to specify how many items will be developed and what behavior each of the items will measure. A two-dimensional specification table is commonly used for this purpose. In the specification table, specific behaviors are labeled on one axis, and types of items are labeled on the other axis. The cell values are number

Conducting extra readings and projects beyond homework

Completing homework

Actively seeking opportunities to participate in science activities

Participating in science activities when invited

Making efforts to attend science class when there is conflict

Attending science class

Knowing that learning science is compulsory

Figure 4.2 A hierarchy of student behaviors related to interest in school science. Higher behaviors represent higher interest in school science.

of items for specified behaviors. For example, a specification table for measuring interest in school science is as follows (Table 4.1).

In the above example, the instrument for measuring student interest in school science will consist of 21 items, with three items for each of the seven identified behavior types. Also, there will be two types of questions, Likert scale and rating scale. Of course, a measurement instrument may use only one item format, such as Likert-scale type questions. The guidelines suggested in Chapter 3 for deciding test length are also applicable here when developing a test specification for measurement of an affective construct.

Based on the table of specification, items are then created. Many item formats are available for measuring affective variables, depending on the construct to be measured. Most commonly used item types are Likert-scale, rating scale, semantic differential, and checklist or inventory.

TABLE 4.1 A Test Specification for Measurement of Interest in School Science

	Likert-Scale Type Questions	Rating Scale
Conducting extra-science readings and activities beyond homework	0	3
Completing homework	0	3
Actively participating in science activities	3	0
Participating in science activities when asked	0	3
Making efforts to attend science classes	3	0
Attending science classes	0	3
Science is compulsory	3	0

A typical Likert-scale type question contains a statement followed by a list of five-category choices from *strongly agree* (SA), *agree* (A), *undecided* (U), *disagree* (D), and *strongly disagree* (SD). The respondent selects one of the five categories as his/her agreement to the statement. Guidelines for constructing Likert-scale items are available in Chapter 1.

Rating scale questions are very similar to Likert-scale questions. Both contain a statement and a set of hierarchical categories. Subjects respond to both rating scale and Likert-scale questions by choosing one of the categories. However, one difference between rating scale questions and Likert-scale questions is how categories are phrased. Compare the following two statements:

Item 1: School science is exciting.
SA A U D SD

Item 2: School science is exciting.
always most times sometimes occasionally never

Item 1 expresses a value judgment about school science, and respondents will decide to what degree they will agree or disagree with the judgment. Item 1 is typical for Likert-scale questions. Item 2 presents a phenomenon—school science is exciting, and respondents will decide how to respond by indicating the frequency of the phenomenon. Further, rating scale categories can be any descriptions that represent a quantitatively or qualitatively different hierarchy (e.g., 1, 2, 3, 4, 5; or *excellent, very good, good, fine,* and *bad*), while Likert-scale categories are always different degrees of agreement (e.g., SA, A, U, D, SD). Because of the above differences, rating scale questions are more appropriate for measuring constructs related to behaviors, while Likert-scale questions are more appropriate for measuring constructs related to values and judgments.

Semantic differential was first proposed by Osgood and colleagues (Osgood et al., 1971). It is intended to probe both cognitive and affective aspects of a construct. A construct is assumed to have three dimensions of meanings: evaluation (e.g., good–bad), potency (e.g., strong–weak), and activity (e.g., fast–slow). Evaluation pertains to feeling or value associated with the construct; potency pertains to the usefulness of the construct; and activity pertains to the dynamic nature of the construct. Each dimension typically consists of 3–15 bipolar adjectives. Scores on each of the dimensions are then averaged, and the three averaged scores are used to describe a respondent's cognitive and affective meanings assigned to the construct. The three dimensions together define a semantic space of the construct.

The following is an example of a semantic differential for school science:

What Does School Science Mean to You?

For each of the following bipolar pairs of adjectives, decide your agreement to each of the bipolar adjectives in relation to "School Science" by placing a √ mark on a dash. The closer the dash to an adjective, the more agreement you have on the adjective.

School Science is

Good	___: ___: ___: ___: ___: ___: ___:	Bad
Positive	___: ___: ___: ___: ___: ___: ___:	Negative
Valuable	___: ___: ___: ___: ___: ___: ___:	Little use
Useful	___: ___: ___: ___: ___: ___: ___:	Useless
Powerful	___: ___: ___: ___: ___: ___: ___:	Powerless
Helpful	___: ___: ___: ___: ___: ___: ___:	Helpless
Fun	___: ___: ___: ___: ___: ___: ___:	Boring
Busy	___: ___: ___: ___: ___: ___: ___:	Light
Engaged	___: ___: ___: ___: ___: ___: ___:	Inattentive

When writing semantic differential items, it is important to keep three dimensions in mind and try to write items accordingly. As the case with any other item type, more items than specified in the specification table need to be created to give room for item deletions and revisions after pilot-testing.

Inventory or checklist questions contain a set of descriptive scenarios or facts; respondents are asked to check if they apply to them or not. Choices are typically binary; that is, they are either "Yes" or "No." Inventory and checklist are more appropriate for measuring affective variables closely related to behaviors.

Here is a set of sample inventory or checklist items that measure science teachers' awareness of professional development activities.

Example: For each of the following activities, please circle *Yes* if you think it is beneficial for your professional development related to teaching, and *No* if you think it is not beneficial for your professional development related to teaching.

Attending workshops offered by the school district	Yes	No
Attending the annual state science teacher conference	Yes	No
Reading popular science magazines	Yes	No
Reading newspapers	Yes	No
Reading science education research journals	Yes	No

In the above example, selecting *Yes* to an activity indicates a teacher's awareness of the potential of the activity for professional development.

After a pool of items has been created, it is necessary to have items reviewed by a panel of experts consisting of both content experts and methodologists. The panelists should be briefed on the purpose of the measurement and the construct to be measured. Test specification may also be presented to the panelists to facilitate their review. Specific questions pertaining to the item content accuracy, language clarity, and alignment with the intended construct may also be prepared to help panelist review items.

After items have been reviewed by the panel and necessary revisions have been made to the items, the items should then be tried out by a small number of respondents. The respondents will not only respond to the items, but also state their impressions or comments on the clarity of the items. Interviews with a few respondents may also be helpful.

Conduct Pilot-Test/Field-Test

The resulting items from the previous development steps will then go for pilot testing. The sample chosen for the pilot testing should represent the intended population, particularly in the range of performance on the construct to be measured. For example, if the construct is attitude toward school science, then the variation in attitude toward school science among the respondents in the sample should be expected to be similar to that of the population. Sample size should also be adequate. For rating scales including the Likert scale, Linacre (2004) suggests that at least 10 observations per category in the item should be used. A larger sample size, such as 25 to 50 observations per category, can help maintain the stability of item parameter estimates during Rasch modeling.

Conduct Rasch Analysis

Figure 4.3 shows a sample data file for an attitude toward science. This survey contains 27 items; all of them have a five-response category format: *strongly disagree* (1), *disagree* (2), *undecided* (3), *agree* (4), and *strongly agree* (5). Person IDs are not contained in the data file. A few missing data points are in blank.

Figure 4.4 shows the Winsteps control file for analyzing the data shown in Figure 4.3. It can be seen that this control file includes an additional command specifically for analyzing data for measuring affective variables. When an instrument contains items with different response categories (e.g., 5 *strongly agree*, 4 *agree*, 3 *undecided*, 2 *disagree*, and 1 *strongly disagree*),

```
File  Edit  Format  View  Help
44324342411312323325543424
242333244422 42244244344334
55554545211114214444454555
33434434324434424443444344
44444444453434535454544433
44443432222222324444442323
54445444222222334454443443
54455555221311215354554455
54354231221423555514351455
544444444332243243444444224
55235344342244434545245444
44444414222 23425454444455
44444324244234344444244434
52435433321315424542554554
34233432222323424443443434
42442435422322225452444244
44332442145225534433313414
53334444232313424323443444
44344444442222244444334424
44344444422332324443443434
444544454444343143344 454254
44344444241215314544354413
44444442222423424444444422
54443425242322425543444522
44433324222222224444454424
44344443344344334444444344
55555555111111115555551555
4  33444332123432444444434434
44234233434224424434444444
54544454424224552445454444
43445435241412415121251155
44245434221222454454452454
44444234342224444424343422
43344343321232234445454555
24244434422224222444542255
44444434151414412444344445
44534524232442345445544433
44334444222222424324443424
55535555555111111155555555
44344444242223434444444444
55545444254535535545543454
54434434222315514445444544
54445445431224415445533255
54454544221313414432543434
35433433221444424424444442
44434344444444443344434414
43444444142222424422222434
44545455223332445445454445
454435 4243424425441542132
```

Figure 4.3 Shows a data file for measuring student attitude toward science. All items have five response categories: 1 (*strongly disagree*), 2 (*disagree*), 3 (*undecided*), 4 (*agree*), and 5 (*strongly agree*). Student IDs are not included in this data file.

the command CLFILE = * is used. If an instrument contains more than one tem format, for example one format is the 5-category Likert scale, and another is a rating scale in frequencies from *always* (3), *sometimes* (2), and *rarely* (1), then the command ISGRUOPS should be used to specify which items are in which same response format. In this case, CLFILE needs to specify category labels for each group of items by using $a + b$ where a is any item number within a group of items with a same format, and b is the code for one response to all items in the group. For example,

```
File  Edit  Format  View  Help
&INST
  TITLE = "Student Attitude toward Science"
   ITEM1 = 1 ; column of response to first item in data record
     NI = 27 ; number of items
  CODES = 12345 ; valid codes in data file
CLFILE = * ; category label file for category naming
1 Strongly Disagree
2 Disagree
3 Undecided
4 Agree
5 Strongly Agree
* ; end of CLFILE=* list
&END
Item 1
Item 2
Item 3
Item 4
Item 5
Item 6
Item 7
Item 8
Item 9
Item 10
Item 11
Item 12
Item 13
Item 14
Item 15
Item 16
Item 17
Item 18
Item 19
Item 20
Item 21
Item 22
Item 23
Item 24
Item 25
Item 26
Item 27
END LABELS
```

Figure 4.4 Shows the Winsteps control file for the data file in Figure 4.3. The "CLFILE" command specifies the labels for the item response categories: 1 (*strongly disagree)*, 2 (*disagree)*, 3 (*undecided)*, 4 (*agree)*, and 5 (*strongly agree)*.

```
ISGROUPS = 1111122222
CLFILE = *
3+A Strongly disagree
3+B Disagree
3+C Agree
3+D Strongly agree
7+A Never
7+B Sometimes
7+C Often
7+D Always
*
```

In the above example, ISGROUPS indicates that there are 10 items, and the first five (Group 1) are in one format and the other five (Group 2) are in another format. 3+A indicates that A for any item (e.g., Item 3) in Group 1 represents *strongly disagree*, 3+B indicates that B for any item (e.g., Item 3) in Group 1 represents *disagree*, and so forth; 7+A indicates that A for any item (e.g., Item 7) in Group 2 represents *never*, 7+B indicates that B for any item (e.g., Item 7) in Group 2 represents sometimes, and so forth.

For items that need to be reversely coded, Winsteps command pairs IREFER and IVALUE* can be used. For example,

```
IFEFER = AAAAABBBBB
CODES = 12345
IVALUEA = 12345
IVALUEB = 54321
```

In the above example, there are 10 items with the first five items coded as 12345 as entered in the data file, while the last five items are reversed coded as 54321 by changing 1 to 5, 2 to 4, 3 to 3, 4 to 2, and 5 to 1.

Review Item Fit, Wright Map, Unidimensionality, and Reliability

Statistical properties of items will then be studied using the pilot-test data. Rasch analysis produces a variety of output tables and diagrams. As usual, a variety of fit statistics should be reviewed. Particular attention should be paid to the item category structure, because questions used for measuring affective variables typically involve more than 2 categories. Item category structure refers to the hierarchical progression among the categories. For Likert-scale questions, item categories are degrees of agreement from *strongly disagree* to *strongly agree*; for rating scale questions, item categories are various levels of ratings assigned to an aspect; for semantic differential items, item categories are the steps between two bipolar adjectives; and for inventory or checklist questions, item categories are two incidents representing yes or no. Categories should clearly form a progression. Figure 4.5 shows an ideal category structure for a three-category Likert-scale question.

In Figure 4.5, the three categories (agree, neutral, and disagree) clearly form a progression. As a respondent's latent trait (e.g., attitude) increases, which is indicated by the increase in value on the *x* axis, the respondent is becoming less and less likely to be in favor of the construct. Although the three categories overlap to some degree, each of them has its own unique zone of most probable response category. Specifically, when a respondent's latent trait level (e.g., attitude) is smaller than –1, the respondent is most likely to be in favor, that is, agree, with the construct; when a respondent's latent trait level is between –1 and +1, the respondent is most likely to be neutral on the construct; and when the respondent's latent trait level is greater than +1, the

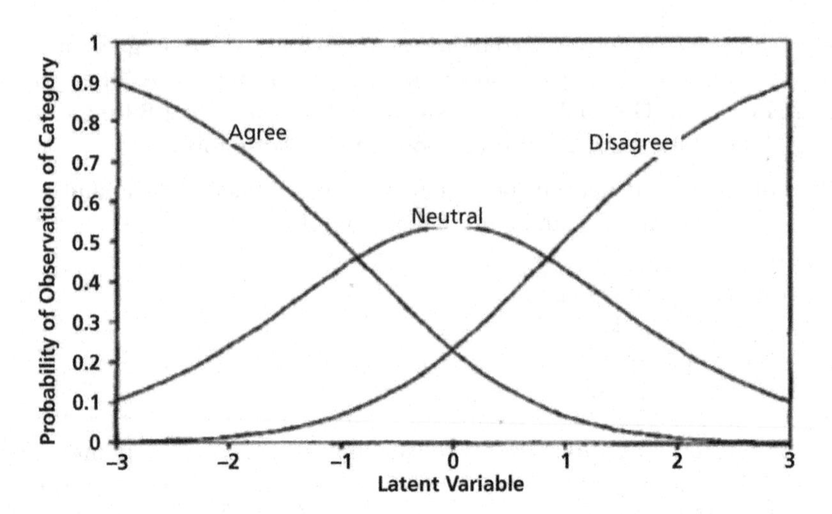

Figure 4.5 Expected category structure for a 3-category Likert-scale question (adapted from Linacre, 2004, p. 264). Each category has its own peak in its characteristic curve.

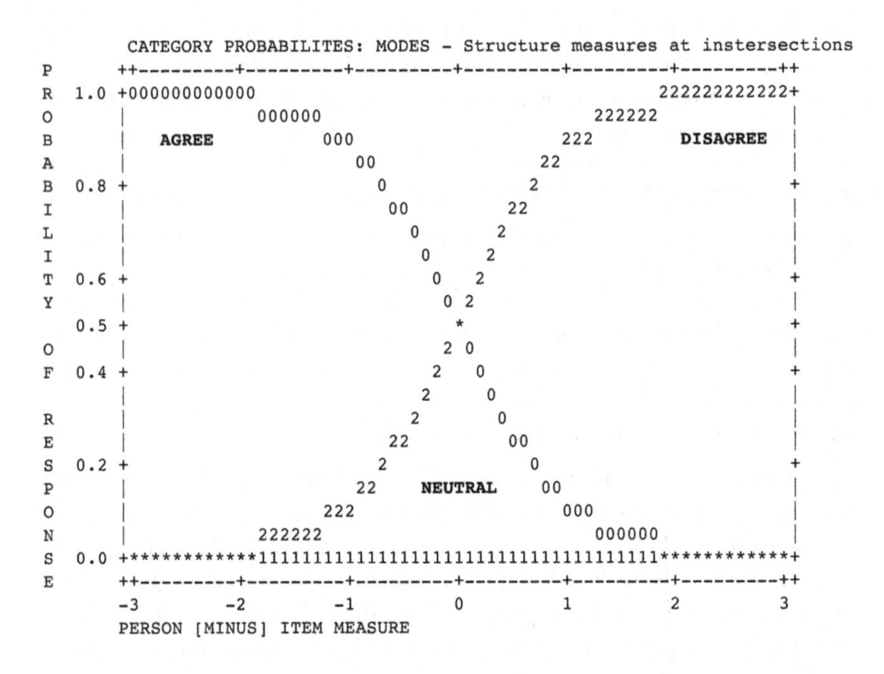

Figure 4.6 A sample inappropriate category structure for a Likert-scale question. Category "Neutral" is subsumed by the other two categories; category "Neutral" has no distinct characteristic curve.

respondent is most likely to be in disfavor, that is, disagree, of the construct. Ideally, step difficulties should increase by at least 1.4 logits but less than 5 logits between two adjacent categories (Linacre, 2004).

Considering a different category structure as shown in Figure 4.6, there are three categories contained in each of the questions, but only two categories have their unique zones of responses; category "neutral" is subsumed by the above two categories. In this case, the three categories do not form a clear progression. We could remove category "neutral" and make the item a binary one, that is, agree or disagree.

Figure 4.7 shows the empirical category structure for a Likert scale concerning university engineering students' attitude toward nanotechnology. In the Likert-scale questions, there are five categories from *strongly agree* (5), *agree* (4), *neutral* (3), *disagree* (2), and *strongly disagree* (1). However, the category structure graph shows only 4 categories; Category 5 is missing. This indicates that no respondent chose category 5 (*strongly agree*). In this

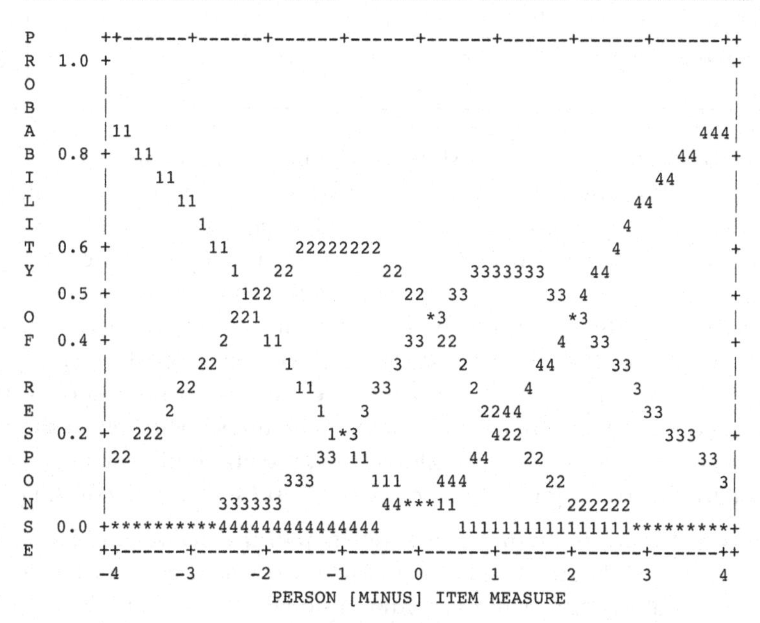

Figure 4.7 Category structure for a Likert scale on attitude toward nanotechnology. Although each category (i.e., 1, 2, 3, and 4) has its own distinct characteristic curve, Category 5 is not shown in the graph, because none selected the response (i.e., *strongly agree*) for any item.

case, categories 1 to 4 form a clear progression; Category 5 is not needed, thus should be removed from the items.

Figure 4.8 shows the Wright map for a 7-item Likert scale measuring university engineering students' attitude toward nanotechnology. The seven items are:

	SA (5)	A (4)	U (3)	D (2)	SD (1)
1. All engineering students should know about nanotechnology.					
2. More funding should be invested in developing nanotechnology.					
3. Nanotechnology can stimulate economic growth.					
4. All engineering majors should take nanotechnology courses.					
5. Nanotechnology courses should be offered in high school.					
6. Nanotechnology is critical for national defense.					
7. Nanotechnology can solve important social problems.					

Although item fit statistics showed adequate fit between data and the rating scale Rasch model, the range of item difficulties and person abilities shown in Figure 4.8 indicates a few noticeable gaps. For example, the difference in difficulty or endorsability between Item 1 (Q1) and Item 3 (Q3) is almost 1 logit, and there are very few students whose attitudes fall within this range. Thus the measurement will not be able to differentiate students within this range. Also, we see that the range of item difficulty is not large enough to cover the student attitude range, because at both ends of the attitude continuum, there are no items to measure students. Even though the sample size is small for this pilot-testing, the above information is still informative for revising the instrument for the subsequent field-testing.

Figure 4.9 shows the dimensionality map of the nanotechnology sale. Figure 4.9 shows that the first factor in the unexplained variance had a size of 2.3 eigenvalues, suggesting multidimensionality. Only Items 2, 3, and 7 fall along one dimension, and the other four items do not conform with the same dimension because their residuals correlate highly with one potential additional construct. The content of items 1, 4, 5 and 6 seems quite different from each other. In fact, the Rasch measures explained only 43%

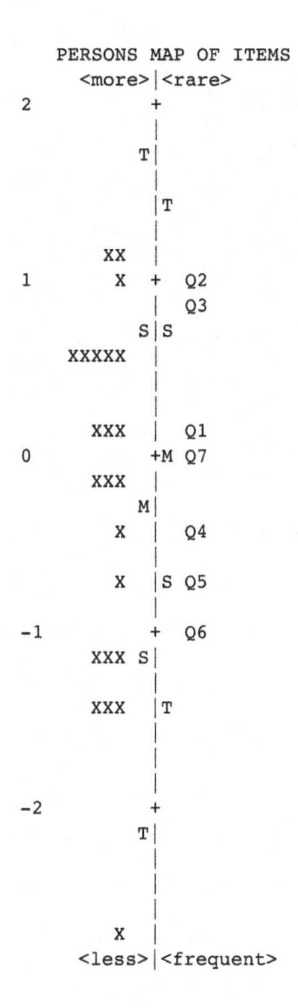

Figure 4.8 Wright map for the measurement instrument of university engineer students' attitude toward nanotechnology. The difficulty range of items does not match the ability range of students, and there are noticeable gaps in the distribution of items.

of the total variance in observations, leaving 57% unexplained. This lack of unidimensionality would call the construct validity of the measures into question.

Reliability evidence of affective measurement is reviewed based on item and subject measures and the construct measure as a whole. After item and person parameters are calibrated, standard errors of measurements for each item and person should be reviewed. Person separation index and its equivalent Cronbach's alpha should be reviewed for their adequacy of reliability for person measures as a whole.

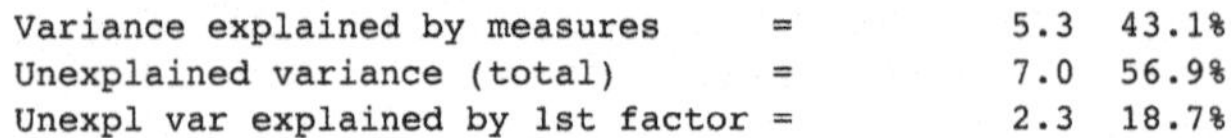

```
Variance explained by measures     =        5.3  43.1%
Unexplained variance (total)       =        7.0  56.9%
Unexpl var explained by 1st factor =        2.3  18.7%
```

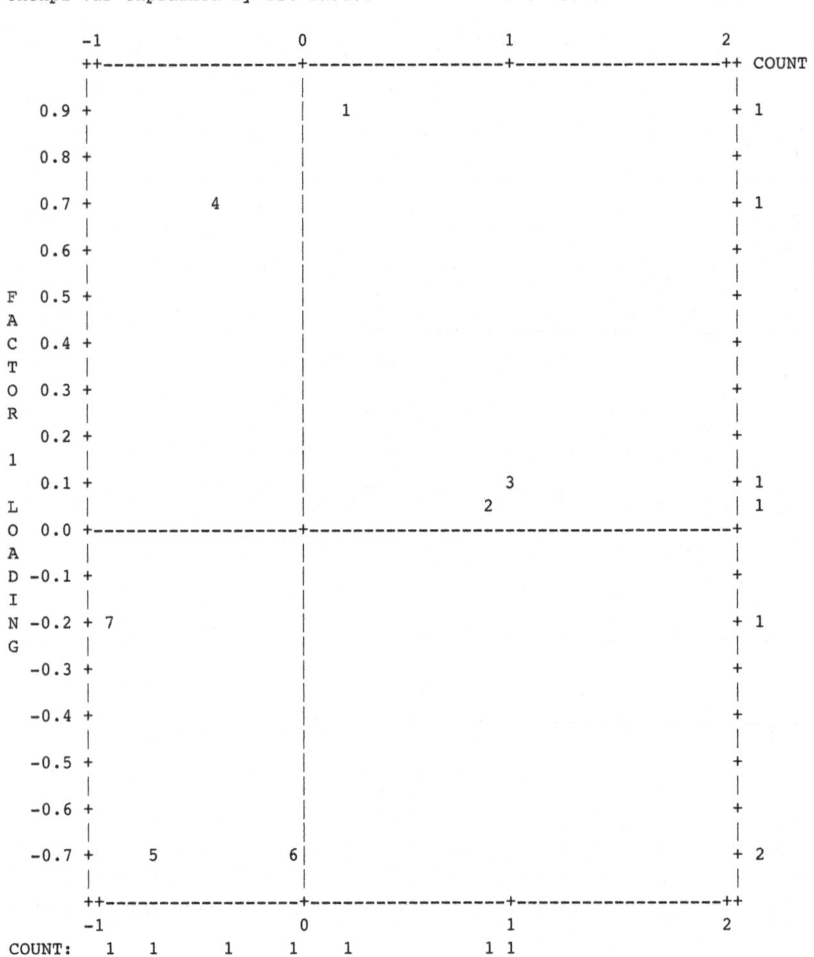

Figure 4.9 Dimensionality map for the measurement of university engineer students' attitude toward nanotechnology. Items 1, 4, 5, and 6 have loadings beyond ±0.4, suggesting multidimensionality.

Examine Invariance Properties

After items have been found to fit the model and the range of items is aligned with the range of student abilities, a final quality assurance check is to examine item and person measure invariance. As a desired benefit of Rasch modeling, item difficulty estimates should not vary according to student subgroup characteristics (e.g., gender), and student ability estimates should not vary by item formats or delivery modes. Rasch modeling software typically provides various statistics to help examine the above properties; this is commonly called analysis of Differential Item Functioning (DIF; for item invariance) and Differential Person Functioning (DPF; for person invariance). In order to conduct DIF and DPF analyses, information on student and item characteristics must be included in additional columns as variables in the data file.

Establish Validity Claims

Evidence for the validity of measures may be established in terms of test content, response processes, internal structure, external structure, and consequences. Examination of test content may involve review of the definition of the affective construct to be measured and how the test items cover the entire range of variation of the construct. Important questions to ask are: (a) "Is the definition based on commonly accepted theories?"; (b) "Is the defined construct clearly affective, that is, involving feeling, judgment, or values?"; (c) "Is there an underlying linear progression for the construct?"; and (d) "How do the test items represent the variation of specific behaviors indicating the construct?" Answers to the above questions should be based on the stated theoretical framework and the test specification. A negative answer to any of the questions will call the measurement validity into question.

Response processes related validity is established by reviewing examinees' response patterns in terms of the above discussed fit statistics and response category structure. In addition, qualitative studies of item functioning during item tryout, and observation, interview, and artifacts of respondents about the measurement instruments can also enhance claims about the validity of the measurement instrument in terms of response processes.

Internal structure related validity evidence may be examined in a number of ways. In addition to reviews of various item-fit statistics and item category structure discussed earlier, the Wright map and dimensionality map also provide additional evidence to support the claim for the construct validity.

External structure related validity can be examined by conducting correlation studies between examinees' Rasch scale scores and their measures of other variables. One important consideration is obviously the selection of the criterion variable for the correlation study. Finally, for consequence-related validation, because results from affective measurement are typically not used for important decision-making such as accountability, a decision procedure related validity is thus of less importance for affective measurement validation.

Develop Guidelines for Use of the Instrument

As with any measurement instrument, appropriate documentation should be available to facilitate users to appropriately use the measurement instrument. Documentation should include the construct definition, test specification, scoring, validity and reliability evidence, and score reporting and interpretation. In order to help users to convert raw scores into Rasch scale scores, a raw score to Rasch scale score conversion table should be included. Table 4.2 is a Winsteps output table for the engineering students' attitude toward nanotechnology instrument described earlier. It gives the logit equivalent score for each of the possible raw scores ranging from 7 to 28. As noted earlier, although each question has five categories, only four of five categories were functional, because no one selected *strongly agree* (SA) to any of the questions. This explains why the total possible raw score in the table is 28. The logit scale scores are set with a mean of 50 and a standard deviation of 10.

Chapter Summary

The affective domain includes various variables related to interests, attitudes, valuing, appreciation, and adjusting; it is closely related to the cognitive and psychomotor domains. One influential conceptualization of affective variables is the taxonomy of affective education objectives (Krathwohl et al., 1964). This taxonomy perceives affective variables as a continuum from external influences to internal driving forces. This continuum is also

TABLE 4.2 Raw Score to Rasch Scale Score Conversion Table for the Nanotechnology Attitude Scale

Score	Measure	SE	Score	Measure	SE	Score	Measure	SE
7	–5.22E	18.70	15	42.47	5.80	23	68.40	6.06
8	7.98	10.81	16	45.77	5.70	24	72.28	6.42
9	16.66	8.24	17	48.98	5.63	25	76.76	7.01
10	22.55	7.22	18	52.13	5.59	26	82.38	8.09
11	27.35	6.67	19	55.25	5.59	27	90.85	10.73
12	31.57	6.33	20	58.38	5.61	28	103.94E	18.66
13	35.42	6.10	21	61.57	5.69			
14	39.03	5.93	22	64.88	5.83			

hierarchical, that is, a higher category subsumes all its lower categories. The Krathworhl's affective taxonomy includes the following five levels or categories: receiving, responding, valuing, organization, and characterization. Each level also contains a few subcategories. Interest is primarily related to categories of receiving, responding, and valuing; attitude is primarily related to categories of valuing and organization.

Specifically, in science education, various definitions of attitude are available. Attitude contains three dimensions: feeling, cognition, and behavior. No matter what attitude is to be measured, it is important to note that attitude cannot be separated from its context and the underlying body of influences that determine its real significance. Motivation is another affective construct commonly studied in science education. Although there is no universally-agreed upon definition of motivation, motivation relates to one's desire to act or not to act. Motivation is also contextualized, such as motivation to learn science, motivation to teach science, motivation to implement a new science curriculum, to name a few.

There is a large collection of measurement instruments of affective variables in science education; some are for students and others for teachers. The validity and reliability of measures of these instruments vary greatly. In particular, the construct validity of measures of many instruments may be questioned because our understanding of the constructs has changed over the years. Continuously validating and expanding the variety of evidence for validity and reliability claims of published instruments is necessary.

Another issue in using current measurement instruments is associated with the ordinal nature of raw total scores. It is common for attitude measurement instruments to adopt a Likert scale. One serious issue associated with the Likert scale is the use of a total score by adding individual item scores. Values such as 1–5 assigned to five choices of a statement do not have the same origin and interval unit because they are not on a ratio or interval scale. The consequence of being non-interval or ratio is that we cannot meaningfully add individual item scores into a total score. Also, different items may have different degrees of endorsability. Rasch modeling addresses the above issues.

Developing measurement instruments for affective variables based on Rasch modeling starts with a clear definition of the affective construct to be measured. The defined construct must be consistent with an accepted theory. The construct should also suggest a hierarchy of subjects' behaviors along the construct. Affective measurement instruments typically use question formats of the Likert scale, a rating scale, semantic differential, and checklist/inventory. Rasch modeling can help identify best category structure of items, and establish evidence for the validity and reliability of the measurement instrument.

Exercises

1. Read Blalock et al.'s (2008) review of attitude toward science instruments. Apply the validity and reliability standards described in Chapter 1 to evaluate the appropriateness of their scoring rubric used to evaluate the measurement instruments, is the rubric reasonable? How would you revise, if you deem necessary?
2. The development of the instrument, *Changes in Attitudes About the Relevance of Science* (CARS; Siegel & Ranney, 2003), was based on Rasch modeling. Read the article carefully and critique the adequacy of the application of Rasch modeling in the development of the instrument. Based on your critique, summarize the instrument's strengths and weaknesses and recommend appropriate uses of the instrument.
3. Suppose that you are going to develop a measurement instrument on high school students' attitude toward biotechnology careers. Based on your best knowledge of biotechnology careers, complete the following initial steps in developing the measurement instrument: (a) define the construct and (b) identify behaviors of the defined construct and write a few sample items.

References

Abd-El-Khalick, F., Summers, R., Said, Z., Wang, S., & Culbertson, M. (2015). Development and large-scale validation of an instrument to assess Arabic-speaking students' attitudes toward science. *International Journal of Science Education, 37*(16), 2637–2663.

Adams, W. K., Perkins, K. K., Podolefsky, N. S., Dubson, M., Finkelstein, N. D., & Wieman, C. E. (2006). New instrument for measuring student beliefs about physics and learning physics: The Colorado Learning Attitudes about Science Survey. *Physical Review Special Topics—Physics Education Research, 2*(1), 010101.

Ajzen, I., & Fishbein, M. (2005). The influence of attitudes on behavior. In D. Albarracin, B. T. Johnson, & M. P. Zanna (Eds.), *The handbook of attitude* (pp. 173–221). Mahwah, NJ: Erlbaum.

American Association for the Advancement of Science (1993). *Benchmarks for science literacy*. New York, NY: Oxford University Press.

Baldwin, J. A., Ebert-May, D., & Burns, D. J. (1999). The development of a college biology self-efficacy instrument for nonmajors. *Science Education, 83*, 397–408.

Bandura, A. (2006). Guide for constructing self-efficacy scales. *Self-efficacy beliefs of adolescents, 5*, 307–337.

Beghetto, R. A. (2009). Correlates of intellectual risk taking in elementary school science. *Journal of Research in Science Teaching, 46*(2), 210–223.

Blalock, C. L., Lichtenstein, M. J., Owen, S. V., Pruski, L. A., Marshall, C. E., & Toepperwein, M. A. (2008). In pursuit of validity: A comprehensive review of science attitude instruments 1935–2005. *International Journal of Science Education, 30*(7), 961–977.

Britner, S. L. (2008). Motivation in high school science students: A comparison of gender differences in life, physical, and earth science classes. *Journal of Research in Science Teaching, 45*(8), 955–970.

Brown, W. R. (1973). Checklist for assessment of science teachers and its use in a science preservice teacher education project. *Journal of Research in Science Teaching, 10*(3), 243–249.

Button, W. H., & Stephens, L. (1963). Measuring attitudes toward science. *School Science and Mathematics, 63*, 43–49.

Caleon, I., & Subramaniam, R. (2008). Attitudes towards science of intellectually gifted and mainstream upper primary students in Singapore. *Journal of Research in Science Teaching, 45*(8), 940–954.

Campbell, J. R. (1971). Cognitive and affective process development and its relation to a teacher's interaction ratio. *Journal of Research in Science Teaching, 8*(4), 317–323.

Cheung, D. (2009). Developing a scale to measure students' attitudes toward chemistry lessons. *International Journal of Science Education, 31*(16), 2185–2203.

Chiang, W.-W., & Liu, C. (2014). Scale of academic emotion in science education: Development and validation. *International Journal of Science Education, 36*(6), 908–928.

Chu, H.-E., Lee, E. A., Ko, H. R., Hee, S. D., Moon, N., Min, B. M., & Hee, K. K. (2007). Korean year 3 children's environmental literacy: A prerequisite for a Korean environmental education. *International Journal of Science Education, 29*(6), 731–746.

Clarkeburn, H. (2002). A test for ethical sensitivity in science. *Journal of Moral Education, 31*(4), 439–453.

Connelly, F. M., Finegold, M., Wahlstrom, M. W., & Ben-Peretz, M. (1977). TOT-SI matches teacher to curriculum. *Science Teacher, 44*(2), 24–26.

Dalgety, J., Coll, R. K., & Jones, A. (2003). Development of chemistry attitudes and experiences questionnaire (CAEQ). *Journal of Research in Science Teaching, 40*(7), 649–668.

Donnelly, L. A., & Boone, W. J. (2007). Biology teachers' attitudes toward and use of Indiana's evolution standards. *Journal of Research in Science Teaching, 44*(2), 236–257.

Doran, R., Lawrenz, F., & Helgeson, S. (1994). Research on assessment in science. In D. Gabel (Ed.), *Handbook of research on science teaching and learning* (pp. 388–442). New York, NY: Macmillan.

Eagly, A. H. (1992). Uneven progress: Social psychology and the study of attitudes. *Journal of Personality and Social Psychology, 63,* 693–710.

Erb, T. O., & Smith, W. S. (1984). Validation of the attitude toward women in science scale for early adolescents. *Journal of Research in Science Teaching, 21,* 391–397.

Evia, J. R., Peterman, K., Cloyd, E., & Besley, J. (2018). Validating a scale that measures scientists' self-efficacy for public engagement with science. *International Journal of Science Education, Part B, 8*(1), 40–52.

Fortus, D., & Weiss-Wedder, D. (2014). Measuring students' continuing motivation for science learning. *Journal of Research in Science Teaching, 51*(4), 497–522.

Fowler, S. R., Zeidler, D. L., & Sadler, T. D. (2009). Moral sensitivity in the context of socioscientific issues in high school science students. *International Journal of Science Education, 31*(2), 279–296.

Francis, L. J., & Geer, J. E. (1999). Measuring attitudes toward science among secondary school students: The affective domain. *Research in Science & Technology Education, 17,* 219–226.

Fraser, B. J. (1978). Development of a test of science-related attitudes. *Science Education, 62,* 509–515.

Gardner, P. L. (1975). Attitudes to science. *Studies in Science Education, 2*(1), 1–41.

Germann, P. J. (1988). Development of the attitude toward science in school assessment and its use to investigate the relationship between science achievement and attitude toward science in school. *Journal of Research in Science Teaching, 25,* 689–703.

Glynn, S. M., & Kobbala Jr., T. R. (2006). Motivation to learn in college science. In J. Mintzes & W. H. Leonard (Eds.), *Handbook of college science teaching* (pp. 25–32). Arlington, VA: National Science Teachers Association Press.

Glynn, S. M., Brickman, P., Armstrong, N., & Taasoobshirazi, G. (2011). Science Motivation Questionnaire II: Validation with science majors and nonscience majors. *Journal of Research in Science Teaching, 48*(10), 1159–1176.

Glynn, S., Taasoobshirazi, G., & Brickman, P. (2009). Science motivation questionnaire: Construct validation with nonscience majors. *Journal of Research in Science Teaching, 46*(2), 127–146.

Gogolin, L., & Swartz, F. (1992). A quantitative and qualitative inquiry into the attitudes toward science of nonscience college majors. *Journal of Research in Science Teaching,* 29, 487–504.

Good, R. G. (1971). Study of the effects of a 'student-structured' laboratory approach to elementary science education methods courses: Affective domain. *Journal of Research in Science Teaching, 8*(3), 255–262.

Gungor, A., Eryrilmaz, A., & Fakroglu, T. (2007). The relationship of freshmen's physics achievement and their related affective characteristics. *Journal of Research in Science Teaching, 44*(8), 1036–1056.

Guttman, L. (1944). A basis for scaling qualitative data. *American Sociological Review, 9,* 139–150.

Herodotou, C., Kyza, E. A., Nicolaidou, I., Hadjichambis, A., Kafouris, D., & Terzian, F. (2012). The development and validation of the GMOAS, an instrument measuring secondary school students' attitudes towards genetically modified organisms. *International Journal of Science Education, Part B, 2*(2), 131–147.

Hopkins, K. D. (1998). *Educational and psychological measurement and evaluation* (8th ed). Needham Height, MA: Allyn & Bacon.

Hoover, K. H., & Schutz, R. E. (1963). Development of a measure of conservation attitudes. *Science Education, 47*(1), 63–68.

Ibrahim, A., Aulls, M. W., & Shore, B. M. (2016). Development, validation, and factorial comparison of the McGill Self-Efficacy of Learners for Inquiry Engagement (McSELFIE) survey in natural science disciplines. *International Journal of Science Education, 38*(16), 2450–2476.

Jones, M. G., & Carter, G. (2007). Science teacher attitudes and beliefs. In S. K. Abell & N. G. Lederman (Eds.), *Handbook of research on science education* (pp. 1067–1104). New York, NY: Routledge.

Juriševi, M., Glažar, S. A., Puko, C. R., & Devetak, I. (2007). Intrinsic motivation of preservice primary school teachers for learning chemistry in relation to their academic achievement. *International Journal of Science Education, 30*(1), 87–107.

Kesamang, M. E. E., & Taiwo, A. A. (2002). The correlates of the socio-cultural background of Botswana junior secondary school students with their attitudes towards and achievements in science. *International Journal of Science Education, 24*(9), 919–940.

Kier, M. W., Blanchard, M. R., Osborne, J. W., & Albert, J. L. (2013). The development of the STEM Career Interest Survey (STEM-CIS). *Research in Science Education, 44*(3), 461–481.

Kind, P., Jones, K., & Barmby, P. (2007). Developing attitudes towards science measures. *International Journal of Science Education, 29*(7), 871–893.

Kitchen, E., Reeve, S., Bell, J., Sudweeks, R. R., & Bradshaw, W. S. (2007). The development and application of affective assessment in an upper-level cell biology course. *Journal of Research in Science Teaching, 44*(8), 1057–1087.

Koballa, T. R., & Glynn, S. M. (2007). Attitudinal and motivational constructs in science learning. In S. K. Abell & N. G. Lederman (Eds.), *Handbook of research on science education* (pp. 75–102). New York, NY: Routledge.

Konnemann, C., Asshoff, R., & Hammann, M. (2016). Insights into the diversity of attitudes concerning evolution and creation: A multidimensional approach. *Science Education, 100*(4), 673–705.

Krathwohl, D. R., Bloom, B. S., & Masia, B. B. (1964). *Taxonomy of educational objectives: Handbook 2 affective domain.* New York, NY: Longman.

Laforgia, J. (1988). The affective domain related to science education and its evaluation. *Science Education, 72*, 407–421.

Lan, Y.-L. (2012). Development of an attitude scale to assess K–12 teachers' attitudes toward nanotechnology. *International Journal of Science Education, 34*(8), 1189–1210.

Lazarowitz, R., & Lee, A. E. (1976). Measuring inquiry attitudes of secondary science teachers. *Journal of Research in Science Teaching, 13*(5), 455–460.

Lee, G., Kwon, J., Park, S.-S., Kim, J.-W., Kwon, H.-G., & Park, H.-K. (2003). Development of an instrument for measuring cognitive conflict in secondary-level science classes. *Journal of Research in Science Teaching, 40*(6), 585–603.

Lee, M.-H., Johanson, R. E., & Tsai, C.-C. (2008). Exploring Taiwanese high school students' conceptions of and approaches to learning science through a structural equation modeling analysis. *Science Education, 92*(2), 191–220.

Lichtenstein, M. J., Owen, S., Blalock, C., Liu, Y., Ramirez, K. A., Pruski, L. A.,... Toepperwein, M. A. (2008). Psychometric reevaluation of the scientific attitude inventory-revised (SAI-II). *Journal of Research in Science Teaching, 45*(5), 600–616.

Linacre, J. M. (2004). Optimizing rating scale category effectiveness. In E. V. Smith Jr., & R. M. Smith (Eds.), *Introduction to Rasch measurement* (pp. 258–278). Maple Grove, MN: JAP Press.

Lumpe, A., Haney, J., & Czerniak, C. (2000). Assessing teachers' beliefs about their science teaching context. *Journal of Research in Science Teaching, 37,* 275–292.

Martin, M. O., Mullis, I. V. S., Foy, P., & Stanco, G. M. (2012). *TIMSS 2011 international results in science.* TIMSS & PIRLS International Study Center: Lynch School of Education, Boston College.

Mazas, B., Manzanal, M. R. F., Zarza, F. J., & Maria, G. A. (2013). Development and validation of a scale to assess students' attitude towards animal welfare. *International Journal of Science Education, 35*(11), 1775–1799.

Menis, J. (1989). Attitudes towards school, chemistry and science among upper secondary chemistry students in the United States. *Research in Science and Technological Education, 7,* 183–190.

McGinnis, J. R., Kramer, S., Shama, G., Graeber, A. O., Parker, C. A., & Watanabe, T. (2002). *Journal of Research in Science Teaching, 39*(8), 713–737.

Moore, K. D. (1977). Development and validation of a science teacher needs-assessment profile. *Journal of Research in Science Teaching, 14*(2), 145–149.

Moore, R. W., & Sutman, F. X. (1970). The development, field test and validation of an inventory of scientific attitudes. *Journal of Research in Science Teaching, 7*(2), 85–94.

Moore, R. W., & Foy, R. L. H. (1997). The scientific attitude inventory: A revision (SAI-II). *Journal of Research in Science Teaching, 34,* 327–336.

Mulkey, L. (1989). Validation of early childhood attitudes toward women in science scale (ECWiSS): A pilot administration. *Journal of Research in Science Teaching, 26*(8), 737–753.

Munby, H. (1983a). Thirty studies involving the "scientific attitude inventory": What confidence can we have in this instrument? *Journal of Research in Science Teaching, 20,* 141–162.

Munby, H. (1983b). *An investigation into the measurement of attitudes in science education.* Columbus, OH: ERIC.

Munby, H. (1997). Issues of validity in science attitude measurement. *Journal of Research in Science Teaching, 34*(4), 337–341.

National Research Council. (1996). *National science education standards.* Washington, DC: National Academic Press.

Odom, A., Stoddard, E. R., & LaNasa, S. M. (2007). Teacher practices and middle-school science achievements. *International Journal of Science Education, 29*(11), 1329–1346.

Oppermann, E., Brunner, M., & Eccles, J. S. (2018). Uncovering young children's motivational beliefs about learning science. *Journal of Research in Science Teaching, 55,* 399–421.

Osborne, J., Simon, S., & Collins, S. (2003). Attitude towards science: A review of the literature and its implications. *International Journal of Science Education, 25*(9), 1049–1079.

Osgood, C. E., Suci, G. J., & Tannenbaum, P. H. (1971). *The measurement of meaning.* Chicago: University of Illinois Press.

Owen, S. V., Toepperwein, M. A., Pruski, L. A., Blalock, C. L., Liu, Y., Marshall, C. E., Lichtenstein, M. J. (2007). Psychometric reevaluation of the women in science scale (WiSS). *Journal of Research in Science Teaching, 44*(10), 1461–1478.

Owen, S. V., Toepperwein, M. A., Marshall, C. E., Lichtenstein, M. J., Blalock, C. L., Liu, Y., . . . Grimes, K. (2008). Finding pearls: Psychometric reevaluation of Simpson-Troost Attitude Questionnaire (STAQ). *Science Education, 92*(6), 1076–1095.

Parkinson, J., Hendley, D., Tanner, H., & Stables, A. (1998). Pupils' attitudes to science in key stage 3 of the national curriculum: A study of pupils in South Wales. *Research in Science and Technological Education, 16,* 165–177.

Pekrun, R., Goetz, T., Frenzel, A. C., Barchfeld, P., & Perry, R. P. (2011). Measuring emotions in students' learning and performance: The Achievement Emotions Questionnaire (AEQ). *Contemporary Education Psychology, 36*(1), 36–48.

Pell, T., & Jarvis, T. (2001). Developing attitude to science scales for use with children of ages five to eleven years. *International Journal of Science Education, 23,* 847–862.

Pell, A., & Jarvis, T. (2003). Developing attitude to science education scales for use with primary teachers. *International Journal of Science Education, 25*(10), 1273–1295.

Prokop, P., Leskova, A., Kubiatlo, M., & Diran, C. (2007). Slovakian students' knowledge of and attitudes toward biotechnology. *International Journal of Science Education, 29*(7), 895–907.

Raker, J. R., Gibbons, R. E., & Cruz-Ramirez de Arellano, D. (2019). Development and evaluation of the organic chemistry-specific achievement emotions questionnaire (AEQ-OCHEM). *Journal of Research in Science Teaching, 56*(2), 163–183.

Riggs, I. M., & Enochs, L. E. (1990). Toward the development of an elementary teacher's science teaching efficacy belief instrument. *Science Education, 74,* 625–637.

Ritter, J. M., Boone, W. J., & Rubba, P. A. (2001). Development of an instrument to assess prospective elementary teacher self-efficacy beliefs about equitable science teaching and learning (SEBEST). *Journal of Science Teacher Education, 12,* 175–198.

Romine, W. L., Water, E. M., Bosse, E., & Todd, A. N. (2017). Understanding patterns of evolution acceptance – a new implementation of the measure of acceptance of the theory of evolution (MATE) with Midwestern university students. *Journal of Research in Science Teaching, 54*(5), 642–671.

Russell, J., & Hollander, S. (1975). A biology attitude sale. *American Biology Teacher, 37*(5), 270–273.

Rutledge, M. L., & Warden, M. A. (1999). The development and validation of the measure of acceptance of the theory instrument. *School Science and Mathematics, 99*(1), 13–18.

Rutledge, M. L., & Sadler, K. C. (2007). Reliability of the measure of acceptance of the theory of evolution (MATE) instrument with university students. *The American Biology Teacher, 69*(6), 332–335.

Salta, K., & Tzougraki, C. (2004). Attitudes toward chemistry among 11th grade students in high schools in Greece. *Science Education, 88*(4), 535–547.

Schwirian, P. M. (1968). On measuring attitudes toward science. *Science Education, 52,* 172–179.

Shrigley, R. (1974). The attitude of preservice elementary teachers toward science. *School Science and Mathematics, 74,* 243–250.

Siegel, M. A., & Ranney, M. A. (2003). Developing the changes in attitude about the relevance of science (CARS) questionnaire and assessing two high school science classes. *Journal of Research in Science Teaching, 40*(8), 757–775.

Simpson, R. D., Koballa, T. R., Oliver, J. S., & Crawley, F. (1994). Research on the affective dimension of science learning. In D. Gable (Ed.), *Handbook of research on science teaching and learning* (pp. 211–234). New York, NY: Macmillan.

Simpson, R. D., & Troost, K. M. (1982). Influences on commitment to learning of science among adolescent students. *Science education, 66*(5), 763–781.

Sjaastad, J. (2013). Measuring the ways significant persons influence attitudes towards science and mathematics. *International Journal of Science Education, 35*(2), 192–212.

Skinner, E., Saxton, E., Currie, C., & Shusterman, G. (2017). A motivational account of the undergraduate experience in science: Brief measures of students' self-system appraisals, engagement in coursework, and identify as a scientist. *International Journal of Science Education, 39*(17), 2433–2459.

Skinner, R., & Barcikowski, R. S. (1973). Measuring specific interest in biological, physical, and earth sciences in intermediate grade levels. *Journal of Research in Science Teaching, 10*(2), 153–158.

Smith, M. U., Snyder, S. W., & Devereaux, R. S. (2016). The GAENE–Generalized Acceptance of EvolutioN Evaluation: Development of a new measure of evolution acceptance. *Journal of Research in Science Teaching, 53*(9), 1289–1315.

Sondergeld, T. A., & Johnson, C. C. (2014). Using Rasch measurement for the development and use of affective assessments in science education research. *Science Education, 98*(4), 581–613.

Summers, R., & Abd-El-Khalick, F. (2018). Development and validation of an instrument to assess student attitudes toward science across grades 5 through 10. *Journal of Research in Science Teaching, 55*(2), 172–205.

Syed Hassan, S. S. (2018). Measuring attitude towards learning science in Malaysian secondary school context: Implications for teaching. *International Journal of Science Education, 40*(16), 2044–2059.

Tee, O. P., & Subramaniam, R. (2018). Comparative study of middle school students' attitudes toward science: Rasch analysis of entire TIMSS 2011 attitudinal data for England, Singapore, and the USA, as well as psychometric properties of attitudes scale. *International Journal of Science Education, 40*(3), 268–290.

Thompson, T. L., & Mintzes, J. J. (2002). Cognitive structure and the affective domain: On knowing and feeling in biology. *International Journal of Science Education, 24*(6), 645–660.

Thompson, C. L., & Shrigley, R. (1986). What research says: Revising the science attitude scale. *School Science and Mathematics, 86,* 331–343.

Thurston, L. L. (1925). A method of scaling psychological and educational tests. *Journal of Educational Psychology, 16,* 433–451.

Tuan, H.-L., Chin, C.-C., & Shieh, S.-H. (2005). The development of a questionnaire to measure students' motivation towards science learning. *International Journal of Science Education, 27*(6), 639–654.

van Aalderen-Smeets, S., & van der Molen, J. (2013). Measuring primary teachers' attitudes toward teaching science: Development of the dimensions of attitude toward science (DAS). *International Journal of Science Education, 35*(4), 577–600.

Velayutham, S., Aldridge, J., & Fraser, B. (2011). Development and validation of an instrument to measure students' motivation and self-regulation in science learning. *International Journal of Science Education, 33*(15), 2159–2179.

Wang, T.-L., & Berlin, D. (2010). Construction and validation of an instrument to measure Taiwanese elementary students' attitudes toward their science class. *International Journal of Science Education, 32*(18), 2413–2428.

Wang, H. A., & Marsh, D. D. (2002). Science instruction with a humanistic twist: Teachers' perception and practice in using the history of science in their classrooms. *Science & Education, 11,* 169–189.

Weible, J. L., & Zimmerman, H. T. (2016). Science curiosity in learning environments: Developing an attitudinal scale for research in schools, homes, museums, and the community. *International Journal of Science Education, 38*(8), 1235–1255.

Weigel, R., & Weigel, J. (1978). Environmental concern: The development of a measure. *Environment and Behavior, 10*(1), 3–15.

Woelfel, J., & Haller, A. O. (1971). Significant others, the self-reflexive act and the attitude formation process. *American Sociological Review, 36*, 74–87.

Zacharia, Z., & Calabrese Barton, A. C. (2004). Urban middle-school students' attitudes toward a defined science. *Science Education, 88*(2), 197–222.

Zint, M. (2002). Comparing three attitude-behavior theories for predicting science teachers' intentions. *Journal of Research in Science Teaching, 39*(9), 819–844.

Zusho, A., Pintrich, P. R., & Coppola, B. (2003). Skill and will: The role of motivation and cognition in the learning of college chemistry. *International Journal of Science Education, 25*(9), 1081–1094.

5

Using and Developing Instruments for Measuring Science Inquiry and Practices

This chapter is about science inquiry. Science inquiry is concerned with both content of learning and an approach to teaching and learning; it is an important domain of measurement in science education. This chapter will first review various theoretical frameworks about science inquiry. It will then introduce standardized instruments for measuring science inquiry. Finally, this chapter will describe the process for developing new instruments for measuring science inquiry using the Rasch modeling approach.

What Is Science Inquiry?

Science inquiry is a comprehensive construct for organizing science curriculum, instruction, and teacher professional development. Accordingly, science inquiry can be conceptualized in three ways—as curriculum content, as epistemology, and as pedagogy (Anderson, 2007). Science inquiry

Using and Developing Measurement Instruments in Science Education, pages 193–242

as content for science learning, a way of knowing, and an approach to science teaching has been around for almost a century, dating back to Dewey's scientific methods in the 1910s, to Schwab's invitations to science enquiry during 1960s, and to hands-on based science learning promoted in NSF funded science curriculums in the 1960s and 1970s (NRC, 2000). Science inquiry as content requires students as well as teachers to demonstrate the ability to conduct science inquiry.

Science inquiry involves posing meaningful questions, designing appropriate procedures to collect data necessary for answering the questions, and analyzing and interpreting data in order to answer the research questions. The *National Science Education Standards* (NRC, 1996) state that:

> *Scientific inquiry* refers to the diverse ways in which scientists study the natural world and propose explanations based on the evidence derived from their work. Inquiry also refers to the activities of students in which they develop knowledge and understanding of scientific ideas, as well as an understanding of how scientists study the natural world. (NRC, 1996, p. 23)

Although inquiry does not take place in a step-by-step fashion, there are essential aspects common to all inquiry activities. These aspects are: hands-on skills pertaining to data collection and analysis, reasoning skills relating data to theories, and the ability to perform inquiry tasks. Examples of hands-on manipulative skills commonly expected in science curriculums are (Liu, 2009, pp. 85–86):

- *Elementary and Middle School Science:*
  - Use hand lens to view objects.
  - Use stopwatch to measure time.
  - Use ruler (metric) to measure length.
  - Use balances and spring scales to weigh objects.
  - Use thermometers to measure temperature (C°, F°).
  - use measuring cups to measure volumes of liquids.
  - use graduated cylinders to measure volumes of liquids.
- *High School Biology:*
  - Count the growth of microorganism.
  - Determine the size of a microscopic object, using a compound microscope.
  - Dissect a frog or a worm.
  - Germinate seeds.
  - Make a serial dilution.
  - Manipulate a compound microscope to view microscopic objects.

  - Prepare a wet mount slide.
  - Sketch an organism.
  - Slice a tissue for microscope examination.
  - Sterilize instruments.
  - Take a pulse.
  - Use appropriate staining techniques.
  - Use paper chromatography and electrophoresis to separate chemicals.
- *High School Chemistry:*
  - Boil liquid in beakers and test tubes.
  - Conduct an acid-base titration.
  - Cut, bend and polish glass.
  - Dilute strong acids.
  - Pour liquid from a reagent bottle.
  - Prepare solutions of a given concentration.
  - Separate mixtures by filtration.
  - Smell a chemical.
  - Transfer powders and crystals from reagent bottles.
  - Weigh chemicals using an analytic balance.
- *High School Earth Science:*
  - Analyze soils.
  - Classify rocks.
  - Classify fossils.
  - Determine the volume of a regular- and irregular-shaped solid.
  - Grow crystals.
  - Locate the epicenter of an earthquake.
  - Measure weather variables (e.g., barometer, anemometer, etc.).
  - Orient a map with a compass.
  - Plot the sun's path.
  - Test the physical properties of minerals.
  - Use a stereoscope to view objects.
- *High School Physics:*
  - Connect electrical devices in parallel and series circuits.
  - Determine the focal length of mirrors and lenses.
  - Determine the electrical conductivity of a material using a simple circuit.
  - Determine the speed and acceleration of a moving object.
  - Locate images in mirrors.
  - Solder electrical connections.
  - Use electric meters.

Besides manipulative skills listed above, "inquiry is in part a state of mind—that of inquisitiveness" (NRC, 2000, p. xii). These *thinking-oriented inquiry skills* relate to generating testable hypotheses, designing controlled experimentation, making accurate observations, analyzing and interpreting data, and making valid conclusions. Lunetta and Tamir (1979) have identified following thinking-oriented inquiry skills:

- *Planning and Designing*
  - Formulate a question or define a problem to be investigated.
  - Predict experimental result.
  - Formulate hypothesis to be tested.
  - Design observation or measurement procedure.
  - Design experiment.
- *Performance*
  - Carry out qualitative observation.
  - Carry out quantitative observation or measurement.
  - Record results, describe observation.
  - Perform numeric calculation.
  - Explain or make a decision about experimental technique.
  - Work according to own design.
- *Analysis and Interpretation*
  - Transform result into standard form (other than graphs).
  - Graph data.
  - Determine qualitative relationship.
  - Determine quantitative relationship.
  - Determine accuracy of experimental data.
  - Define or discuss limitations and/or assumptions that underlie the experiment.
  - Formulate or propose a generalization or model.
  - Explain a relationship.
  - Formulate new questions or define problem based upon the result of investigation.
- *Application*
  - Predict, based upon result of this investigation
  - Formulate hypothesis based upon result of this investigation.
  - Apply experimental techniques to new problem or variable.

Specific inquiry skills described above focus on individual actions involved in science inquiry. Although they are essential, additional complex processes are necessary. Science inquiry is essentially an integrated practice that involves generating and evaluating scientific evidence and explanations. Generating and evaluating scientific evidence and explanations encompasses the knowledge and skills used for building and refining models and explanations, designing and analyzing empirical investigations and observations, and constructing and defending arguments with empirical evidence (NRC, 2007). In fact, the Next Generation Science Standards (NGSS; NRC, 2012, 2013) does not use the term science inquiry anymore; instead it identifies eight practices that mirror what professional scientists and engineers do. Using the term practices instead of inquiry, NGSS emphasizes the integration of inquiry skills with science and engineering concepts as well as disposition. The eight science practices are: (a) asking questions and defining problems, (b) developing and using models, (c) planning and carrying out investigations, (d) analyzing and interpreting data, (e) using mathematics and computational thinking, (f) constructing explanations and designing solutions, (h) engaging in argument from evidence, and (i) obtaining, evaluating, and communicating information.

The socio-cultural context is also essential in science inquiry activities. Accordingly, science inquiry involves a range of activities that involve cognitive, affective, and psychomotor domains. The above integrated practices may be generally called scientific reasoning. Research in cognitive sciences and science education over the past few decades has shown that school children develop their scientific reasoning with age, but this development is significantly enhanced by their prior knowledge, experience, and instruction (NRC, 2007). That is, scientific reasoning is intimately intertwined with conceptual knowledge of the natural phenomena under investigation.

Besides the ability to demonstrate science inquiry described above, the National Science Education Standards also expect students to demonstrate understanding about science inquiry (NRC, 1996). *Understanding about science inquiry* relates to "how and why scientific knowledge changes in response to new evidence, logical analysis, and modified explanations debated within a community of scientists" (NRC, 2000, p. 21). Examples of understanding about science inquiry are:

- Different kinds of questions suggest different kinds of scientific investigation.
- Current scientific knowledge and understanding guide scientific investigation.
- Technology used to gather data enhances accuracy and allows scientists to analyze and quantify results of investigations.
- Scientific investigations emphasize evidence, have logically consistent arguments, and use scientific principles, models, and theories.

One important aspect of understanding science inquiry is understanding the Nature of Science (NOS). Because science inquiry is an essential component of science, understanding science inquiry is an essential component of understanding the NOS. *Nature of Science* refers to the epistemology of science or science as a way of knowing, including the values and assumptions inherent to science, scientific knowledge, and/or the development of scientific knowledge (Lederman, 1992, 2007), or in brief, the epistemology of science as distinct from science process and content (Lederman, Wade, & Bell, 1998). According to Lederman (2007), important understandings of NOS relevant to K–12 include the distinction between observation and inference, the distinction between scientific laws and theories, the roles of both observation and human imagination and creativity in knowledge construction, the theory-laden nature of scientific knowledge, the cultural context of scientific enterprise, and the tentative nature of scientific knowledge. Similarly, Osborne, Collins, Ratcliffe, Miller, and Duschl (2003) identified the following aspects of NOS by consensus of representatives of the "science expert community" that should be taught in schools: scientific method and critical testing, creativity, historical development of scientific knowledge, science and questioning, diversity of scientific thinking, analysis and interpretation of data, science and certainty, hypothesis and prediction, and cooperation and collaboration. Thus, while understanding inquiry is concerned with inquiry processes, understanding about NOS is concerned with both the processes and products of science inquiry—scientific knowledge. The two constructs are different but overlap significantly.

Science inquiry as related to teaching and learning requires that teachers understand the nature and process of developing scientific knowledge as inquiry, and conduct science teaching through science inquiry. The *National Science Education Standards,* Teaching Standard A, states that teachers of science plan an inquiry-based science program for their students (NRC, 1996). Essential features of inquiry science teaching and learning include:

- Learners are engaged by scientifically oriented questions.
- Learners give priority to evidence, which allows them to develop and evaluate explanations that address scientifically oriented questions.
- Learners formulate explanations from evidence to address scientifically oriented questions.
- Learners evaluate their explanations in light of alternative explanations, particularly those reflecting scientific understanding.
- Learners communicate and justify their proposed explanations. (NRC, 2000, p. 25)

Based on the above conceptualizations of science inquiry, measurement of science inquiry may include three areas: measurement of students' science inquiry/practice abilities, measurement of understanding science inquiry/practice, and measurement of science teachers' understanding and practices of inquiry science teaching. The first area includes measurement of specific inquiry skills and integrated practices. The second area involves measurement of student understanding processes and products of science inquiry and practice. The last area involves measurement of the ways science teachers conceptualize, plan, teach, and evaluate student science learning.

Although this chapter focuses on science inquiry, it is important to note that there are other constructs used in the science education literature that are closely related to science inquiry. For example, the National Research Council's recent report, *Taking Science to School* (NRC, 2007), conceptualizes student proficiency in science to contain four strands: (a) know, use, and interpret scientific explanations of the natural world; (b) generate and evaluate scientific evidence and explanations; (c) understand the nature and development of scientific knowledge; and (d) participate productively in scientific practices and discourse. Strands *b* and *c* are closely related to science inquiry abilities and an understanding of science inquiry discussed above. Given that there are diverse theoretical orientations in science education research, when using and developing instruments for measuring any construct including science inquiry, it is critically important to situate the construct within a particular theoretical framework and define it accordingly. All definitions of construct have a theoretical orientation; they are not applicable universally, regardless of the assessment purpose and context.

Instruments for Measuring Science Inquiry

A variety of standardized measurement instruments related to science inquiry are available in science education. The following summative descriptions introduce instruments published over the past 50 years in major science education research journals, mostly in *Journal of Research in Science Teaching, Science Education*, and *International Journal of Science Education*. The majority of the instruments have been developed based on CTT. They are for various intended uses and based on various theoretical frameworks of science inquiry. Each instrument description is information only, not intended to be a critical review of its strength and weakness. Also, descriptions do not follow a same format or even in similar length; how and how much an instrument is described depend on what is reported in the publication. In general, when available each description contains information about the instrument's intended population, purpose, composition (e.g., number and type of items), validation process and key indices of validity and reliability. Instruments are described in their publication years and in alphabetical order of authors' last names.

Instruments for Measuring Science Inquiry Skills and Practices

Evaluating Laboratory Work in High School Biology (Robinson, 1969)

This performance assessment contains 15 items organized in two groups: identifying objectives and placing objects into designated groups. The test measures high school biology students' lab performances in four areas: measuring, identifying, selecting, and computing. Students rotate around the stations to perform the tasks. Point-biserial correlation for the items ranged from 0.29 to 0.68. Correlation between scores on this test and that on a 63-item multiple choice achievement test was 0.33. Reliability coefficient was 0.63.

The Inquiry Skill Measure (Nelson & Abraham, 1973)

This test measures upper elementary school students' inquiry skills. The test employs a sealed box with a number of different colored sticks protruding from it. The child examines the outside of the box using all of five senses in order to tell as much about the outside of the box as possible. This portion of the test is to obtain a measure of students' ability to observe. The child is then asked what is likely to be inside the box based upon observations. This procedure provides a measure of the child's ability to draw inferences. For each inference made, the child is asked to give the reason, and to describe how to verify the inference without opening the box. The child's ability to classify is measured by presenting nine transparent vials of varying sizes containing different amounts of different colored liquids. The child tries to group the vials using different criteria. Validation of the instrument involved expert panel review and pilot-testing. Factor analysis showed that four factors were present: verifying, inferring, classifying, and frequency. The equivalent form reliability coefficients ranged from 0.441 to 0.707.

Science Process Skills Test (Molitor & George, 1976)

Science Process Skills Test (SPST) measures upper elementary, that is, Grades 4–6, school students' science process skills. It has 20 multiple-choice items, 10 for each of the following two process skills: (a) the ability to make an inference and (b) the ability to verify. Instrument validation involved review of items by an expert panel, and pilot-testing which involved comparison between student responses and their provided reasons. Students who had studied the *Science: A Process Approach* curriculum for 4 years scored significantly higher than students who had not studied the curriculum. KR-20 for the inference scale ranged from 0.54 to 0.59 from Grade 4 to Grade 6, and for the verification scale ranged from 0.72 to 0.84 from Grade 4 to Grade 6.

Test of Enquiry Skills (Fraser, 1980)

Test of Enquiry Skills (TOES) measures nine separate enquiry skills that fall into three major groups: (a) using reference materials such as dictionaries, encyclopedias, library card indexes, book indices, and tables of contents; (b) interpreting and processing information such as reading various scales, calculating averages, percentages and proportions, interpreting charts and tables, using graphical materials; and (c) critical-thinking-in-science skills such as comprehension of science reading materials, design of experimental procedures in science, and the ability to draw valid conclusions and generalizations from data. Each skill forms one scale. Items in TOES are of the multiple-choice format with five alternatives per item. Items are also content-free, that is, they measure general intellectual skills which transcend all science contents.

Development of TOES proceeded in a number of identifiable stages. The first stage was a comprehensive review of the science education literature. The second stage was development of a large pool of items to measure the nine enquiry skills. The third stage involved the scrutiny of the initial pool of items by an expert panel of 15 people. The fourth stage involved pilot-testing the instrument. The fifth stage involved the assembling and pilot-testing a second version of TOES. The sixth and last stage was the assembling and administration of the final version of TOES to 1,158 students from Grades 7–10. The KR-20 coefficients for the nine scales ranged from 0.57 to 0.83 for the Grade 7 sample, from 0.57 to 0.83 for the Grade 8 sample, from 0.53 to 0.77 for the Grade 9 sample, and from 0.50 to 0.75 for the Grade 10 sample. The test–retest reliability ranged from 0.65 to 0.82. Standard error of measures ranged from 1.1 to 1.5. Inter-correlations among the nine scales ranged from 0.30 to 0.56, providing evidence for the discriminant validity.

Test of Integrated Process Skills (Dillashaw & Okey, 1980)

Test of Integrated Process Skills (TIPS) is a multiple-choice question test for measuring 7th–10th grade students' process skills associated with planning, conducting, and interpreting results from investigations. Questions are related to independent variables, dependent variables, controlled variables, hypotheses, experimental designs, graphing of data, and pattern of relationships. Validation involved expert panel review of items and field test with students of diverse backgrounds. The alpha internal consistency reliability was 0.89.

Practical Test Assessment Inventory (Tamir, Nussinovitz, & Friedler, 1982)

Practical Test Assessment Inventory (PTAI) is a categorization and scoring system to grade student practical performances. It was validated by three science educators who read 100 papers and classified questions and answers into categories. Inter-rater reliability was checked by three raters who assessed independently the same 40 tests by assigning marks to each response. The PTAI classifies student performances into the following 21 categories: (a) formulating problems, (b) formulating hypotheses, (c) identifying dependent variables, (d) identifying independent variables, (e) designing control, (f) the fitness of the experiment to the tested problem or hypothesis, (g) completeness of experimental design, (h) understanding the role of the control in the experiment, (i) making and reporting measurements, (j) determining and preparing adequate dilutions, (k) making observations with a microscope, (l) describing observations, (m) making graphs, (n) making tables, (o) interpreting observed data, (p) drawing conclusions, (q) explaining research findings, (r) examining results critically, (s) analyzing knowledge, (t) understanding and interpreting data presented in a graph, and (u) suggesting ideas and ways to continue the investigation.

Environment Literacy Instrument for Korean Children (Chu, Lee, Ko, Hee Moon, Min & Hee, 2007)

Environment Literacy Instrument for Korean Children (ELIKC) measures third-grade (8–9 years old) Korean students' environmental literacy. It includes the following scales: Environmental Knowledge (24 items), Environmental Attitude (22 items), Behaviors to Solve Environmental Problems (16 items), and Skills to Solve Environmental Problems (7 items). The knowledge and skill scales use multiple-choice questions, while attitude and behaviors scales use the five category Likert-scale questions. The draft instrument including both multiple-choice and open-ended questions was pilot-tested with 250 second grade students at the end of the school year. Items were reviewed by four science education professors and three environmental education researchers for content validity. The final version was administered to 969 third-grade students. Cronbach's alphas for the four scales were 0.65 (knowledge), 0.81 (attitude), 0.74 (behaviors), and 0.46 (skills).

Decision-Making Questionnaire (Eggert & Bogeholz, 2010)

The decision-making questionnaire is an instrument that measures students' decision-making competence in situations relating to sustainable development. Final items were separated into two booklets. In the first booklet, students are confronted with two decision-making situations. The first decision-making task is concerned with the problem of overfishing, and the second with the topic of neophyte invasion. The second booklet is designed to analyze how students reflect on the structure of different decision-making strategies that are presented to them, and the context is a consumer choice situation about buying chocolate bars. The initial questionnaire with open-ended questions was based on pertinent curriculum standards. The preliminary test items were discussed by researchers and pre-piloted with 45 junior high school students and university undergraduates. And then, a quantitative pilot study was conducted to analyze the test items' functioning with 291 students. The final validation study involved 436 students from Grade 6 through second year undergraduates. Student responses were scored using a scoring rubric.

The Rasch partial credit model was applied to the whole sample as well as to each subgroup separately. Fit statistics were reviewed, and score category thresholds were examined. Item separation reliability was found to be 0.99; person separation reliability was found to be moderate with 0.66. EAP/PV reliability was of 0.70; inter-rater reliability was high with 95%.

Exploratory Behavior Scale (van Schijndel, Franse, & Raijmakers, 2010)

The Exploratory Behavior Scale (EBS) is a quantitative measure of young children's hands-on behaviors of preschool age children in science museums. The EBS was designed to have three levels of increasingly extensive exploration of the physical environment: (a) passive contact, (b) active manipulation, and (c) exploratory behavior. The EBS items are informed by psychological literature on exploration and play. In the pilot study, two experiments were conducted. Experiment 1 was administered to 71 4- to 6-year-old children. The sample in Experiment 2 consisted of 75 4- to 5-year-olds together with a parent.

Inter-observer reliabilities and Kappas were calculated to assess the consistency of scoring between multiple observers. For the EBS, it was done in different settings. Studies in science museum settings yielded percentage agreements of 81% and 92%, or Kappas 0.63 and 0.81. Studies in child care and development settings, in which an extended version of the EBS was used, yielded average percentage agreements of 78% and 82%, or Kappas 0.56 and 0.64. The test–retest reliability of the extended version of the EBS was established in a child care and development setting by the correlations between two administrations, which were 0.53 and 0.74. In a child care and development setting, a correlation of 0.43 had been found between the mean EBS level and age.

Students' Socioscientific Reasoning and Decision-Making on Energy-Related Issues (Sakschewski, Eggert, Schneider, & Bögeholz, 2014)

The measurement of socioscientific reasoning and decision-making on energy-related issues is located within the framework of education for sustainable development. While working on these issues, students need to acknowledge the ecological, economic as well as the social dimension by integrating needs of the present as well as future generation. The measurement instrument is composed of a set of open-ended questions and multiple-choice questions embedded in three main contexts: generating electric power from wind, energy storage, and energy usage. Three main steps were followed to develop the final instrument. First, initial items were created based on review of state curricula for physics. Second, a qualitative study was conducted with 14 students of Grades 6, 8, 10, and 12 using think-aloud protocols. Student answers to open-ended questions were based on a scoring guide; the degree of accordance between raters was 0.84. The third step was the validation study. Validation involved 850 students from Grades 6 through 12 in Germany. Rasch modeling was used to establish evidence for construct validity. Weighted Likelihood Estimate (WLE) person reliability and Cronbach's alphas were 0.63 and 0.65.

Simulation-Based Assessment of Inquiry Ability (Wu, Wu, & Hsu, 2014)

The Simulation-Based Assessment of Inquiry Ability (SAIA) was developed to estimate how high-school students actually demonstrated their inquiry abilities in the SAIA tasks. The SAIA included four simulation tasks involving a camera (Task A), viscosity (Task B), buoyancy (Task C), and a population distribution (Task D), and 14 items to evaluate the inquiry abilities. Task A provided a simulated camera that students could manipulate, take photos, and view the resulted photos. Task B presented a virtual laboratory that allowed students to investigate the relationships among the submersion velocity, viscosity of the liquid, and density of the liquid. Task C presented a virtual laboratory that allowed students to carry out experiments involving the selection of appropriate instruments for determining the relationship between the density of a liquid and the buoyancy of objects therein. Task D was set in a simulated farm and students were asked to investigate what color of flypapers could catch the most fruit flies. For the inquiry abilities, it was developed to focus on two types: experimental and explaining; each has three sub-abilities. Experimental abilities include: identifying and choosing variables (CV), planning an experiment (PE), and selecting appropriate measurements (SAM). Explaining abilities include: interpreting the testing data by making a claim (MC), using evidence (UE), and evaluating alternative explanations (EAE). The pilot study was conducted with 48 12th graders from a local high school, and they were categorized into three groups based on their program majors with different degrees of science intensity.

Three experts were invited to evaluate the representativeness and the kappa coefficient was used to calculate the degree of agreement among the experts. The agreement analysis showed that the kappa coefficients between the three experts were within the range of 0.88–0.96, which indicated that the quality of SAIA content in evaluating the inquiry abilities was high. Criterion-related validity was established by exploring the degree of association between the SAIA and the Classroom Test of Scientific Reasoning (CTSR; Lawson, 1978). Pearson's correlation between the SAIA and CTSR was 0.40, which is positively significant.

Advancing Science by Enhancing Learning in the Laboratory Survey (Barrie et al., 2015)

The Advancing Science by Enhancing Learning in the Laboratory (ASELL) Survey includes 14 Likert scale items for evaluating individual laboratory exercises. The purpose of ASELL survey is to provide feedback for instructors to iteratively improve experiments. The development of the instrument went through three stages. Lab instructors in biology and physics reviewed the items.

A total of 3,153 student responses from 19 universities in three countries (Australia, New Zealand, and the United States) was collected and analyzed. Three factors, motivation, assessment, and resources have been identified from the data. All items fell clearly into a single factor with factor scores in the range of 0.58–0.77. No significant cross-loadings were evident; all secondary loadings had eigenvalues at least 0.25 smaller than the dominant loading.

Multimedia-Based Assessment of Scientific Inquiry Abilities (Kuo, Wu, Jen, & Hsu, 2015)

The development of Multimedia-Based Assessment of Scientific Inquiry Abilities (MASIA) followed the construct modeling approach in five steps: developing an assessment framework, designing tasks and items, developing scoring rubrics, conducting pilot testing, and applying the Rasch modeling. Inquiry abilities consist of questioning, experimenting, analyzing, and explaining, and each of the abilities is defined in three levels. Following the framework, scenario-based tasks in four content areas (chemistry, biology, physics, and earth science) across four inquiry abilities were developed to engage students in meaningful and authentic inquiry situations. A total of 114 items with an equal number in each inquiry ability was developed. Content validity was established through expert reviews of the items. Student think-aloud protocols were also used to illuminate the consistencies between the cognitive demands of the simulation-based tasks and the complexity of the demands of the assessment framework. A two-stage stratified cluster sampling method was used to obtain a representative sample across different proficiency levels of the target population. The Balanced Incomplete Block Design (BIB) was applied to divide the tasks and items into several exclusive item blocks that were assembled into six booklets. The sample included 1,066 students from 8th grade to 11th grade in Taiwan. Four-dimensional Rasch modeling was applied to the data; personal reliabilities for the four dimensions ranged from 0.83 to 0.88. Correlation between the four dimensions ranged from 0.87 to 0.96.

Quantitative Assessment of Socio-Scientific Reasoning (Romine, Sadler, & Kinslow, 2017)

Quantitative Assessment of Socio-Scientific Reasoning (QuASSR) presents students with a brief scenario describing a particular socio-scientific issue (SSI) and a series of two-tiered ordered multiple choice questions mapped to one of the four dimensions of SSR. Social scientific reasoning (SSR) is conceptualized as consisting of four dimensions: complexity, perspective-taking, inquiry, and skepticism. The instrument is designed for online administration that allows for adaptive testing. Qualtrics software was used to implement the online survey. In the first tier, students respond to a yes/no question, and in the second tier, students select the reason that best corresponds to the response to the first tier question. One hundred thirty-two (132) students from five environmental science classes at a small private college participated in the study. Rasch analysis suggested that the above four dimensions were representative of a single construct. Cronbach's alpha was 0.79.

Instruments for Measuring Cognitive and Metacognitive Reasoning

A Test of Science Comprehension (Nelson & Mason, 1963)

Test of Science Comprehension (TSC) is a 60-item test of Grades 4–6 students' critical thinking skills. Items are in blocks, with each block centering around the interpretation of a situation, the application of scientific principles or theory in accounting for what has happened, and analysis of the situation as a basis for arriving at a solution to the problem. Validation of the instrument involved classes that explicitly taught critical thinking and classes that did not. Students took both pre- and post-tests. The gains were statistically significant for classes that had received explicit instruction. Reliability coefficients ranged from 0.63 to 0.76 on the pretest and from 0.72 to 0.82 on the post-tests for different classes.

X-35 Test of Problem Solving (Forms A and B; Butts, 1964)

The Test of Problem Solving (TPS) assesses secondary school students' ways of knowing in terms of hypothesis formation, experimental design, independent and dependent variables, and hypothesis verification. It has only two problems and each problem has three parts: Part I presenting a problem, Part II presenting data, and Part III presenting solutions. Professors of child psychology and science education reviewed the problems and scoring guides. The overall agreement between the investigator's evaluation and the experts' was 0.75. The correlation between students' scores on the two problems was 0.54.

Cognitive Preference Test: High School Chemistry (Marks, 1967)

This test contains 100 items. Each item has four options corresponding to four cognitive preferences: memory or recall of chemical facts, practical application, critical questioning of information, and identification of a fundamental principle. The 100 items are divided equally into four subtests, one for each cognitive style. The mean biserial correlations of items for the four cognitive preference subtests ranged from 0.44 to 0.82, and Cronbach's alphas ranged from 0.28 to 0.70 for different student samples.

The Classification-Seriation Test (Bridgham, 1969)

The Classification-Seriation Test (CST) measures third grade students on two Piagetian constructs, classification and seriation. The assessment tasks involve students in classifying various geometric shapes that are made of plywood, and painted into one of seven colors. There are seven items on hierarchical classification and four on multiplicative classification; there are also 3 items on seriation. The correlation between the hierarchical and additive classification test scores was 0.31, and the correlation between the seriation test scores and classification test scores was $< = 0.35$.

The Mathematical Skill Test in Chemistry (Denny, 1971)

The Mathematical Skill Test (MAST) measures mathematical skills identified in popular chemistry texts from 1960–1970. Examples of math skills are logarithm, percentage, manipulation of one-variable equations, and interpreting *x*, *y* graphs. Items were reviewed by experts as being essential. Pilot-testing was conducted to ensure appropriate item discrimination and difficulty. Three levels of math skills were tested; Level I had 14 items, Level II 32 items, and Level III 14 items. Correlation between MAST scores and the American Chemical Society's (ACS) chemistry test was 0.633, and between MAST scores and ACS chemistry calculation subtest was 0.823 ($df = 200$, $p = 0.01$). The KR-20 was 0.963, with a standard error of 0.78.

Test for Formal Reasoning (Lawson, 1978)

The TFR is a 15-item test for assessing Grades 8-12 students' concrete and formal operational reasoning. Each item involves a demonstration using some physical materials and/or apparatus, and student responses in writing on test booklets. Students respond by checking the box next to the best answer and explaining why they choose the answer. Most items were based on interview tasks used by Piaget and his associates. Face validity was established by a panel who were Piagetian research experts. The concurrent validity based on correlation between student test scores and scores on two Piagetian interview tasks was found to be 0.76 ($p < 0.001$). Factor analysis of combined test scores and interview scores showed that all the test items and interview tasks loaded on the same factors (i.e., formal reasoning, early formal reasoning, and concrete reasoning). The KR-20 was 0.78.

Piagetian Task Instrument (Walker, Hendrix, & Mertens, 1979)

The Piagetian Task Instrument (PTI) measures university students' formal operational thinking. The six problems are based on the published tasks Piaget and his colleagues used. Adaptations were made to make them as a paper-and-pencil test. Construct validity was determined by comparing the interview results and the written responses. Validation of the instruments also involved students responding to the problems both verbally and in writing. There was no statistically significant correlation between student ages and their test scores. A biology genetic course was found to increase students' scores from 72% correct to 91%.

A Scientific Creativity Structure Model (Hu & Adey, 2002)

The Scientific Creativity Structure Model (SCSM) is an open-ended, paper-and-pencil test consisting of 7 tasks to be completed within 60 minutes, or one class period. It measures secondary school students' scientific creativity. Scientific creativity is defined "as a kind of intellectual trait or ability producing or potentially producing a certain product that is original and has social or personal value, designed with a certain purpose in mind, using given information" (p. 392). It involves three dimensions: process (i.e., imagination and thinking), trait (i.e., fluency, flexibility, and originality), and product (i.e., technical product, science knowledge, science phenomena, and science problems). An initial set of 48 items, two for each of the 24 cells of the three dimensions' interaction, were developed. Fifty science education researchers and teachers in China reviewed the items and found 9 to be appropriate for measuring secondary school students' creativity. The 9 items were then tried out by 60 13-years old girls in London, which resulted in the deletion of 2 more items. The final instrument consists of 7 items, with a scoring rubric for each.

Field testing of the instrument took place with 160 students from Year 7 to Year 10 (ages 12, 13, and 15) in a suburban school in England. Item discrimination was found to be high, ranging from 0.654–0.892. Cronbach's alpha internal consistency coefficient was found to be 0.893. Inter-rater reliability measured by correlation coefficients between two sets of scores was found to be between 0.793–0.916 for the 7 items. Face validity was established by 35 science education researchers and teachers in China and England who stated that the items were measuring creativity. The construct validity was established by principal component factor analysis which found only one dominant factor. Significant difference in creativity was found between Year 7 students and Year 8 students, and between Year 8 students and Year 10 students. Also, significant difference in creativity was found between high-ability students and low-ability students.

Well-Structured Problem-Solving Process Inventory and Ill-Structured Problem-Solving Process Inventory (Shin, Jonassen, & McGee, 2003)

The Well-Structured Problem-Solving Process Inventory (WPSPI) and the Ill-Structured Problem-Solving Process Inventory (IPSPI) is a 2-item essay question instrument measuring high school students' problem-solving skills in the context of astronomy. Students may use any resources including computers to solve the problems. Validation of the instruments involved extensive expert review of the content, pilot-testing, and scoring based on data from both expert problem solvers and novice problem solvers. Students who took the reformed astronomy course designed specifically for improving problem-solving abilities scored significantly higher than those who did not take the course. Inter-rater reliability was 0.83.

Metacognition Baseline Questionnaire (Anderson & Nashon, 2007)

The Metacognition Baseline Questionnaire (MBQ) is a 53-item Likert scale survey of students' self-reported engagement in metacognition learning situations within both formal and informal learning settings. Metacognition is conceptualized to consist of six dimensions: awareness, control, evaluation planning, monitoring, and self-efficacy. Each item asks students to self-assess their degrees of agreement with propositions related to the above six dimensions by selecting *always or almost always true of me* (5), *frequently true of me* (4), *half of the time true of me* (3), *sometimes true of me* (2), and *only rarely true of me* (1). The items were derived from a wide consultation of experienced high school teachers and research partners. Extensive review of items by teachers and science education researchers resulted in deletions and revisions of initial items. The instrument was then pilot-tested with 40 Grade 11 students. The Cronbach's alpha coefficients for the six dimensions were: 0.798 (control), 0.717 (monitoring), 0.671 (awareness), 0.765 (evaluation), 0.842 (planning), and 0.894 (self-efficacy).

Self-Efficacy and Metacognition Learning Inventory—Science (Thomas, Anderson, & Nashon, 2008)

Self-Efficacy and Metacognition Learning Inventory (SEMLI-S) is a 30-item rating scale survey of high school students' meta-cognitive science learning orientations. Meta-cognitive science learning orientations are based on students' self-perceptions of elements of their meta-cognition, self-efficacy, and science learning processes. An extensive literature review on meta-cognition, self-regulation, and constructivist science learning processes helped generate an initial list of 72 items. An iterative factor analysis and Rasch modeling resulted in reduction of the items into the final set of 30 items. The 30 items are related to the following five subscales: (a) constructivist connectivity (7 items), (b) monitoring, evaluation, and planning (9 items), (c) science learning self-efficacy (6 items), (d) learning risks awareness (5 items), and (e) control of concentration (3 items). The Rasch fit statistics for the above 30 items indicated a good model-data-fit. Cronbach's alpha coefficients for the above five subscales were 0.84, 0.84, 0.85, 0.77, and 0.68. Correlation coefficients among the subscales ranged from 0.29 to 0.58. The instrument is available in both English and Chinese.

Approaches to Learning Science Questionnaire (Lee, Johanson, & Tsai, 2008)

The Approaches to Learning Science Questionnaire (ALS) is a 24-item survey on high school students' approaches to learning. The ALS questions were adapted from a similar published instrument to make them science-specific. Approaches to learning were conceptualized as a hierarchical construct with two elements: motivation and strategy. The ALS includes four subscales; they are deep motive, deep strategy, surface motive, and surface strategy. Initially, each subscale had six to nine items presented as bipolar always/never statements on a five-point scale. An expert reviewed the initial items to ensure its content validity. Both exploratory and confirmatory factor analyses were conducted to establish the construct validity. Cronbach's alpha coefficients for the four subscales were 0.90, 0.89, 0.84, and 0.84, and the alpha for the entire instrument was 0.89.

Physics Metacognition Inventory (Taasoobshirazi & Farley, 2013)

The 24-item Physics Metacognition Inventory (PMI) was developed to measure physics students' metacognition for problem solving. Items were classified into eight subcomponents subsumed under two broader themes: knowledge of cognition and regulation of cognition. A total of 505 introductory-level college physics students from six classes at four universities in Nevada and Georgia participated in the validity study. The students' total PMI scores correlated significantly with their reported physics course grade ($r = 0.27$, $p < 0.001$) and physics motivation ($r = 0.56$, $p < 0.001$), providing evidence for criterion-related validity. Principal component analysis suggested six factors accounting for 69.46% of the total variance. All but one of the items met the criterion of loading at least 0.35 on their respective factor. Reliability for Factor 1, knowledge of cognition, was 0.90; Factor 2, information management 0.91; Factor 3, monitoring 0.87; Factor 4, evaluation 0.78; Factor 5, debugging 0.92; and Factor 6, planning 0.68. The reliability of the 24 items was 0.90. No significant differences were found between women and men or among different ethnic groups in total scores on the PMI.

Taasoobshirazi, Bailey, and Farley (2015) further validated PMI using confirmatory factor analysis and Rasch modeling. Participants included 285 introductory-level college physics students. Confirmatory factor analysis consisting of both the measurement model and structure model suggested that all of the items loaded significantly on their respective factors. Rasch rating scale analysis results also showed that there was a good model-data fit and match between item difficulty range and student ability range. Person reliability was 0.86. Again, student total scores on PMI were statistically significantly correlated with their reported physics course grades ($r = 0.33$, $p < 0.01$) and physics motivation ($r = 0.52$, $p < 0.01$).

Global Scientific Literacy Questionnaire (Mun et al., 2015)

Global Scientific Literacy Questionnaire (GSLQ) measures student scientific literacy in four dimensions: (a) habits of mind, (b) character and values, (c) science as human endeavor, and (d) metacognition and self-direction. The initial item pool included 114 statements representing thoughts and behaviors of students who possess global scientific literacy competencies. Items were in a five-point scale format: *never, seldom/rarely, sometimes, often/frequently,* and *always.* Items were mostly from published instruments. A panel of eight experts reviewed the items for content validity. The initial items were then piloted with 541 8th and 9th grade students from schools in South Korea. Fifty-eight (58) items were revised as the result of the pilot test. The revised 58 items were given to 3,784 secondary school students. The sample was divided into two samples, one for exploratory factor analysis and the other for confirmatory factor analysis. Excluding 10 items not having appropriate loadings, the remaining 48 items resulted in an eight factor solution from exploratory factor analysis accounting for 50.22% of total variance. Two factors corresponded to each of the habits of mind and character and values dimensions; three factors corresponded to the science as human endeavor dimension; and one factor corresponded to the metacognition and self-direction dimension. The internal consistency reliability coefficients were 0.87 for the habits of mind, 0.80 for character and values, 0.85 for science as human endeavor, and 0.88 for metacognition and self-direction. Confirmatory factor analysis was conducted to test the structure of the above four dimensions eight factors; there was a satisfactory goodness-of-fit.

Graphical Interpretation Assessment (Peterman, Cranston, Pryor, & Kermish-Allen, 2015)

Graphical Interpretation Assessment (GIA) is a standards-aligned interview-based performance assessment to measure primary students' (6–10 years olds) interpretation skills in relation to three types of graphs: bar graphs, line graphs, and dot plots. Nine items were related to bar graphs, seven to dot plots; and three to line graphs. Data were from 55 students who completed both pre–post performance assessments. The graphical interpretation scoring rubric was used to create a second set of scores in addition to *in situ* scores. Results showed that students became more sophisticated in their understanding of data and graphs from ages 9 to 10. Also differences in scores on bar graphs, line graphs, and dot graphs provided additional evidence for discriminant validity. There was a statistically significant correlation between the two sets of scores: GIA scores based on scoring rubrics and that based on interview observations (*in situ*). GIA scores were also sensitive to detect effect of curriculum interventions focusing on graphical interpretations.

Metaconceptual Awareness and Regulation Scale (Kirbulut, Uzuntiryaki-Kondakci, & Beeth, 2016)

Metaconceptual Awareness and Regulation Scale (MARS) consists of 10 items scored on a six-point Likert scale to measure degrees to which high school students realize, monitor, and evaluate their ideas. It includes two scales: metaconceptual awareness and metaconceptual regulation. Metacognition was defined as (a) knowledge of cognition and (b) regulation of cognition. An item pool of 17 items was developed. The pool of items was then examined by three experts, resulting in 12 items. The 12-item instrument was given to 349 10th grade students as a pilot study. Exploratory factor analysis was conducted on the pilot study sample. The metaconceptual regulation factor explained 33% of total variance and the metaconceptual awareness factor accounted for 11.9% of total variance. Two items were removed due to poor loadings. The remaining 10-item instrument was then given to 338 11th grade students in the validation study. Confirmatory factor analysis supported the two factor solution. Cronbach's alpha for the metaconceptual regulation scale was 0.80 and for the metaconceptual awareness scale was 0.72. There was a statistically significant canonical correlation between students' metaconceptual awareness and regulation and their use of learning strategies.

Instrument for Measuring Understanding About Science Inquiry and Nature of Science

Given the considerable overlap between understanding about science inquiry and understanding about NOS, and there is no separate chapter on using measurement instruments for understanding of NOS, this section includes instruments for measuring both constructs.

Nature of Science Scale (Kimball, 1967)

The Nature of Science Scale (NOSS) is a 29-item Likert-scale survey of university students' and science teachers' conceptions of science. Nature of science was defined by eight assertions such as "the fundamental driving force in science is curiosity concerning the physical universe" (p. 111). Development of the instrument involved expert reviews of the initial items, and pilot-testing with university science and nonscience majors. Items included in the final version had few neutral responses by students, discriminated in favor of science majors, and presented a logical progression either from agreeing to disagreeing or from disagreeing to agreeing. The split-half reliability was 0.72. Philosophy students scored significantly better than those in other fields including science education. Science teachers did not score significantly different from practicing scientists.

Nature of Science Test (Billeh & Hasan, 1975)

The Nature of Science Test (NOST) is a 60-item multiple-choice test assessing students' understanding of the NOS in the following aspects: (a) assumptions of science (8 items), products of science (22 items), processes of science (25 items), and ethics of science (5 items). The test consists of two types of items. The first type measures the individual's knowledge of assumptions and processes of science, and the characteristics of scientific knowledge; the second type presents situations which require the individual to make judgments in view of his/her understanding of NOS. The validation took place in Jordan. The content validity of the instrument was established by a panel of 25 scientists, science educators, and science supervisors. Correlation of NOST scores with general GPA was 0.58, and with mathematics GPA was 0.60. The instrument was also found to discriminate significantly between comparable groups of secondary school students majoring in science and arts. The split-half reliability for different sample of students ranged from 0.58 to 0.82.

Conceptions of Scientific Theories Test (Cothan & Smith, 1981)

The Conceptions of Scientific Theories Test (COST) is a 40-item Likert scale instrument measuring preservice elementary teachers' conceptions of science. The items are presented in five contexts: Bohr's theory of the atom, Darwin's theory of evolution, Oparin's theory of abiogenesis, theory of plate tectonics, and a nontheoretical context. Items selected through pilot-testing had high item-subscale correlation. Results showed that elementary teachers scored significantly differently than university chemistry major and philosophy students. Standard error of measurement for the instrument was 0.3.

Views on Science–Technology–Society (Aikenhead & Ryan, 1992)

Views of Science–Technology–Society (VOSTS) assesses high school students' viewpoints of science and its relations to technology and society. It contains 114 items. There is no need to administer 114 items entirely; users can select the most appropriate items to form a survey. Each item presents an extreme viewpoint about science that students have to determine how strongly they agree or disagree with the viewpoint. Underneath each statement is a list of positions about that statement. The student is asked to read each and to select which statement seems to fit their own position. The development of the instrument began with thousands of Grade 12 students writing a paragraph about their views of science, technology, and society. Common themes were identified from student writings as "student positions." To determine how well these VOSTS "student positions" fit with the students' true beliefs, further interviews were conducted. Strictly speaking, VOSTS is not an instrument ready to use; no scoring schemes were provided. VOSTS items can be used to develop specific measurement instruments.

Views About Sciences Survey (Halloun & Hestenes, 1996, 1998)

Views About Sciences Survey (VASS) contains 30 items probing high school and college students' personal beliefs about the NOS and learning science in specific science disciplines. There is one scale in each of physics, chemistry, and biology. Beliefs about science involve three scientific dimensions: structure, methodology, and validity. Beliefs about learning science involve three cognitive dimensions: learnability, reflective thinking, and personal relevance. Each dimension is framed by pairs of contrasting views that are primarily held by scientists/experts and novices. When responding to an item, the student chooses from an eight-point scale representing a continuum from the two contrasting views. Expert reviews and interviews with students validated the items.

Scientific Epistemological Views (Tsai & Liu, 2005)

Scientific Epistemological Views (SEVs) include the following dimensions:

1. The role of social negotiation (SN): The development of science relies on communications and negotiations among scientists.
2. The inventive and creative NOS (IC): Scientific reality is invented rather than discovered.
3. The theory-laden exploration (TL): Scientists' personal assumptions, values, and research agendas influence the scientific explorations they conduct.
4. The cultural impacts (CU): Scientific knowledge is culture-dependent.
5. The changing and tentative nature of scientific knowledge (CT): Scientific knowledge is always changing and its status is tentative.

Questions for SEVs were based on previous interview studies with students. They are presented in a 1–5 Likert-scale format. Validation took place in Taiwan with 613 high school students. Initially, 35 items were developed, but only 19 were retained based on principal component factor analysis. The SN subscale contains 6 items with an alpha of 0.71; the IC subscale contains 4 items with an alpha of 0.60; the TL subscale contains 3 items with an alpha of 0.68; the CU subscale contains 3 items with an alpha of 0.71; and the CT subscale contains 3 items with an alpha of 0.60. The alpha for the entire instrument was 0.67. Correlation among the five subscales ranged from 0.09 to 0.27. Four students from each of three different levels of understanding of scientific epistemology based on SEVs scores were interviewed to further study the construct validity.

Views on Science and Education Questionnaire (Chen, 2006)

The Views on Science and Education (VOSE) questionnaire is a written survey of university students' and preservice science teachers' conceptions of the NOS and their related teaching attitude. Conceptions of NOS is differentiated as views on what NOS is (actual), and views on what NOS ought to be (ought). The VOSE is in both Chinese and English. It has 15 questions, each followed by several items representing different philosophical positions and asking respondents to rate on a five-point scale their degrees of agreement. Altogether, there are 85 items. Items are based on previously published both qualitative and quantitative surveys, particularly VOSTS and VNOS. The development of VOSE went through three stages. The first stage involved a comprehensive literature review and a pilot study to collect college students' NOS views and attitudes. Results of the first stage helped determine the content and format of the questionnaire. The second stage involved actual item development, pilot testing, and a review of the items by a panel consisting of six experts. The initial draft instrument contained 102 items organized around 19 questions. Interviews of seven students also helped develop the items. The third stage was the actual field testing for establishing evidence for validity and reliability. Field testing took place at two research universities in Taiwan. The clarity of the instrument was checked again through interviews after the field testing.

Views on Science and Education (VOSE) questionnaire focuses on seven aspects of NOS relevant to K–12 science education, they are: tentativeness of scientific knowledge, nature of observation, scientific methods, hypotheses, laws and theories, imagination, validation of scientific knowledge, and objectivity and subjectivity in science. The Cronbach's alpha coefficients for the above aspects/subscales ranged from 0.34 to 0.81. The attitudes toward teaching the NOS issues contain the following aspects/subscales: tentativeness, nature of observations, scientific methods, theories and laws, and subjectivity and objectivity; the Cronbach's alpha coefficients for the above subscales ranged from 0.59 to 0.81. Test–retest reliability was 0.82.

Students' Ideas About Nature of Science (Chen et al., 2013)

Students' Ideas About Nature of Science (SINOS) is a 47-item questionnaire to measure young learners' conceptions of NOS. The items are of the Likert-type with response options represented by sad faces, smiley faces to symbolize the degree of dis-/agreement and do not understand; they relate to seven aspects in NOS: theory-ladenness, creativity and imagination, tentativeness of scientific knowledge, durability of scientific knowledge, coherence and objectivity in science, the science for girls stereotype, and the science for boys stereotype. Students' Ideas About Nature of Science (SINOS) was first constructed based on the written responses of 431 sixth graders, elementary students' quotations in theses and books related to NOS, and student interviews. The face validity and coding categories were verified by a team of five science educators. The draft questionnaire was administered to 1,139 third to sixth graders in Taiwan, and then 47 out of 62 items organized by seven aspects were selected for the final version. It was further validated by a sample of 1,091 5th and 6th graders.

The first test showed that the Cronbach's alphas of the subscales ranged between 0.67 and 0.84; with the overall alpha to be 0.85. In the second test, the Cronbach's alphas of the subscales ranged between 0.70 and 0.87. Confirmatory factor analysis revealed a seven factor solution as expected. The correlations between seven factors were moderate, ranging from 0.48 to 0.58, and loadings on the intended factors were positive and high.

Views About Scientific Inquiry (Lederman et al., 2014)

Views About Scientific Inquiry (VASI) questionnaire is a set of 10 open-ended questions designed to solicit middle and high school students' understanding of scientific inquiry aspects. Based on major reform documents including Next Generation Science Standards, the following aspects of scientific inquiry of which students should develop an informed understanding were identified: (a) scientific investigations all begin with a question and do not necessarily test a hypothesis; (b) there is no single set of steps followed in all investigations; (c) inquiry procedures are guided by the question asked; (d) all scientists performing the same procedures may not get the same results; (e) inquiry procedures can influence results; (f) research conclusions must be consistent with the data collected; (g) scientific data are not the same as scientific evidence; and (h) explanations are developed from a combination of collected data and what is already known. The VASI questions were vetted by a panel of committee for fully addressing the eight aspects. The face validity was established by a group of middle school teachers. The VASI was then piloted with a group of 60 eighth grade students. The VASI was also shown to be sensitive to instructional effect of explicit scientific inquiry instruction. Student VASI responses were also compared to interview data with 20% of the questionnaire respondents; results showed agreement between the two. Inter-rater reliability was determined by agreement between raters, which was > 90%. Administration of VASI typically took 30–45 minutes. Student responses to each scientific inquiry aspect were rated as informed, mixed, naïve, and unclear based on a rubric.

Scientific Epistemic Beliefs Instrument and the Goal Orientations in Learning Science Instrument (Lin & Tsai, 2017)

Scientific Epistemic Beliefs Instrument (SEBI) includes multiple scales for high school students: source of knowledge, certainty of knowledge, development of knowledge, justification for knowing, purpose of knowledge, and purpose of knowing. The initial sample included 600 senior high school students in Taiwan. The initial item pool was reviewed by experienced science education professors and senior high school science teachers and students. A total of 36 items were adopted from a published instrument or created new; each item was stated in a 5-point Likert scale from *strongly agree* to *strongly disagree.* The Goal Orientations in Learning Science Instrument (GOLSI) was adapted from a published instrument. The GOLSI contains 12 items, with three items for each of the four dimensions: mastery-approach goal, mastery-avoidance goal, performance-approach goal, and performance-avoidance goal. The GOLSI items were also stated in a five-point Likert scale from *strongly agree* to *strongly disagree.* Exploratory factor analysis and confirmatory factor analysis were conducted to establish construct validity of the measures. Cronbach's alphas for the six dimensions of SEBI ranged from 0.80 to 0.88, and for the four dimensions of GOLSI ranged from 0.83 to 0.90.

What Is Technology (Lachapelle, Cunningham, & Oh, 2019)

What is Technology (WT) is a 21-item instrument to measure conceptions of technology of children of ages 8–11. Technology is defined as all types of human-made systems and processes. It contains the following scales: electrical technologies, mechanical technologies, simple technologies, and natural items (non-technologies). Items are in picture grid formats. Initial items were revised based on 26 interviews with young children aged 6–11. A sample of 508 children completed the pre- and post-surveys of WT, and more than 6,000 more students completed the pre- and post-surveys of WT in 2013 during engineering education units in 2011. Both exploratory and confirmatory factor analyses were conducted to establish validity evidence. Students' responses to open-ended questions asking them to explain their understanding of technology were coded and there was a significant correlation between the codes and confirmatory factor analysis scale scores. The instrument was able to detect changes in students' conceptions after an intervention.

Instruments for Measuring Science Teachers' Inquiry Conceptions and Practices

Thinking About Science Instrument (Cobern & Loving, 2002)

The Thinking About Science Instrument (TSI) is a 35-item Likert scale survey of preservice elementary teachers' views about science. It covers the following aspects of science: epistemology, science and the economy, science and the environment, public regulation of science, science and public health, science and religion, science and aesthetics, science race and gender, and science for all. Items were validated by a group of experts, and verified by reviews of a small group of elementary teachers. The alpha for the entire instrument was 0.82, and alphas for the subscales ranged from 0.41 to 0.80. Survey results showed that elementary teachers had different views on different aspects of science, but overall viewed science positively.

Views of Nature of Science Questionnaire Form B and Form C (Lederman, Abd-El-Khalick, Bell, & Schwartz, 2002)

The Views of Nature of Science (VNOS) questionnaire is a constructed-response question survey of preservice and inservice science teachers' views about the NOS. It is administered in a controlled condition with no time limit (but typically taking about 1.5 hours). After the survey, a sample of respondents (e.g., 10%) is interviewed. The VNOS questionnaire assesses the following aspects of the NOS: scientific knowledge is tentative, empirical, theory-laden, partly the product of human inference, imagination, and creativity; and socially and culturally embedded. Three additional important aspects are the distinction between observation and inference, the lack of a universal recipe-like method for doing science, and the functions of and relationships between scientific theories and laws. The VNOS-B contains 7 items, and VNOS-C contains 10 items; both are the results of many rounds of interviews and analyses in addition to expert panel reviews. According to Lederman (2007), there is a shorter version, VNOS-D, producing almost identical results in a much shorter time frame (i.e., less than one hour), and another version, VNOS-E, has been specifically developed for younger age students (K–3) or those who cannot read or write. Responses to the survey and interview questions were then analyzed qualitatively, including coding for different levels of understanding and constructing different profiles of understanding.

Scientific Habits of Mind Survey (Çalik & Coll, 2012)

The 59-item Scientific Habits of Mind Survey (SHOMS) instrument was developed to explore preservice science teachers', and scientists' understanding of habits of mind (HoM). The SHOMS includes seven subscales: mistrust of arguments from authority, open-mindedness, skepticism, rationality, objectivity, suspension of belief, and curiosity. Items followed a four-point Likert scale, with items scored in two ways: positive (1–4) or negative (4–1). The sample used to validate the instrument consisted of 145 third-year preservice primary school teachers and 145 third-year preservice science teachers in Turkey. Face validity was established by the use of a panel of experts, including five science educators and five nonscience educators, and semi-structured interviews. Three teachers and six non-English-speaking-background graduate students were employed to confirm readability and comprehension of items. The survey was translated from English into Turkish. To ensure construct, three bilingual educators checked the translation. Confirmatory factor analysis showed that the items defined the expected seven factors; Pearson correlation between the seven factors was below 0.3, suggesting that the seven factors were distinct. There was also a statistically significant difference in scores of SHOMS between preservice teachers with and without science backgrounds. The internal reliability of the instrument based on Cronbach's alpha was 0.73.

Inquiry-Based Tasks Analysis Inventory (Yang & Liu, 2016)

Inquiry-Based Tasks Analysis Inventory (ITAI)) was developed to gather evidence to evaluate how well inquiry-based tasks in science textbooks perform intended functions. A review of literature suggests that inquiry-based tasks in science textbooks should serve: (a) to assist in the construction of understandings about scientific concepts, (b) to provide students opportunities to use inquiry process skills, (c) to contribute to the establishment of understandings about scientific inquiry, and (d) to provide students with opportunities to develop higher order thinking skills. Accordingly, items covered the following areas: (a) consistency with curriculum knowledge objectives, (b) opportunities for students to use inquiry process skills, (c) reflection of understanding of scientific inquiry, (d) opportunities to develop higher order thinking skills. There was a total of 22 items. A sample of 53 inquiry-based tasks from three most widely used senior secondary school biology textbooks in China were selected for pilot-testing. Four experts were recruited to use ITAI to rate each of the inquiry tasks according to the scoring rubrics. The average concordance rate between raters was 87.7%. Rasch analysis was conducted to examine item properties and scale properties. Results suggest that there was a good model-data-fit; Cronbach's internal consistency based on raw scores was 0.79.

Commentary

A number of measurement instruments related to various aspects of science inquiry are available. Selecting an instrument to use should first clearly define the construct related to the science inquiry to be measured. This definition will then guide the selection of an available instrument to use. In most situations, a perfect match between the defined construct and the construct measured by an instrument is unlikely. One approach to using current measurement instruments is to modify a best available instrument and collect data to further validate it. Even if there is a good match between the defined construct and the measured construct of an instrument, continuing to validate the instrument is still necessary because the sample involved in the original validation study may not represent the sample in your study.

One trend in developing measurement instruments related to science inquiry is that more attention is being given to measuring integrated and complex science inquiry abilities or practices as compared to specific isolated inquiry skills. Examples include the ability to deal with socio-scientific issues, authentic science inquiry abilities, and scientific decision-making ability. Similarly, more complex scientific reasoning such as habit of mind has also been measured in recent years. This trend is consistent with the direction the Next Generation Science Standards is driving toward.

In terms of measuring NOS, many instruments still adopt a Likert-scale or rating scale that is often accompanied by some kind of scoring (such as scores 1 to 5 for *strongly agree* to *strongly disagree*). Two potential problems are associated with this practice. One problem, as Aikenhead (1973) pointed out, is the ambiguity of total scores: What does a total score on an instrument mean and what does a difference between two students or between two tests means in terms of student understanding of the NOS? Many instruments do not have a clear scale to facilitate qualitative interpretation. Another potential problem is bias or privilege assigned to a particular version of the NOS. This problem is pointed out by Lederman et al. (1998) in their review of measurement instruments of NOS. Because there is no universally agreed-upon version of NOS, any selected response or close-ended response question format, such as the Likert scale, is likely to force students to think in terms of one version of NOS, and it remains unclear exactly what are students' true understandings of the NOS. In order to address the above two problems, VOSTS adopts the no-scoring approach and VNOS adopts the interview and open-ended response question format. As Lederman et al. (1998) pointed out, a forced response format like a Likert scale can still play a role in assessing a specific version of NOS, but a more comprehensive and accurate assessment of students' and teachers' understandings of NOS may require a combination of both quantitative and qualitative methods.

Developing Instruments for Measuring Science Inquiry and Practices

State the Purpose and Intended Population

The first step in developing an instrument for measuring inquiry is to consider the uses of scores derived from the instrument. There are various reasons for which we need to measure science inquiry. Given that science inquiry and practices are key science learning standards in national and state science content standards, it is necessary to measure student mastery of science inquiry and practices during the instructional processes as formative assessment and at the end of instructional processes as summative assessment. Assessing teachers' practices of inquiry science teaching and understanding of science inquiry and practices may also be necessary for development and evaluation of effective preservice and inservice teacher education programs. Different intended uses of test scores will have implications on the process and technical quality requirements of the measurement instrument development.

Define the Construct to Be Measured

Science inquiry and practices include many constructs; different constructs may have quite different theoretical underpinnings and content domains. For example, a science modeling ability is very different than an inquiry ability to conduct a collaborative scientific investigation. Similarly, understanding science inquiry is fundamentally different in both theoretical frameworks and content from conducting science inquiry. Thus, it is very important to clearly define the science inquiry construct and its content domains.

One requirement for the construct to be measured is its linearity. That is, the construct should imply a linear progression along which various levels of performance on inquiry or practice may be differentiated. This unidimensional progression is assumed to underlie both subjects' inquiry performances and items' difficulties; it will guide the subsequent development of the measurement instrument. The linearity of the construct should not simply be assumed; it should be derived from a commonly accepted theory or theories. Clearly stating the theory or theories is necessary and important for the subsequent development stages.

When defining the science inquiry or practice construct, it is important to keep in mind that science inquiry (and practice) is both context and content dependent (Wong & Hodson, 2008). Researchers have pointed out that general science inquiry abilities and understanding of science inquiry applicable to all science content domains and contexts are neither real nor informative for improving science teaching and learning (Elby & Hammer, 2001; Rudolph, 2000). "There is no single set of NOS elements, static with time and fitting all disciplines and contexts" (Wong & Hodson, 2008, p. 123). Thus, when defining the construct of science inquiry or practice to measure, it is very important to situate the construct in a specific science context (e.g., everyday, industrial, research laboratory) and identify subject performances and behaviors that are related to specific science content. For example, understanding of the nature of observation may be situated within an everyday context and be related to specific topics of chemistry, physics, or biology. Similarly, the ability for science argumentation may be situated around a socio-scientific issue or within a school science subject such as chemistry.

Identify Performances of the Defined Construct

Once the construct has been defined, the specific behaviors that describe different levels of subjects' performance on the construct should then be identified. These behaviors are related to specific science contents and contexts. For example, if the construct is on secondary school students' conceptions of science and technology, then the behaviors that represent the construct may include the following:

1. Define science and technology.
2. Recognize similarities and differences between science and technology.
3. Explain how science and technology interact with each other.
4. Analyze how science and technology are involved in society.
5. Appreciate the historical and socio-cultural dimensions of science and technology.

The above behaviors are related to science subjects taught in secondary schools and represent different levels of subjects' conceptions of science and technology; they are derived from a defined construct of conceptions of science and technology. Once these behaviors are identified, they will then be used to develop a test specification.

The most important consideration in developing a test specification is the number of questions. The number of questions should be determined by the defined construct and its content domain, specifically the range of expected variation of performances or behaviors within the target population. There should be questions for every possible level of performance, and the difficulty range of test items should well cover the range of the performances. Based on the above considerations, there will be no fixed number of items for all measurement instruments; different constructs and target populations require different number of items. However, the general rule that the more items in a measurement instrument, the more reliable the measurement scores are, is still applicable. Measurement instruments should also strive for efficiency—achieving maximum reliability using the least number of items.

Depending on the context and content domain of the construct to be measured, items can be in various formats and media such as paper-and-pencil tests using multiple-choice and constructed response questions, performance assessment using objects and materials, observation using a checklist, paper-and-pencil Likert-scale questions asking subjects to agree or disagree, to name a few.

Laboratory Skills

Although laboratory skills in using tools may be assessed using paper-and-pencil test questions, the most effective assessment of using tools involves observation of students performing the skills. In order to observe student mastery of laboratory tools during labs, a systematic plan must be in place. An observation plan can be a checklist or a rating scale. A sample rating scale for a laboratory skill assessment on using a microscope is presented below.

A Rating Scale for Using a Microscope

Name ______________________________ Date ______________

Lab Skills	Developing	Mastery	Proficient
1. Carry a microscope with two hands.			
2. Clean up microscope lenses.			
3. Identify parts of a microscope.			
4. Prepare a microscope slide.			
5. Mount a slide on the microscope stage.			
6. Set the microscope light control.			
7. Focus the view using coarse and fine adjustments.			

Paper-and-pencil tests using multiple-choice and constructed response questions may also be used to assess laboratory skills involving reasoning. For example, the instrument *Test of Integrated Process Skills* (TIPS; Dillashaw & Okey, 1980) is a multiple-choice question test for measuring 7th–10th grade students' laboratory skills associated with planning, conducting, and interpreting results from investigations. Questions are related to independent variables, dependent variables, controlled variables, hypotheses, experimental designs, graphing data, and pattern of relationships. Table 5.1 shows some sample questions from TIPS.

Integrated Science Inquiry Performances

Integrated science inquiry performances are best measured by performance assessment. Performance assessment is a hands-on test that requires students to perform a task by carrying out a scientific procedure, conducting a scientific investigation, or producing a useful product. Performance

TABLE 5.1 Sample TIPS Questions

Assessment Objective	Sample Item
A. Given a description of an investigation, identify the independent, dependent, and controlled variables and the hypothesis being tested	Sarah wanted to find out if temperature has an effect of the growth of bread mold. She grew the mold in nine containers containing the same amount and type of nutrients. Three containers were kept at 00C, three were kept at 900C, and three were kept at room temperature (about 270C). The containers were examined and the growth of the bread mold was recorded at the end of four days. The dependent or responding variable is: 1. growth of bread mold 2. amount of nutrients in each container 3. temperature of the containers 4. number of containers at each temperature.
B. Given a description of an investigation, identify how the variables are operationally defined	The superintendent is concerned about the accidents in schools. He makes the hypothesis that safety advertising will reduce school accidents. He decides to test the hypothesis in four middle schools. Each school will use a different number of safety posters to see if the number of accidents are reduced. Each school nurse will keep a record of students that come to the office because of an accident. How is safety advertising measured in this study? 1. number of accidents reported to the nurse 2. number of middle schools involved 3. number of safety posters in each school 4. number of accidents in the school
C. Given a problem with a dependent variable specified, identify variables which may affect it	Sue wants to find out what might affect the length of bean seedlings. She places a bean wrapped in moist tissue paper in each of ten identical test tubes. She puts five tubes in a rack in a sunny window. She puts the other five tubes on a rack in a dark refrigerator. She measures the lengths of the bean seedlings in each group after each week. Which of the following variables might affect the length of the bean seedlings? 1. temperature and moisture 2. moisture and length of test tubes 3. light and temperature 4. light and amount of time

(continued)

TABLE 5.1 Sample TIPS Questions (Continued)

Assessment Objective	Sample Item
D. Given a problem with dependent variables specified and a list of possible independent variables, identify a testable hypothesis	A student has been playing with a water rocket. He can change the amount of water in the rocket and the angle at which he releases the rocket. He can also change the weight of the rocket by adding sand in the nose cone. He wants to see what might affect the height to which the rocket will rise. Which of the following is a hypothesis he could test? 1. Rockets with warm water will rise higher than rockets with cold water 2. Rockets with four tail fins will rise higher than rockets with two tail fins 3. Rockets with pointed nose cones will rise higher than rockets with rounded nose cones 4. Rockets with more water will rise higher than rockets with less water.
E. Given a verbally described variable, select a suitable operational definition for it	The effect of exercise on pulse rate is studied by a science class. Students do different numbers of jumping jacks and then measure their pulse rates. One group does jumping jacks for one minute. A second group does these for two minutes. A third group jumps for three minutes. A fourth group does not jump. How would you measure the pulse rate in this study? 1. by counting the number of jumping jacks for one minute 2. by counting the number of heart beats in one minute 3. by counting the number of jumping jacks done by each group 4. by counting the number of exercises for each group
F. Given a problem with dependent variable specified, identify a testable hypothesis	Some chickens lay an egg almost every day. Other chickens produce few eggs. A study is planned to examine factors that might affect the number of eggs produced by chickens. Which of the following is NOT a suitable hypothesis for the study? 1. More eggs are produced by chickens that receive more hours of light 2. The more eggs produced by chickens the more weight they lose 3. The larger the cage for chickens the more eggs they will produce 4. The more proteins there are in the feed the more eggs produced

assessment usually takes place as an investigation, which can be both structured investigation and open-ended or extended investigation.

Performance assessment in the forms of investigations and extended investigations consists of three components: performance task, response format, and scoring rubric (Brown & Shavelson, 1996). Liu (2009, pp. 74–76) provides an elaboration of the above three components. Specifically, a performance task should be around a problem and meet the following requirements:

1. The problem invites students to solve a problem or conduct an investigation. The problem can be well-structured or ill-structured. Well-structured problems have all elements of the problem and a known solution. The procedures required to solve the problem are also routine and predictable. On the other hand, ill-structured problems do not necessary have all elements of the problem presented. More importantly, ill-structured problems may possess multiple solutions, solution paths, or no solutions at all. Because ill-structured problems are open-ended, personal judgment is constantly needed in the process of solving the problem. Well-structured problems are more appropriate for short investigations, and ill-structured problems are more appropriate for extended investigations.
2. The problem requires the use of concrete materials. This requirement is to differentiate performance assessment tasks from paper-and-pencil based problems. Performance tasks must involve hands-on operations.
3. Provide a concrete contextualization of the problem or investigation. Context makes the problem authentic. Performance assessment tasks approximate real-world problems, thus presenting the problem in a meaningful context is important.

Once a performance task is defined, the next component is to decide on the response format. The response format is to gather evidence of student performance for scoring. The response format must meet the following requirements: (a) provide opportunities for students to record processes and solutions; (b) prompt students for specific information; (c) allow students to decide how to summarize findings, and (d) ask students to justify solutions.

The last component of performance assessment is a scoring scheme or rubric. A rubric is a description of different levels of competence in performing a task. Given a task, there is always a variation in student performances; some are more novice-like while others more expert-like. The rubric must capture the "right" answer and reasonableness of procedures. It should also provide students useful feedback plus numerical scores. Given the number of students to be scored, the scoring rubric should also be easy and quick to use.

Here is a sample performance assessment with its scoring rubric for assessing students' integrated inquiry performance involving energy transfer and conservation for 8th grade students (Liu & Collard, 2005).

Task: Working in groups of 3, you will design a simple machine system to lift a 20-gram weight by 1 meter. You will use the materials provided in the materials box. You will have a total of 4 weeks to complete this task. Class time will be provided to complete the task. After you have completed the task, you will write a report describing the process you followed (including changes in your plans), identifying what forms of energy are involved, and how different forms of energy are transferred in your final design.

Materials: omitted

Scoring Rubric:

Level	Quantitative Reflection (QR)	Design Capacity (DC)	Relational Thinking (RT)	Design Consistency and Success (DCS)	Score
Energy Conservation	The heat energy generated by pulley accounts for the imperfect efficiency	Task + four types of simple machines	Characteristics (nature of energy such as state or property of matter)	Plans changed systematically and task completed successfully	4
Energy Degradation	The efficiency can never be 100%	Task + three types of simple machines	Evidence (energy loss during transformation due to friction)	Plans changed systematically and task completed unsuccessfully	3
Energy Transfer	Quantitative relationships among three or more variables	Task + two types of simple machines	Leads-to (sequential energy transformation)	Plans change irregularly and task completed unsuccessfully	2
Energy does work	Quantitative relationships between two variables	Task + one simple machine	Type-of or kind-of (e.g., energy types, sources, etc.)	Plans changed irregularly and task not completed	1
Energy as activity	Listing pulley, weight, incline plane etc. separately/no connection between them	Task	No explicit discussion of the nature of energy	No attempt to complete the task	0

Understanding About Science Inquiry

Various item formats are available for measuring understanding science inquiry. The most commonly used item format remains paper-and-pencil multiple-choice questions (e.g., Chen, 2006). Open-ended questions followed by in-depth interviews are also viable (e.g., Lederman et al., 2002). The following question is from VOSTS (Aikenhead & Ryan, 1992):

Item 10411

Science and technology are closely related to each other:

Your position, basically: (please read from A to H, and then choose one)
They are closely related to each other.

A. Because science is the basis of all technological advances; though it is hard to see how technology could aid science.
B. Because scientific research leads to practical applications in technology, and technological developments increase the ability to do scientific research.
C. Because, although they are different, they are linked so closely that it's hard to tell them apart.
D. Because technology is the basis of all scientific advances, though it's hard to see how science could aid technology.
E. Because science and technology are more or less the same thing.
F. I don't understand.
G. I don't know enough about this subject to make a choice.
H. None of these choices fits my basic viewpoint.

Items should be reviewed for content validity. This is typically accomplished by an expert panel. "Experts" for the panel should be selected based on their demonstrated expertise instead of assumed expertise. For example, if the items are intended to measure understanding about science inquiry, science teachers may not necessarily be considered experts because research has shown repeatedly that many science teachers do not possess an adequate understanding of the NOS. Practicing scientists may not have the highest level of understanding of NOS either. Besides content validity, items may also be reviewed by assessment experts on the appropriateness of item formats and scoring. Item bias in terms of cultures, gender, and socioeconomic status should also be reviewed and avoided.

Conduct Pilot-Test/Field-Test

Items should also be tried with a small number of selected subjects before field-testing. Pilot-testing is typically accompanied by interviewing or think aloud with a small number of subjects. Rasch analysis of pilot-test data (described below) can help expose problems and issues with the items so that necessary revisions may be made before they go to field-testing. Pilot-test also provides an opportunity to understand the process of measurement in order to better plan for field-testing.

After items have been tried out and necessary revisions have been made, the instrument can now go through a field-testing. The sample for the field-testing should be representative of the intended population. Representativeness of the sample does not necessarily have to be achieved by random sampling—a major challenge in most science education research situations. One important consideration of representativeness is the range of performance variations in the target population. Thus, field-testing samples may be conveniently selected based on the variation in subject performances that matches the variation of subject performances in the target population.

Conduct Rasch Analysis

Data collected from pilot-testing/field-testing will then be submitted to Rasch analysis. When performance assessment is used to measure student science inquiry performances, it is common that more than one rater will score each student's performance. In Rasch modeling, the many-facet Rasch model is applicable to this situation. In many-facet Rasch modeling, raters are considered to be one additional facet, in addition to the two facets, that is, students and items, in a typical paper-pencil test situation. When including raters as one additional facet, each rater's degree of severity in scoring is estimated as a measure placed on the same scale as item difficulty and subject ability measures. This way, we can directly compare a rater's bias or scoring severity against item difficulties and subject abilities.

Figure 5.1 shows an integrated control and data file for many-facets Rasch modeling. In the file, the upper half before "Data=" presents commands describing the three facets and how they are arranged in the data file. The lower half starting from "Data=" are student scores on the three traits given by the two independent coders. Feeding this integrated control and data file into many-facets Rasch analysis software (Linacre, 1997), a series of output tables and diagrams will be produced in an output file.

```
File  Edit  Format  View  Help
Title = Energy Progression
Facets = 3               ; three facets: judges (two coders), students, items (traits)
Inter-rater = 1 ; facet 1 (judege) is the rater facet
Positive = 2    ; students have greater understanding with greater score
Non-centered = 1         ; students and items are centered on 0 logits, judges are allowed to float
Model = ?B,?B,?,R4       ; judges, students and items produce ratings with maximum rating of 5.
    ; A bias/interaction analysis, ?B,?B will report interactions between facets 1 (judges) and 2 (students)
Labels =
1, Judges   ; name of first facet: judges
 1 = Coder1
 2 = Coder2
*
2, Students   ; name of second facet: students
 1 =1a
 2 =1b
 3 =1c
 4 =1d
 5 =2a
 6 =2b
 7 =2c
 8 =2d
 9 =3a
 10 =3b
 11 =3c
 12 =4a
 13 =4b
 14 =4c
 15 =5a
 16 =5b
 17 =5c
 18 =6a
 19 =6b
 20 =7a
 21 =7b
 22 =7c
 23 =8a
 24 =8b
 25 =8c
 26 =8d
*
3, Traits        ; name of third facet: items
 1 = QT
 2 = AC
 3 = QL
*
Data=
1,1,1-3,4,4,2
1,2,1-3,2,2,2
1,3,1-3,3,3,2
1,4,1-3,3,2,4
1,5,1-3,2,2,2
1,6,1-3,3,3,3
1,7,1-3,3,3,2
1,8,1-3,2,3,2
1,9,1-3,3,3,4
1,10,1-3,3,4,2
1,11,1-3,3,2,2
1,12,1-3,3,3,2
1,13,1-3,2,3,3
1,14,1-3,2,3,2
1,15,1-3,2,3,4
1,16,1-3,4,3,4
1,17,1-3,3,3,2
1,18,1-3,2,3,4
1,19,1-3,1,2,2
1,20,1-3,3,4,4
1,21,1-3,3,4,4
1,22,1-3,3,4,4
1,23,1-3,3,3,2
1,24,1-3,3,2,2
```

Figure 5.1 Shows a combined control and data file. Lines before "Data=" are Rasch analysis commands describing the three facets and how they are arranged in the data file. Lines after "Data=" are student scores on the three traits given by two independent coders.

Review Fit Statistics, Wright Map, Unidimensionality and Reliability

Table 5.2 presents the fit statistics obtained for the middle school performance assessment task on energy transfer described earlier based on many-facet Rasch modeling.

Like item analysis for developing other types of measurement instruments, various statistics related to items based on Rasch analysis outputs

TABLE 5.2 Fit Statistics for Items on a Middle School Performance Assessment on Energy Transfer

			INFIT		OUTFIT	
Item	**Measure**	**SE**	**MNSQ**	**ZSTD**	**MNSQ**	**ZSTD**
Capacity	–5.59	0.36	0.9	0.0	1.0	0.0
Qualitative	2.19	0.32	0.6	–1.0	0.3	–2.0
Quantitative	3.40	0.34	0.6	–2.0	0.7	–1.0
Judge 1	–0.17	0.24	0.7	–1.0	0.8	–1.0
Judge 2	0.17	0.25	0.6	–1.0	0.6	–2.0

Source: Liu & Collard, 2005

should be reviewed. In particular, item fit statistics, point-biserial correlation, misfitting responses, and category structures should be reviewed. From Table 5.2, we see that the fit statistics suggest that all three items/traits seem to have a good fit with the model. As for the two coders/judges, Judge 1 is less severe than Judge 2 in scoring; the difference between the two judges' severity measures is 0.34. Statistical testing for the difference by considering their standard errors of measures suggested that the difference was statistically insignificant. Thus, there was overall a good inter-rater agreement. The two judges' fit statistics are also within the acceptable ranges, thus the judges' scoring performances fit the Rasch model well.

Figure 5.2 shows the scoring category functions for five levels of performance specified in the scoring rubric for the same middle school performance assessment. Only categories 0, 2, 3, and 4 are distinct, that is, having their own peaks, suggesting that Category 1 is not distinct and may be combined with either Category 0 or Category 2. In the performance assessment on energy transfer, categories correspond to different stages of energy concept development. Category 0 is about energy responsible for work; Category 1 is about energy sources and forms; Category 2 is about energy transfer; Category 3 is about energy degradation; and Category 4 is about energy conservation.

Figure 5.3 shows the Wright map for the performance assessment on energy transfer based on the same middle school group of students.

In Figure 5.3, measures are in logit; two coders are labeled Judge 1 and Judge 2; students are marked by *x*; *trait* corresponds to the three aspects specified in the scoring rubric and is considered as *item*; finally *stage* is the category in the scoring rubric corresponding to different stages of energy concept development. We can see from Figure 5.3 that the two judges' severity degrees are very close and at about the average of students' ability levels, thus the two judges are unlikely to be biased against most students.

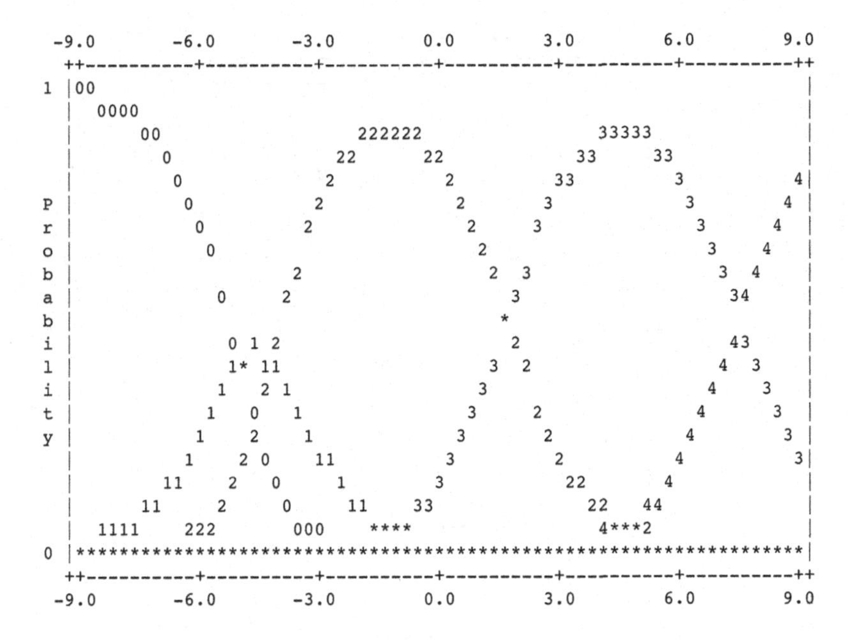

Figure 5.2 Category structure probability curves for the middle school energy transfer performance assessment. Although Categories 0, 2, 3, and 4 have distinct characteristic curves, Category 1 does not possess a distinct characteristic curve.

This Wright map also shows that *ql* and *qt* are above most students' ability levels, and there is a need for more items at the average level of most students' abilities.

Rasch modeling also provides a variety of ways to examine the reliability of the instrument. First, item and person standard errors of measurement provide an indication on the precision of item and person measures. Lower standard errors of measurement relative to the item and person measures are desirable. Second, the person separation index and its Cronbach's alpha equivalent provide an indication of how the items are internally spread along the construct and are able to differentiate subjects. The higher the separation index and the alpha, the more reliable the instrument. Third, for measurement instruments that involve more than one rater, many-facet Rasch modeling also provides information on inter-rater reliability. The difficulty estimates for different raters should be statistically insignificant.

Examine Invariance Properties

After items have been found to fit the model and the range of items is aligned with the range of student abilities, a final quality assurance check

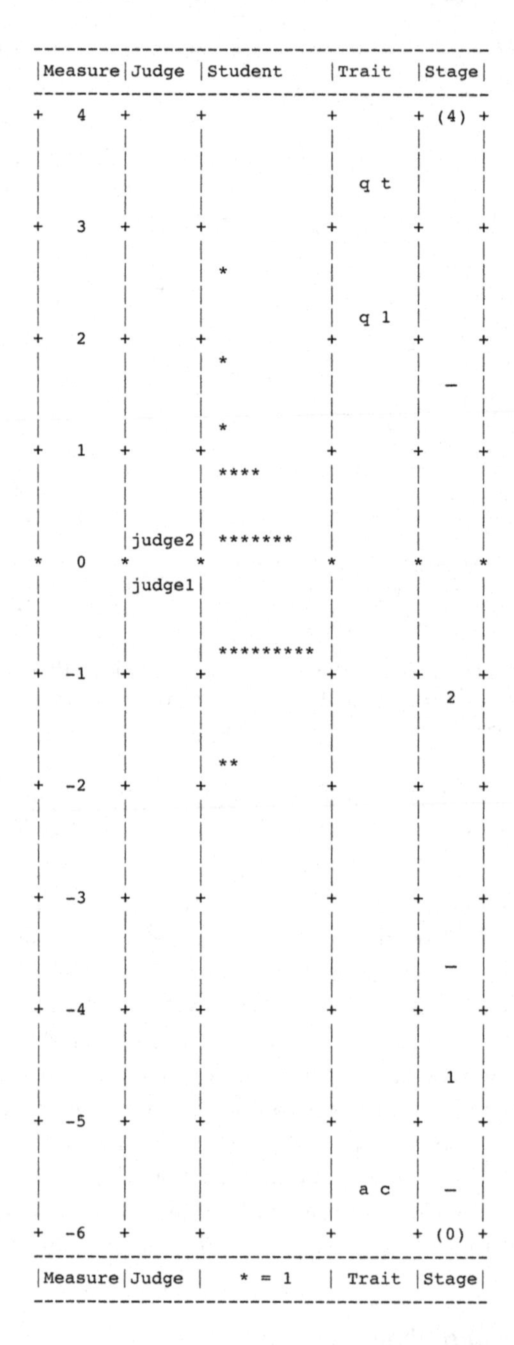

Figure 5.3 Wright map for the middle school energy transfer performance assessment. Two judges showed similar severity in scoring; different traits possess different difficulties.

is to examine item and person measure invariance. As a desired benefit for Rasch modeling, item difficulty estimates should not vary according to student subgroup characteristics (e.g., gender), and student ability estimates should not vary by item formats or delivery modes. Rasch modeling software typically provides various statistics to help examine the above properties; this is commonly called analysis Differential Item Functioning (DIF; for item invariance) and Differential Person Functioning (DPF; for person invariance). In the case of many-facet Rasch modeling, additional facets to person and item can also be considered necessary for invariance. For example, if significant difference is found between different coders, then neither item difficulty nor person ability estimates may be expected to be invariant. Thus, in many-facet Rasch modeling, there should not be significant differences found among elements of additional facets.

Establish Validity Claims

Rasch modeling provides various types of evidence to help establish validity claims. First and foremost, the entire measurement development process is guided by the theory of science inquiry and practices on the construct, and the Rasch modeling is a tool to test the agreement between data and the theory. If there is an overall good model-data-fit based on item fit statistics, then there is evidence for construct validity. Construct validity can also be demonstrated by examining the interaction between persons and items. A good match in range of measures between persons and items, and an even distribution of items along the construct demonstrates construct validity. Construct validity can further be demonstrated by examining the dimensionality of the items. If the factor analysis of residuals shows no additional dominant factors, then there is evidence for the construct validity, as the instrument is unidimensional. Validity evidence related to the test content, response processes, and consequences may also be established following such procedures as expert item review, think aloud of respondents when answering items, and examination of impacts of Rasch score uses.

The Rasch modeling primarily facilitates validity claims related to internal structure of the measurement instrument; additional validity studies may also be conducted to establish other validity related evidence. For example, if it is known that understanding about science inquiry should be statistically significantly correlated with students' science achievement, then a criterion-related validation may be conducted to establish the prediction validity—external structure. Similarly, if there is another valid and reliable instrument available, administering the available instrument to the same sample may be conducted to establish concurrent validity, that is, external structure, of the measurement instrument under development.

Develop Guidelines for Use of the Instrument

After the instrument has been finalized through cycles of revisions and pilot-testing/ field-testing, and the validity and reliability of measures of the instrument have been established, the instrument is ready for dissemination and adoption by others. Detailed documentation should then be developed to facilitate users' administration of the instrument. The documentation should clearly state the intended uses of the test scores, the construct measured and the correspondent domains of performance and behaviors. The documentation should also include the procedures for administering the instrument, scoring responses, and reporting measures of subjects. Information on the validity and reliability of the instrument should also be provided in the documentation.

Chapter Summary

Science inquiry and practices are comprehensive constructs for organizing science curriculum, instruction, and teacher professional development. Science inquiry and practices can be conceptualized as content for science learning, a way of knowing, and an approach to science teaching. Although inquiry does not take place in a step-by-step fashion, there are essential aspects common to all inquiry activities. These aspects are: hands-on skills pertaining to data collection and analysis, reasoning skills in relating data to theories, and abilities to perform inquiry tasks. Science inquiry is essentially an integrated practice that involves generating and evaluating scientific evidence and explanations. Generating and evaluating scientific evidence and explanations encompasses the knowledge and skills used for building and refining models and explanations, designing and analyzing empirical investigations and observations, and constructing and defending arguments with empirical evidence. The socio-cultural context is also essential in the above inquiry activities.

Besides the ability to demonstrate science inquiry and practices described above, the *National Science Education Standards* and *Next Generation Science Standards* also expect students to demonstrate understanding about science inquiry and practices. Understanding about science inquiry and practices relates to how and why scientific knowledge changes in response to new evidence, logical analysis, and modified explanations debated within a community of scientists. Related to understanding about science inquiry and practices is understanding about the NOS. Because science inquiry and practices are essential components of science, understanding about science inquiry and practices is an essential component of understanding about the NOS. In addition, science inquiry as teaching and learning requires that teachers understand the nature and process of students' developing scientific knowledge as inquiry, and conduct science teaching through science inquiry.

Many measurement instruments are available for science inquiry and practices. These instruments use a variety of question formats, including multiple-choice, true–false, rating scale, Likert scale, checklist or inventory, and performance assessment. When selecting a measurement instrument to use, one most important consideration is the validity of the construct the instrument is intended to measure. Theories about science inquiry and practices have changed considerably over the years. As a result, many theories that underlie the constructs of previous instruments may be outdated, which makes the construct validity of those instruments questionable. Consideration should also focus on other types of evidence of validity and reliability for the instruments. Even if a measurement instrument developed in the past is reported to be valid and reliable, it is still necessary to further validate the instrument because the targeted subject population (e.g., students, teachers) may have changed considerably.

Developing measurement instruments for science inquiry and practices follows a systematic process. This process starts with identifying a valid and commonly accepted theory about science inquiry and practices. Based on the theory, the intended construct of measurement is defined and the subject behaviors corresponding to different levels of performance on the construct are identified. The measurement instrument development process continues with the development of a test specification, item constructions, item review, item try-out and field testing. Data collected from pilot-testing and field-testing are submitted to Rasch modeling. If more than one rater is involved in scoring, a many-facet Rasch model is applied. Review of fit statistics, the Wright map, dimensionality map, as well as separation indices and alpha reliability should be conducted according to established conventions. Using Rasch modeling to develop measurement instruments helps establish evidence for validity and reliability claims.

Exercises

1. There are at least three approaches to developing measurement instruments on understanding of the nature of science. One approach is using selected response question formats such as the Likert scale questions; another approach is using non-scoring items

such as that in VOSTS; and the third approach is using open-ended questions such as items in VNOS. Compare the above three approaches, discuss the pros and cons of each of the approaches to measuring understanding of the nature of science in the context of Rasch modeling.

2. One important characteristic of inquiry science teaching is authentic science investigation activities. Suppose that you are interested in developing a measurement instrument to measure the degree of authenticity of science investigation activities performed by students. Please complete the following initial steps of developing such an instrument: (a) define the construct of authenticity, (b) identify student behaviors that represent different degrees of authenticity, and (c) write five sample items corresponding to five different levels of authenticity.
3. The development of *Self-Efficacy and Metacognition Learning Inventory—Science* (SEMLI-S; Thomas, Anderson, & Nashon, 2008) was based on Rasch modeling. Read the report to critique the adequacy of application of Rasch modeling in the development of the measurement instrument.
4. Hands-on performance tasks are claimed to be one of the best ways to measure student science inquiry abilities. Develop a plan to conduct validity and reliability studies for the development of a measurement instrument that is based on hands-on performance tasks.

References

Aikenhead, G. (1973). The measurement of high school students' knowledge about science and scientists. *Science Education, 57*(4), 539–549.

Aikenhead, G. S., & Ryan, A. G. (1992). The development of a new instrument: Views on science–technology–society (VOSTS). *Science Education, 76*(5), 477–491.

Anderson, R. D. (2007). Inquiry as an organizing theme for science curricula. In S. K. Abell & N. G. Lederman (Eds.), *Handbook of research on science education* (pp. 807–830). New York, NY: Routledge.

Anderson, D., & Nashon, S. (2007). Predators of knowledge construction: Interpreting students' metacognition in an amusement park physics program. *Science Education, 91*(2), 298–320.

Barrie, S. C., Bucat, R. B., Buntine, M. A., Silva, K. B., Crisp, G. T., George, A. V., . . . Yeung, A. (2015). Development, evaluation and use of a student experience survey in undergraduate science laboratories: The advancing science by enhancing learning in the laboratory student laboratory learning experience survey. *International Journal of Science Education, 37*(11), 1795–1814.

Billeh, V. Y., & Hasan, O. E. (1975). Factors influencing teachers' gain in understanding the nature of science. *Journal of research in Science Teaching, 12*(3), 209–219.

Bridgham, R. G. (1969). Classification, serration, and the learning of electrostatics. *Journal of Research in Science Teaching, 6,* 118–127.

Brown, J. H., & Shavelson, R. J. (1996). *Assessing hands-on science: A teacher's guide to performance assessment*: Thousand Oaks, CA: Corwin Press.

Butts, D. P. (1964). The evaluation of problem solving in science. *Journal of Research in Science Teaching, 2,* 116–122.

Calik, M., & Coll, R. K. (2012). Investigating socioscientific issues via scientific habits of mind: Development and validation of the scientific habits of mind survey. *International Journal of Science Education, 34*(12), 1909–1930.

Chen, S. (2006). Development of an instrument to assess views on nature of science and attitudes toward teaching science. *Science Education, 90*(5), 803–819.

Chen, S., Chang, W.-H., Lieu, S.-C., Kao, H.-L., Huang, M.-T., & Lin, S.-F. (2013). Development of an empirically based questionnaire to investigate young students' ideas about nature of science. *Journal of Research in Science Teaching, 50*(4), 408–430.

Chu, H.-E., Lee, E. A., Ko, H. R., Hee, S. D., Moon, N., Min, B. M., & Hee, K. K. (2007). Korean year 3 children's environmental literacy: A prerequisite for a Korean environmental education. *International Journal of Science Education, 29*(6), 731–746.

Cobern, W. W., & Loving, C. C. (2002). Investigation of preservice elementary teachers' thinking about science. *Journal of Research in Science Teaching, 39*(10), 1016–1031.

Cothan, J., & Smith, E. (1981). Development and validation of the conceptions of scientific theories test. *Journal of Research in Science Teaching, 18*(5), 387–396.

Denny, R. T. (1971). The mathematical skill test (MAST) in chemistry. *Journal of Chemical Education, 48*(12), 845–646.

Dillashaw, F. G., & Okey, J. R. (1980). Test of the integrated science process skills for secondary science students. *Science Education, 64*(5), 601–608.

Eggert, S., & Bogeholz, S. (2010). Students' use of decision-making strategies with regard to socioscientific issues: An application of the Rasch partial credit model. *Science Education, 94*(2), 230–258.

Elby, A., & Hammer, D. (2001). On the substance of a sophisticated epistemology. *Science Education, 85*(5), 554–567.

Fraser, B. J. (1980). Development and validation of a test of enquiry skills. *Journal of Research in Science Teaching, 17,* 7–16.

Halloun, I., & Hestenes, D. (1996, March). *Views about sciences survey: VASS.* Paper presented at the annual meeting of the National Association for Research in science Teaching, St. Louise, MO.

Halloun, I., & Hestenes, D. (1998). Interpreting VASS dimensions and profiles for physics students. *Science & Education, 7,* 533–577.

Hu, W., & Adey, P. (2002). A scientific creativity test for secondary school students. *International Journal of Science Education, 24*(4), 389–403.

Kimball, M. E. (1967). Understanding of the nature of science: A comparison of scientists and science teachers. *Journal of Research in Science Teaching, 5*, 110–120.

Kirbulut, Z. D., Uzuntiryaki-Kondakci, E., & Beeth, M. E. (2016). Development of a metaconceptual awareness and regulation scale. *International Journal of Science Education, 38*(13), 2152–2173.

Kuo, C.-Y., Wu, H.-K., Jen, T.-H., & Hsu, Y.-S. (2015). Development and validation of a multimedia-based assessment of scientific inquiry abilities. *International Journal of Science Education, 37*(14), 2326–2357.

Lachapelle, C. P., Cunningham, C. M., & Oh, Y. (2019). What is technology? Development and evaluation of a simple instrument for measuring children's conceptions of technology. *International Journal of Science Education, 41*(2), 188–209.

Lawson, A. E. (1978). The development and validation of a classroom test for formal reasoning. *Journal of Research in Science Teaching, 15*(1), 11–24.

Lederman, J. S., Lederman, N. G., Bartos, S. A., Bartels, S. L., Meyer, A. A., & Schwartz, R. S. (2014). Meaningful assessment of learners' understandings about scientific inquiry—The Views about Scientific Inquiry (VASI) Questionnaire. *Journal of Research in Science Teaching, 51*(1), 65–83.

Lederman, N. G. (1992). Students' and teachers' conceptions of the nature of science: A review of the research. *Journal of Research in Science Teaching, 29*(4), 331–359.

Lederman, N. G. (2007). Nature of science: Past, present, and future. In S. K. Abell & N. G. Lederman (Eds.), *Handbook of research on science education.* New York, NY: Routledge.

Lederman, N. G., Abd-El-Khalick, F., Bell, R. L., & Schwartz, R. S. (2002). Views of nature of science questionnaire: Toward valid and meaningful assessment of learners' conceptions of nature of science. *Journal of Research in Science Teaching, 39*(6), 497–521.

Lederman, N. G., Wade, P. D., & Bell, R. L. (1998). Assessing understanding of the nature of science: A historical perspective. In W. McComas (Ed.), *The nature of science and science education: Rationales and strategies* (pp. 331–350). Dordrecht, The Netherlands: Kluwer Academic.

Lee, M.-H., Johanson, R. E., & Tsai, C.-C. (2008). Exploring Taiwanese high school students' conceptions of and approaches to learning science through a structural equation modeling analysis. *Science Education, 92*(2), 191–220.

Linacre, J. M. (1997). *Facets* (version 3.08). Chicago, IL: MESA Press.

Lin, T.-J., & Tsai, C.-C. (2017). Developing instruments concerning scientific epistemic beliefs and goal orientations in learning science: A validation study. *International Journal of Science Education, 39*(17), 2382–2401.

Liu, X. (2009). *Essentials of science classroom assessment.* Thousands, CA: SAGE.

Liu, X., & Collard, S. (2005). Using Rasch model to validate stages of understanding the energy concept. *Journal of Applied Measurement, 6*(2), 224–241.

Lunetta, V. N., & Tamir, P. (1979). Matching lab activities with teaching goals. *The Science Teacher, 46*(5), 22–24.

Marks, R. L. (1967). CBA high school chemistry and concept formation. *Journal of Chemical Education, 44*(8), 471–474.

Molitor, L. L., & George, K. D. (1976). Development of a test for science process skills. *Journal of Research in Science Teaching, 13*(5), 405–412.

Mun, K., Shin, N., Lee, H., Kim, S.-W., Choi, K., Choi, S.-Y., & Krajcik, J. S. (2015). Korean secondary students' perception of scientific literacy as global citizens: Using global scientific literacy questionnaire. *International Journal of Science Education, 37*(11), 1739–1766.

National Research Council. (1996). *National science education standards.* Washington, DC: National Academy Press.

National Research Council. (2000). *Inquiry and the national science education standards: A guide for teaching and learning.* Washington, DC: National Academic Press.

National Research Council. (2007). *Taking science to school: Learning and teaching science in Grades K–8.* Washington, DC: The National Academies Press.

National Research Council. (2012). *A framework for K–12 science education: Practices, crosscutting concepts, and core ideas.* Washington, DC: The National Academies Press.

National Research Council. (2013). *Next generation science standards: For states, by states.* Washington, DC: The National Academies Press.

Nelson, M., & Abraham, E. C. (1973). Inquiry skill measures. *Journal of Research in Science Teaching, 10*(4), 291–297.

Nelson, C. H., & Mason, J. M. (1963). A test of science comprehension for upper elementary grades. *Science Education, 47*(4), 319–330.

Osborne, J., Collins, S., Ratcliffe, M., Millar, R., & Duschl, R. (2003). What "ideas-about-science" should be taught in school science? A Delphi study of the expert community. *Journal of Research in Science Teaching, 40*(7), 692–720.

Peterman, K., Cranston, K. A., Pryor, M., & Kermish-Allen, R. (2015). Measuring primary students' graph interpretation skills via a performance assessment: A case study in instrument development. *International Journal of Science Education, 37*(17), 2787–2808.

Robinson, J. T. (1969). Evaluating laboratory work in high school biology. *American Biology Teacher, 31*(4), 236–240.

Romine, W. L., Sadler, T. D., & Kinslow, A. T. (2017). Assessment of scientific literacy: Development and validation of the Quantitative Assessment of Socio-Scientific Reasoning (QuASSR). *Journal of Research in Science Teaching, 54*(2), 274–295.

Rudolph, J. L. (2000). Reconsidering the "nature of science" as a curriculum component. *Journal of Curriculum Studies, 32*(3), 403–419.

Sakschewski, M., Eggert, S., Schneider, S., & Bögeholz, S. (2014). Students' socioscientific reasoning and decision-making on energy-related issues—Development of a measurement instrument. *International Journal of Science Education, 36*(14), 2291–2313.

Shin, N., Jonassen, D. H., & McGee, S. (2003). Predictors of well-structured and ill-structured problem solving in an astronomy simulation. *Journal of Research in Science Teaching, 40*(1), 6–33.

Taasoobshirazi, G., & Farley, J. (2013). Construct validity of the physics metacognition inventory. *International Journal of Science Education, 35*(3), 447–459.

Taasoobshirazi, G., Bailey, M., & Farley, J. (2015). Physics metacognition inventory Part II: Confirmatory factor analysis and Rasch analysis. *International Journal of Science Education, 37*(17), 2769–2786.

Tamir, P., Nussinovitz, R., & Friedler, Y. (1982). The design and use of practical tests assessment inventory. *Journal of Biological Education, 16*(1), 42–50.

Thomas, G., Anderson, D., & Nashon, S. (2008). Development of an instrument designed to investigate elements of science students' metacognition, self-efficacy and learning processes: The SEMLI-S. *International Journal of Science Education, 30*(13), 1701–1724.

Tsai, C.-C., & Liu, S.-Y. (2005). Developing a multi-dimensional instrument for assessing students' epistemological views toward science. *International Journal of Science Education, 27*(13), 1621–1638.

van Schijndel, T. J. P. V., Franse, R. K., & Raijmakers, M. E. J. (2010). The exploratory behavior scale: Assessing young visitors' hands-on behavior in science museums. *Science Education, 94*(5), 794–809.

Walker, R., Hendrix, J. R., & Mertens, T. R. (1979). Written Piagetian task instrument: Its development and use. *Science Education, 63*(2), 211–220.

Wong, S. L., & Hodson, D. (2008). From the horse's mouth: What scientists say about scientific investigation and scientific knowledge. *Science Education, 93*(1), 109–130.

Wu, P.-S., Wu, H.-K., & Hsu, Y.-S. (2014). Establishing the criterion-related, construct, and content validities of a simulation-based assessment of inquiry abilities. *International Journal of Science Education, 36*(10), 1630–1650.

Yang, W., & Liu, E. (2016). Development and validation of an instrument for evaluating inquiry-based tasks in science textbooks. *International Journal of Science Education, 38*(18), 2688–2711.

6

Using and Developing Instruments for Measuring Learning Progression

This chapter is concerned with learning progression. Learning progression is an emerging field of research in science education; it is gaining attention rapidly. For example, The *Canadian Journal of Science, Mathematics, and Technology Education* devoted an entire issue (i.e., Volume 4, Issue 1, 2004) to the topic of learning progression of children's long-term science concept development. The *Journal of Research in Science Teaching* devoted its entire Issue 6 of Volume 46 (2009) to learning progressions in science. More recently, the *Science Education* journal published a special issue on learning progression in May 2015 (Volume 99, Issue 3). With growing recognition of the importance of studying learning progression in science, learning progression is becoming an important measurement domain. This chapter will first review current theoretical frameworks about learning progression. It will then introduce standardized instruments for measuring learning progression. Finally, this chapter will describe the process for developing new instruments for measuring learning progression using the Rasch modeling approach.

Using and Developing Measurement Instruments in Science Education, pages 243–277

What Is Learning Progression in Science?

One key difference between school science learning and informal science learning is the organization of content. School science learning is characterized by a long-term systematic organization of content so that student learning is progressive. National and state science content standards are examples of systematic organization of content to ensure that learning takes place progressively over time across grades. The progressive nature of school science learning requires us to assess student learning over a long time span, often across multiple grades. This type of assessment is important not only to ensure that students have reached certain levels of learning proficiency or standards at a given time, but also to provide information about trajectories of student learning in order to inform ongoing teaching and learning. In order to meet this requirement, it is necessary to develop measurement instruments for assessing learning progression.

Learning progression can be defined as

> descriptions of the successively more sophisticated ways of thinking about a topic that can follow one another as children learn about and investigate a topic over a broad span of time (e.g., 6–8 years). They are crucially dependent on instructional practices if they are to occur. That is, traditional instruction does not enable most children to attain a good understanding of scientific frameworks and practices. (NRC, 2007, p. 219)

Similarly, Smith, and Wiser (2015) define learning progressions to be "proposals about how students' knowledge in major science domains (e.g., matter, energy, genetics) can evolve, over multiple grades, from initial ideas (lower anchor) to more scientifically informed explanations (upper anchor), *with appropriate instruction*" (p. 417). A learning progression is characterized by the following features:

1. *It is domain-specific.* Learning progression must pertain to science content instead of general reasoning or performance skills. Duschl, Maeng, and Sezen (2011) go even further to propose that a learning progress should integrate disciplinary ideas with science practices. This is in contrast to the cognitive development (e.g., logico-mathematical reasoning) in cognitive psychology. The *Next Generation Science Standards* (NRC, 2013) presents hypothetical learning progression for core disciplinary ideas, crosscutting concepts and science and engineering practices across grades.
2. *It describes different levels of competence.* A learning progression should map out a hierarchy of student learning outcomes from least competent to most competent. This hierarchy is also called a learning

trajectory. A learning progression starts with what students typically know and can do without formal school science learning; this starting point represents the level of competence students typically develop in informal settings before formal science learning takes place (the lower anchor). A learning progression ends with what is expected of students as the result of systematic science learning over a period of time. National and state curriculum standards typically describe what is expected of students by the end of a certain grade (the upper anchor). However, what is specified in the standards is not necessarily the end point of a learning progression, because standards usually specify only a minimal competence. The highest level of a learning progression should be a desired level of learning competence, even although it may not be achievable by all students. Between the starting and desired levels in the learning trajectory, there can be multiple interim levels of competence. No matter how many interim levels a learning progression contains, the levels must be distinct from each other, that is, qualitatively different and are in increasing sophistication.

3. *It is driven by a way of thinking.* Stages are not sufficient for a learning progression. Learning progressions "are a way to rethink and restructure the content and/or sequence of the subject matter that is taught (Lehrer & Schauble, 2015, p. 433); they are cognitively based, which means that there is a close interaction between epistemology and disciplinary core ideas to make possible student progression from one level to another possible (Smith & Wiser, 2015). Progression over different levels resembles a conceptual change, thus major epistemological commitments from the learners are essential. There are different perspectives of conceptual change in the literature and they have major implications to learning progression development and validation (Duschl et al., 2011).
4. *It is tied to curriculum and instruction.* A learning progression should capture the effects of different types of curriculum and instruction. Specifically, a learning progression represents what learning is possible with a certain instruction. This is called instruction-assisted learning (Duschl et al., 2011). Changes in student thinking without instruction is not enough for a learning progression. Lehrer and Schauble call learning progressions to seek transformative changes in student thinking instead of conservative changes as the result of conventional instruction (Lehrer & Schauble, 2015). Thus, a learning progression is typically a part of an innovative curriculum or new teaching materials; it seeks to improve student learning through transformative curriculum, instruction, and assessment.

5. *It is a model.* A learning progression does not necessarily account for all sources of variability in instruction (Lehrer & Schauble, 2015); thus a learning progression is not the same as individual student learning trajectories or pathways. For example, a learning progression cannot be the same as a series of lessons or learning units for students to complete. As a model, a learning progression is developed to explain and inform instruction; it may not be complete and is accurate only to some degree.

There are many ways to realize learning progression in curriculum and instruction. One way is to use a spiral curriculum to organize content (Bruner, 1960). According to Bruner, each discipline consists of a few central ideas that form the structure of the discipline. These structures vary in complexity, and cannot be introduced to students completely at once. As the result, it is necessary to teach the central ideas in a progressive way from lower grades through higher grades. A spiral curriculum has the organization of content around central ideas so that they are continually learned in a progressively complex form. Such an organization helps ensure the continuity, integration, articulation, and balance of the curriculum (AAAS, 2001). Continuity of the curriculum means that major ideas and skills are continuously practiced and developed; integration of curriculum means that all types of knowledge and experiences are connected and unified at a higher level; articulation of curriculum means that various aspects of the curriculum are interrelated; and finally balance of curriculum means that appropriate weights are given to each related basic ideas of central topics.

As an example of learning progression, Cately, Lehrer, and Reiser (2005) have developed a learning progression on evolution for K–8. They conceptualize that evolution accounts for life's diversity at all levels—its variation at genetic, habitat, and species levels. The core concepts involved for a complete understanding of evolution are:

1. *Diversity:* Biodiversity occurs at three levels—diversity of species, diversity within species and among individuals, and diversity of habitats. Species are the units of evolutionary change. The genetic diversity is the engine of evolution; and habitat is the agent of natural selection.
2. *Structure-Function:* Structures perform functions that allow individuals to survive. Structure-function relations are the cornerstone of adaptation.
3. *Ecology/Interrelationships:* Living things populate a particular habitat and are embedded within a complex system. Changes in a habitat

affect the functioning of the ecology and the chance for individual organisms to survive and replicate.

4. *Variation:* Random variation results from genetic recombination and genetic mutation; directed variation or natural selection results mostly from habitat variables. The interplay between random and directed variation is the foundation of life's diversity.
5. *Change:* Change occurs at different scales of time and organization.
6. *Geologic processes:* Geologic processes are related to past environments and the life history of the planet.
7. *Forms of argument:* Model-based reasoning and historic interpretation are primary forms of argument for development of the evaluation theory.
8. *Mathematical tools:* Mathematical tools such as measurement, data creation, distribution, Venn diagrams, and cladograms are critical in representing evolutionary processes.

Based on the above core concepts, Catley et al. (2005) propose the following learning progression on evolution by grade span:

- **Grades K–2**
  - Diversity: Diversity may exist in attribute/character, and can be understood by comparative analysis.
  - Structure-Function: Attributes often serve specific functions.
  - Ecology: There are relationships between habitat and the kinds of organisms that live there.
  - Variation: There are individual differences among everyday "kinds" of life.
  - Change: Individual organisms change over the course of their life span.
  - Geologic record: Fossils are artifacts of some of the processes that created them.
  - Forms of argument: Argumentation may take place by using models and comparative analysis.
  - Mathematical tools: Measure, Venn diagram, arithmetic (difference) are used to study evolution.
- **Grades 3–5**
  - Diversity: Diversity is understood by classification and comparative analysis.
  - Structure-Function: Characters often serve specific functions.
  - Ecology: There is a relationship between qualities of the environment and characters of the organism.
  - Variation: There is both within and between species variation.

- Change: Individual change helps with survival.
- Geologic record: There are a wide range of fossils.
- Forms of argument: Argumentation may take place using models and comparative analysis.
- Mathematics tools: Measures, Venn diagram, arithmetic (ratios) are used to study evolution.

- **Grades 6–8**
  - Diversity: Diversity is a result of mechanisms involving change, variation, and ecology.
  - Structure-Function: External and internal structures help organisms survive and reproduce.
  - Ecology: What affects one species may affect others in the system.
  - Variation: Organisms that are of the same species can also be different from one another.
  - Change: Change occurs continuously at both the individual and population levels.
  - Forms of argument: Argumentation may take place using models and comparative analysis.
  - Mathematical tools: Distribution and cladograms are used to study evolution.

We can see from the above description, learning progression on evolution is around a big idea and domain-specific, because it involves a number of core concepts or dimensions including diversity, structure and function, ecology and interrelation, variation, change, and geological processes. The learning progression also contains a hierarchy of competency levels, organized by three grade spans—Grades K–2, Grades 3–5, and Grades 6–8. The learning progression also implies a new way of thinking, that is, an interdisciplinary thinking, because the above core concepts involve not only domains of biology, geology, and chemistry, but also mathematical tools and logical reasoning. The above learning progression also expects a different curriculum and instruction. Because it does not prescribe how specific lessons and units should be designed, although different curriculums may have different effects on students' learning pathways along the learning progression, the learning progression should be considered as a model.

Assessment is an important component of learning progression, but assessment alone is not learning progression. Duschl et al. (2011) differentiate validation learning progression from evolutional learning progression. Validation learning progression focuses on developing instruments to measure and validate a sequence or levels of student performance with an implicit misconception replacement or "fix it" conceptual change model, while evolutionary learning progression focuses on defining and refining "the developmental pathway(s) through identification of mid-levels or stepping stones that are then used to bolster meaning making and reasoning employing crafted instructional interventions" (p. 157). This chapter is about validation learning progression; it focuses on using and developing measurement instruments that facilitate or are part of development of a learning progression.

Instruments for Measuring Student Learning Progression

Because learning progression is a relatively new research area in science education, only a few standard instruments have been published. This section introduces 10 measurement instruments related to learning progression. Although they are for different intended uses and are based on different conceptions of learning progression and development of understanding, one common element among the instruments is that Rasch modeling was used to develop them. Please note, the following descriptions of the instruments are for readers' information only; whether or not a science education researcher will choose an instrument for a particular research question requires a critical review of the instrument by the researcher, which is beyond the scope of this book.

Descriptions

Progression of Understanding Matter (Liu, 2007)

Progression of Understanding Matter (PUM) is a paper-and-pencil test assessing Grades 3 to 12 students' knowledge and understanding of matter. It consists of both multiple-choice and constructed response questions. It has three forms: elementary (Grades 3–6), middle school (7–9), and high school (10–12). The three forms are linked by common items. There are 20 questions in the elementary form, 29 in the middle school form and 29 in the high school form. The questions are from publically released items from TIMSS, NAEP and the NY State Regents exams. One common scale was constructed for all three forms through Rasch modeling so that students' knowledge and understanding at different grades can be directly compared. Validity is claimed by adequate model-data fit, the Wright map and unidimensionality, and reliability is claimed by person and item separation indices (person separation index was 1.94 or alpha equivalent of 0.79; item separation index was 9.68 or alpha equivalent of 0.99).

ChemQuery Assessment System (Claesgens, Scalise, Wilson, & Stacy, 2009)

The ChemQuery is a criterion-referenced assessment system to measure the paths of student understanding in chemistry. It includes assessment questions, a scoring rubric, item exemplars, a framework to describe the paths of student understanding that emerge, and criterion referenced analysis using item response theory (IRT) to map student progress. Understanding of chemistry is called chemistry perspectives and is conceptualized to consist of three "big ideas" or constructs: matter, change, and energy. Each construct requires students to continuously develop understanding about it; thus each construct is a progress variable. Each progress variable is defined by five hierarchical levels of understanding with increasing complexity: (a) Level 1—notions (e.g., What do we know about matter?), (b) Level 2—recognition (e.g., How do chemists describe matter?), (c) Level 3—formulation (e.g., How can we think about interactions between atoms?), (d) Level 4—construction (e.g., How can we understand composition, structure, properties, and amounts?), (e) Level 5—generation (e.g., What new experiments can we design to gain a deeper understanding of matter?). All questions are open-ended constructed response questions, and scored by a rubric. Each rubric describes specific student responses for each of the above performance levels.

Validation of the instrument followed an iterative process and was based on Rasch modeling. The iterative process involves:

1. Use a well-balanced expert group and pay attention to the research literature in the area to be measured to propose some "big ideas" and general outlines for an assessment framework.
2. On the basis of the current framework ideas, examine the curriculum to determine appropriate topic areas and contexts to consider for measuring developing conceptual understanding of progress variables.
3. Explore item-type possibilities and scoring approaches with curriculum developers, content experts, instructors, and project participants, and develop sets of items and rubrics.
4. Test items on a small informant population of students.
5. Collect feedback on items during an item paneling and scoring review session, and attempt to develop a body of exemplars, or examples of student work, at each level of scoring.
6. Qualitatively analyze the results of scoring.
7. Refine scoring rubrics and engage in or adjust approaches to rater training for the larger body of data collection to come.

8. Collect and analyze a larger set of diverse student data through both qualitative and quantitative approaches that can be analyzed with the use of measurement models.
9. Use the results of this larger data collection to refine the framework, improve items, evaluate and update the scoring approaches, and assess the appropriate use of measurement models.
10. Repeat the process using the updated approaches, and iteratively engage in further rounds of student data collection.

Examination of item fit statistics and the Wright map shows a good model-data fit. The expected *a posteriori* estimation based upon plausible values (EAP/PV) was 0.85 for the matter progress variable and 0.83 for the change variable, with person separation reliability estimated at 0.82 and 0.80, respectively. Difference in Rasch scale scores and their correspondent levels of understanding between high school and university students were found to be significant. ChemQuery may be used for both formative and summative assessment.

Connected Chemistry as Formative Assessment (Liu, Waight, Gregorius, Smith, & Park, 2012; Park, Liu, & Waight, 2017)

Connected Chemistry as Formative Assessment (CCFA) is a series of 10 linked end-of-unit tests measuring high school students' conceptual understanding of matter and energy and of computer models over the entire year of a general chemistry course. The progression of matter and energy understanding was defined as three qualitatively different stages from a systems thinking perspective; similarly the progression of understanding computer models was defined as three stages from using models as literal illustration to representing a range of phenomena to tools to make predictions and testing hypotheses. Ten tests were developed for the following ten commonly taught topics in high school chemistry: atomic structure, acids and bases, periodic table, solutions, chemical equilibrium, state of matter, stoichiometry, redox, ideal gases, and chemical bonding. Questions were in the ordered multiple-choice format; all questions were based on a set of computer models developed specifically for each of the topics. During a chemistry course, students learn each unit by using the computer models and then complete a test at the end of each unit, and their understanding of matter and energy and of models is measured and the change trajectory is mapped continuously over the course.

Content validity of the items was established by a multidisciplinary project team and an external advisory board. Pilot-testing was conducted with students from different classes ranging from 10 to 15 students per test.

Cognitive interviews were also conducted with select students. Field-testing of revised tests was conducted in three chemistry classes ($n = 22–26$) in three schools over the entire chemistry course. Multidimensional Rasch modeling was conducted to help improve items and tests. The revised items and tests were then field-tested and field-tested again, that is, extended field-testing, with 10 chemistry classes in 10 different schools during the entire chemistry course. Numbers of students completing each of the 10 tests ranged from 103 to 154; total number of students completing the tests was 1,230. Multidimensional Rasch modeling was conducted on this extended field-testing data set to establish construct validity. Results showed that item fit statistics were overall satisfactory; there was a reasonable alignment between student ability variation and item difficulty variation for both scales. The correlation between student understanding of matter-energy and understanding of models ranged from 0.625 to 0.889 for the 10 topics. The overall correlation between matter-energy and models for all 10 topics together was 0.764. The correlation between student understanding of matter-energy and NY State Regents exam scores was 0.189, and between student understanding of models and NY State Regents exam scores was 0.277. The EAP/PV reliabilities for the 10 tests for the matter-energy scale ranged from 0.258 to 0.785, and for the models scale ranged from 0.221 to 0.506.

Energy Concept Assessment (Neumann, Viering, Boone, & Fischer, 2013)

The Energy Concept Assessment (ECA) instrument was composed of a set of 102 multiple-choice items developed to measure the complexity of students' understanding with respect to four energy conceptions: (a) forms and sources, (b) transfer and transformation, (c) degradation, and (d) conservation. The instrument development went through three two-stages of pilot-testing and item revisions before the final validation study. The final version of ECA was administered to 1,856 students of Grades 6–10 in Germany. Rasch analysis was conducted to establish evidence of validity and reliability.

All the items had an acceptable model-data fit based on infit and outfit MNSQs and ZSEDs. The Wright map suggested a reasonable targeting of student abilities by items. The overall item reliability was 0.98 and person reliability was 0.61. The mean standard error of item difficulty estimates was 0.11 logits, while the item difficulty had a span between –1.78 and 1.38. Mean person ability was –0.66 logits with a minimum of –4.13 logits and a maximum of 3.81 logits.

Park and Liu (2016) developed the Inter-Disciplinary Energy concept Assessment (IDEA) for first year university students. The IDEA consists of four linked forms, one each in environmental science, biology, chemistry, and physics. In addition, IDEA was linked to the Energy Concept Assessment (ECA) for middle and high school students. The initial instrument included 47 items; most of them came from various published items of large-scale assessments including AAAS, NAEP, and TIMSS; 27 items were developed by the authors. Open-ended questions were scored based on rubrics. Five experts reviewed the items for content validity. The initial instrument was then given to 164 students taking the first-year introductory courses in one of the four science disciplines. Unidimensional partial credit Rasch model was applied to pilot data. The revised items were then given to 356 students in 12 first-year introductory science courses at six universities. Four-dimensional partial-credit Rasch analysis was conducted with the field test sample. There was a good model-data fit. The EAP/PV reliability coefficients for environmental science form was 0.772, biology 0.761, chemistry 0.758, and physics 0.756. Correlation between student scores on the four forms ranged from 0.772 to 0.872.

Learning Progression of Scientific Argumentation Assessment (Osborne et al., 2016)

Argumentation is a competency whose goal is the resolution of one or more claims, where resolution is only possible by engaging in a process of critique. A competency for scientific argumentation demands a complex orchestration of construction and critique of claims, warrants, and evidence in situations that require scientific knowledge to resolve. The learning progression for argumentation consists of two dimensions, constructing and critiquing, differentiated by intrinsic cognitive load for different levels requiring different degrees of coordination to be made between claims and pieces of evidence. Validation of the hypothetical learning progression was conducted with Grades 6–8 students in the San Francisco Bay Area. Development of the assessment was based on the BEAR Assessment System operationalized through four building blocks: the construct map, items design, outcome spaces, and the measurement model. Validation of the instrument involved 803 eighth grade students for the written test and 16 eighth grade students for the think-aloud interviews in Phase I, and 119 eighth grade and 159 tenth grade students for the written test and 15 eighth grade students in Phase II. Assessment items were contextualized in the physical behavior of matter. Items sharing a same scenario were grouped into bundles. Twenty-one (21) item bundles were developed, tested, and analyzed psychometrically. The 21 bundles contained 93 individual items. A further three general argumentation item bundles without specific demand for science content knowledge (15 individual items) were also developed. Two dimensional Rasch analysis of the four scientific argumentation bundles and three general argumentation bundles suggested a good model-data fit with EAP/PV reliability for the scientific dimension to be 0.87 and for the general dimension to be 0.84. Tenth grade students demonstrated a significant higher level of scientific argumentation than eighth grade students.

Measuring Concept Progressions in Acid-Base Chemistry (Romine, Todd, & Clark, 2016)

Measuring Concept Progressions in Acid-Base Chemistry (MCAB) is a 33-item instrument to measure undergraduate students' understanding of acids and bases. Student understanding of acids and bases is hypothesized to progress in three levels. The most simplistic way is called the phenomenon model; the next level is the use of character-symbol model or inference model characterized by memorization; the highest level utilizes a scientific model going beyond memorization. Validation of MCAB went through three rounds. The first version of the MCAB containing 33 ordered multiple-choice items was administered to 85 undergraduate students. Content validity was established through textbook content analysis and expert review. The second round of validation used a revised MCAB instrument containing 29 items that was administered to 181 science students. In the third and final round of validation, MCAB contained again 33 items and was administered to 311 students taking the two-course introductory general chemistry courses for science, health/medicine, and engineering majors. Rasch modeling was used to establish the construct validity and latent class analysis was used to cross-validate student progression levels. Person reliability of the 33-item version was 0.76. The measure was unidimensional and all items displayed satisfactory fit with the Rasch rating scale model.

Learning Progression-Based Assessment of Modern Genetics (Todd & Romine, 2016; Todd, Romine & Whitt, 2017)

Learning Progression-Based Assessment of Modern Genetics (LPA-MG) is a 34-item instrument covering 12 modern genetics ideas such as genes code for proteins and genetic information is hierarchically organized. With a few new items created, most items were adapted from previously existing assessment items to fit the learning progression framework. Three items were constructed for each of the major ideas using the ordered multiple-choice question format. There was also a certainty of response index (CRI) for each item to control for the confounding effects of guessing. The CRI was based on student responses to indicate their confidence after responding to each question from *guessing* (1), *uncertain* (2), *certain* (3), and *very confident* (4). The LPA-MG was administered to 65 tenth grade high school students at three time points over 23 weeks of instruction. Rasch analysis indicated that LPA-MG measures demonstrated high reliability and construct validity. Person reliability was 0.68 without adjusting for guessing and 0.91 after adjusting for guessing. The LPA-MG can provide both quantitative and qualitative information about how students progress along each main idea.

Todd and Romine (2016) then created a parallel instrument to measure university introductory biology students' learning progression on modern genetics; the instrument is called Version 2 of LPA-MG 2. Only modification of items was made to make them appropriate for university students. The validation sample contained 316 students of both biology majors and non-biology majors. The assessment was administered using Qualtrics survey software. Rasch rating scale and partial credit models were applied to the data. Results suggest that the measures were both valid and reliable for assessing students' unidimensional learning progression of modern genetics. Person reliability was 0.86.

The Energy Concept Progression Assessment (Yao, Guo, & Neumann, 2017)

The Energy Concept Progression Assessment (ECPA) is based on published learning progressions of energy; it proposes a revised learning progression of energy to consist of two progress variables: key ideas of energy (form, transfer and transform, dissipation and conservation) and conceptual development levels (fact, mapping, relation, and systematic). Initial items for ECPA were identified and modified if necessary from various published and public instruments and items were categorized according to the two progress variables. Invited experts reviewed the items for content validity, which resulted in 56 items to form the initial version of the instrument. Two pilot studies ($n1 = 735$, $n2 = 1{,}033$) were conducted before the final validation study. Two rounds of interviews with students from Grade 8 to Grade 12 (10 and 13 students, respectively) were also conducted. The final version contained 52 items. Vertical linking was used to connect booklets for different grades. The main validation study involved 4,550 students from Grades 8 to 12 in a major city in China. The partial credit Rasch model was applied to the data. Results suggest that there was a good model-data fit; unidimensionality was acceptable; person reliability was 0.82.

Scientific Explanation Progression Assessment (Yao & Guo, 2018)

Scientific Explanation Progression Assessment (SEPA) is based on a theoretical framework of scientific explanation called Phenomenon–Theory–Data–Reasoning (PTDR). The hypothetical learning progression of scientific explanation consists of two progress variables: Complete degree that describes the completeness of the language structure in an explanation using PTDR, and component levels that describe different levels of sophistication. The two progress variables together define a learning progression from Stage 1 through Stage 7. After two rounds of pilot tests, the final validation study involved a sample of 4,554 students in Grades 8–12 in a large city in China. A set of open-ended tasks was created and was piloted with students in two rounds involving 735 and 1,033 students. A scoring rubric was created for each task. Selected students were also interviewed (10 and 13 students, respectively). Pilot test data were analyzed by Rasch modeling and diagnostic information from Rasch modeling was used to revise assessment tasks. Twelve tasks were used in the final validation study. In the validation study, a vertical linking design was used so that different grades of students only responded to a subset of tasks. Rasch analysis was conducted to establish evidence for construct validity. Results show that there was a good model-data fit. The person reliability was 0.88.

University Students' Learning Progression in Quantum Mechanics (Testa et al., 2019)

The instrument for measuring learning progression in quantum mechanics contains 10 ordered-multiple-choice items. The initial hypothetical learning progression was based on the four big ideas: (a) wave-particle duality, (b) wave function, (c) atoms, and (d) complex quantum behavior. A questionnaire with 25 open items adapted from the literature was given to 30 third-year physics undergraduate students and 50 experienced physics teachers including university professors and teaching assistants. Responses to the open questions were used to define levels of learning progression and to develop 10 ordered multiple-choice questions. The validation sample included 244 students taking the bachelor in physics degree. Students were in four groups: no course, introductory course, introductory and upper-level course, and non-physics major students. Data analysis included exploratory factor analysis and Rasch modeling. Exploratory factor analysis found that the 10 items loaded evenly onto two factors: atomic description and measurement, and wave function and its properties. Factor analysis of Rasch residuals suggested unidimensionality with 47% variance explained by measures and the first contrast of unexplained variance had an eigenvalue 1.7. Items had an overall good fit with the partial credit Rasch model. Average difficulties of items increased gradually from lower levels to higher levels. The Wright map showed adequate targeting. The differences in student Rasch ability estimates among the four groups were statistically significant. Only one item exhibited potential DIF.

Commentary

Developing measurement instruments on learning progression is still in its early stage; published measurement instruments are few. Most published instruments are about matter and energy. This is likely due to the fact that well-established learning progressions are limited to such big ideas as matter and energy. Using a measurement instrument of learning progression must be accompanied with well-established learning progression, because a well-established learning progression requires a measurement and developing a measurement instrument requires a well-developed learning progression.

However, learning progressions on vast constructs (e.g., disciplinary ideas, crosscutting concepts, and science and engineering practices) are still under development; universally accepted learning progressions are rare. In this situation, we can only assume that the learning progression to guide instrument development is hypothetical, which creates a dilemma when there is not a sufficient model-data fit in Rasch analysis. This is because it is possible that the misfit could be due to the deficiency in the learning progression, or deficiency in items, or both. Researchers using Rasch modeling to develop measurement instruments of learning progression typically take a pragmatic approach to addressing the above dilemma. Because construct validity can only be claimed when there is good model-data fit including adequate targeting of items on student abilities, initially a coarse hypothetical learning progression is used as the starting point to develop items and instrument. After pilot-testing, Rasch analysis results are then used to refine both items and hypothetical learning progression. Such fine-tuning is iterative and continues until there is sufficient agreement between the hypothetical learning progression and the performances of items and the instrument as a whole. Applied this way, Rasch modeling is used simultaneously for the development of both the learning progression and measurement instrument.

All the published learning progression instruments have been developed using Rasch modeling. This is not surprising because measurement of learning progressions typically involves multiple grades and Rasch modeling allows measures from different grades to be placed on the same scale even when students at different grades respond to different items. Another important reason is that Rasch modeling is based on a rationale that is consistent with learning progression. Specifically, Rasch modeling assumes that there is a linear hierarchy in student performances of a construct and a learning progression provides exactly this. Further, because learning progressions are typically multidimensional (e.g., involving disciplinary ideas, crosscutting concepts, and science and engineering practices), multi-dimensional Rasch modeling is uniquely suited for modeling student performances of this nature.

Developing Instruments for Measuring Learning Progression

State the Purpose and Intended Population

Assessment of learning progression can be conducted for either formative or summative evaluation, or both. A formative use of test scores of learning progression assessment is to inform teachers in planning ongoing instruction and for students to reflect on the ongoing progress toward learning goals. This type of learning progression assessment typically takes place during a course of study. An example of such uses is CCFA (Liu et al., 2012). On the other hand, a summative use of learning progression is to identify the level of student achievement on a defined learning progression. This type of learning progression assessment typically takes place at multiple grade levels. An example of such uses is the PUM (Liu, 2007). Formative uses of learning progression assessment require that data collection takes place at multiple time points over a time span such as a course, while summative uses of learning progression assessment require data collection to be conducted only once—at the end of a learning cycle. Of course, assessment of learning progression may be used for both formative and summative purposes. In this last scenario, test scores of learning progression assessment at the end of a learning cycle are used for summative purposes, while test scores of learning progression assessment before the end of a learning cycle are used for formative purposes. Whether used for formative or summative use, a learning progression assessment must clearly define the intended populations, because a learning progression often takes place over multiple grades or years.

Define the Construct To Be Measured

One key task in developing assessment of a learning progression is to decide on the construct to be assessed. Constructs for learning progression assessment are domain specific; they typically pertain to the integrated learning of core disciplinary ideas, crosscutting concepts, and science and engineering practices. For example, the BEAR system, applied to the Issues, Evidence, and You (IEY) curriculum units, identifies five IEY progress variables as its measurement constructs. The five progress variables are defined as follows (Wilson & Sloane, 2000):

1. *Understanding concepts* (UC): Understanding scientific concepts (e.g., properties and interactions of materials, energy, or thresholds) to apply the relevant scientific concepts to the solution of problems.
2. *Designing and conducting investigations* (DCI): Designing a scientific experiment, carrying through a complete scientific investigation, performing laboratory procedures to collect data, recording and organizing data, and analyzing and interpreting results of an experiment.
3. *Evidence and tradeoffs* (ET): Identifying objective scientific evidence as well as evaluating the advantages and disadvantages of different possible solutions to a problem based on the variable evidence.
4. *Communicating scientific information* (CM): Organizing and presenting results in a way that is free of technical errors and effectively communicates with the chosen audience.
5. *Group interaction* (GI): Developing skills in working with teammates to complete a task (such as a lab experiment) and in sharing the work of the activity.

As can be seen in the BEAR assessment system example, the constructs defined for learning progression assessment are broad and comprehensive; they need to be continuously developed over a long time period. There can be more than one construct to be measured within a measurement instrument; but for each defined construct, there must be only one dimension of competence that describes the progression of the subjects' performance and the increasing demand of measurement items.

Identify Performances of the Defined Construct

Once the construct is defined, then it is necessary to identify subject behaviors or performances that represent the construct. This collection of behaviors defines the domain of assessment. Behaviors should involve both content and reasoning skills. Typically, a learning progression describes student performances at various levels. For example, for the core concept or idea of diversity which can be a construct of measurement, it may have the following behaviors for Grades 6–8 according to Cately et al. (2005):

- *Develop* attributes.
- *Identify* attributes/characters and use them to *classify* an organism.
- *Predict* an attribute given the presence of another attribute.
- *Determine* similarities and differences at the individual species level.
- *Compare and Contrast* species.

- *Construct, revise, present/defend, and critique* explanations of the contrasts observed.
- *Identify* characters which only evolved once and support large radiations of species that is, metamorphosis.
- *Distinguish* between characters that evolved independently on several occasions (convergent evolution), and those that appeared only once in the history of life.
- *Explain* that convergent evolution is driven by local ecological requirements and does not typically lead to large radiations of species.

The above behaviors are defined by both content and reasoning skills. Since a learning progression describes hierarchical levels of students' performances, it is common to define students' performances along a continuum from a lower level to a higher level. For example, the above behaviors related to understanding of diversity in living things may be represented in Table 6.1.

The above identified student behaviors form a hierarchy, similar to a scoring rubric. However, one key difference between the above-defined hierarchical student behaviors and a scoring rubric for an assessment task is that hierarchical student behaviors are not specific to any one assessment task; they apply to all assessment tasks, which will be discussed next.

Depending on the intended uses of learning progression measurement discussed earlier, a test specification based on intended student performances of measurement may take different formats. Taking the above core

TABLE 6.1 Student Behaviors of Diversity of Living Things

Student Behaviors	Performance Level
• *Identify* characters which only evolved once and support large radiations of species (i.e., metamorphosis). • *Distinguish* between characters that evolved independently on several occasions (convergent evolution), and those that appeared only once in the history of life. • *Explain* that convergent evolution is driven by local ecological requirements and does not typically lead to large radiations of species.	3 (Competent)
• *Identify* attributes/characters and use them to classify an organism. • *Predict* an attribute given the presence of another attribute. • *Compare and Contrast* species. • *Construct, revise, present/defend,* and *critique* explanations of the contrasts observed.	2 (Developing)
• *Develop* attributes. • *Determine* similarities and differences at the individual species level.	1 (Basic)

TABLE 6.2 Test Specification With Values for Assessment of Learning Progression of Diversity in Living Things

Grade Span	Student Behaviors	Weight	Linking Points
Grade 6	• *Identify* characters which only evolved once and support large radiations of species (i.e., metamorphosis). • *Distinguish* between characters that evolved independently on several occasions (convergent evolution), and those that appeared only once in the history of life. • *Explain* that convergent evolution is driven by local ecological requirements and does not typically lead to large radiations of species.	30 points	
Grade 7	• *Identify* attributes/characters and use them to classify an organism. • *Predict* an attribute given the presence of another attribute. • *Compare and Contrast* species. • *Construct, revise, present/defend,* and *critique* explanations of the contrasts observed.	30 points	10 points
Grade 8	• *Develop* attributes. • *Determine* similarities and differences at the individual species level.	30 points	

concept of diversity in living things as an example once again, a possible test specification for the summative use of learning progression assessment to be used by three grade levels is in Table 6.2.

In the above test specification, three forms of the test will be developed, with each form worth 30 points. In addition to the 30 points unique for each form, there will also be 10 additional points of test items from another form to provide a linkage between test forms. Thus, each test form will actually have 40 points.

Once a test specification with values such as that in Table 6.2 is developed, the number of items may then be decided. Multiple-choice items are typically scored as *right* (1) or *wrong* (0), and constructed and performance items are typically scored using a scoring rubric by awarding scores from 0 to a maximal score (e.g., 5). Once the total number of items is decided for each of the forms of the instrument, a linking scheme among the different forms of the instrument must be created. A sample linking design is represented graphically in Table 6.3.

In the sample linking design in Table 6.3, cells with *x* indicate that students respond to the corresponding questions in the specified forms, and blank cells indicate that students do not respond to the corresponding

TABLE 6.3 Test of Specification With Values for the Three Forms of Assessment Instrument for Learning Progression of Diversity in Living Things

	Form 1			Form 2			Form 3		
Student	Item 1–10	Item 11–20	Item 21–30	Item 1–10	Item 11–20	Item 21–30	Item 1–10	Item 11–20	Item 21–30
Grade 6	x	x	x	x					
Grade 7				x	x	x	x		
Grade 8			x				x	x	x

questions in the specified forms. Form 1 is indented for Grade 6, Form 2 for Grade 7, and Form 3 for Grade 8. Based on the design, students from each grade will respond to a total of 40 items, with 30 from the intended form and 10 additional from another form. This common item linking enables measures from Forms 1–3 to be placed on the same scale and will be directly comparable. Although there are no absolute rules available for deciding the required number of common items, a rule of thumb is that linking items should be about 20 items, or 20% of the total items in each form (Wolfe, 2004).

In addition to using items from different forms to be linked, which is called internal anchoring, items may also be from other instruments. This latter design is called external anchoring (Wolfe, 2004). The external anchoring uses a common set of items from other sources not to be part of the two forms to be linked as linking items. When using external anchoring, it is important to make sure that the external anchoring items measure the same construct of the learning progression.

If assessment of learning progression is for formative uses, that is, informing teachers and students on students' learning trajectories during a course, a test specification will involve scheduling the time points at which repeated tests will be given to students. A general consideration is to give tests repeatedly throughout the course plus giving a pretest at the beginning and a posttest at the end of the course. Although the number of questions for each of the tests may vary according to specific content, all the tests must be linked by a common scoring rubric along the defined dimension of progression on the construct. Specific indicators for different levels in the scoring rubric can vary depending on the assessment questions. Table 6.4 shows a sample scoring rubric that is used for all formative tests during the course.

When using a common scoring rubric for all formative assessment tests, it is possible to graph students' learning trajectories individually. Figure 6.1 shows sample learning trajectory.

TABLE 6.4 A Sample Scoring for Items on Using Evidence (Wilson & Sloan, 2000)

Score	Outcome
4	Response accomplishes Level 3 AND goes beyond in some significant way, such as questioning or justifying the source, validity, and/or quantity of evidence.
3	Response provides major objective reasons AND supports each with relevant and accurate evidence.
2	Response provides *some* objective reasons AND some supporting evidence, BUT at least one reason is missing and/or part of the evidence is incomplete.
1	Response provides only subjective reasons (opinions) for choice and/or uses inaccurate or irrelevant evidence from the activity.
0	No response; illegible response; response offers no reasons AND no evidence to support choice made.
X	Student had no opportunity to respond.

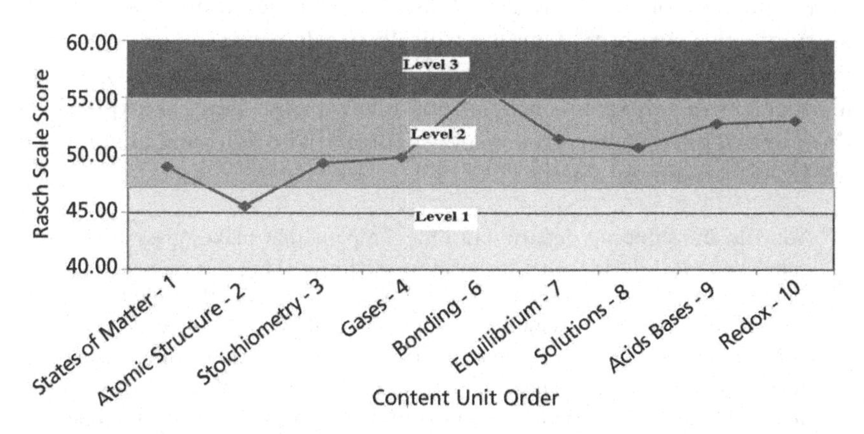

Figure 6.1 A student's learning trajectory on developing energy-matter understanding during a chemistry course. No assessment was given for Topic 5—Periodic Table.

Figure 6.1 shows that there are 9 formative assessment tests during the chemistry course, and there are 3 levels of students' understanding of matter-energy. Changes in their performance from Assessment 1 to Assessment 10 are represented by growth curves. These types of trajectories should be very informative for teachers in planning their ongoing instruction during the course.

One issue in designing a learning progression assessment for formative uses is the changing state of students' abilities. It should be expected that students' abilities on the defined construct will increase as learning progresses. As students' abilities increase, each time a new formative assessment test is given, we have a new group of students, which creates a potential problem for later Rasch modeling because of the lack of linkage in both items and students. Internal and external anchoring designs described earlier specifically address this problem. For example, in the BEAR assessment system applied to the EYE curriculum, in addition to the assessment tasks that are part of the teaching and learning activities, a series of link tests that are not specific to the teaching and learning activities and use conventional item types (e.g., multiple-choice) were also given at major junctures (e.g., end of a unit) along with formative assessment tests. In this way, the formative assessment tests are linked by common linking tests—external anchoring. In fact, the BEAR assessment system for EYE includes pretest, linking tests, and post-test, in addition to the formative assessment tests during various curriculum units.

Items for assessment of learning progression vary in format depending on the purpose of the assessment. If assessment of learning progression is for summative purposes, items are typically in selected response (e.g., multiple-choice) and short constructed response question formats. If assessment of learning progression is for formative purpose, items are typically in short or extended-constructed response format. The following item is from the BEAR assessment system:

> You run the shipping department of a company that makes glass kitchenware. You must decide what material to use for packing the glass so that it does not break when shipped to stores. You have narrowed the field to three materials: Shredded newspaper, Styrofoam® pellets, and cornstarch foam pellets. Styrofoam® springs back to its original shape when squeezed, but newspaper and cornstarch foam do not. Both Styrofoam® and cornstarch foam float in water. Although Styrofoam® can be reused as a packing material, it will not break down in landfills. Newspaper can be recycled easily, and cornstarch easily dissolves in water.
>
> Which material would you use? Discuss the advantages and disadvantages of each material. Be sure to describe the trade-offs made in your decision. (Wilson and Sloan, 2000, p. 190)

Items for assessment of learning progression should be reviewed by a panel of experts. In addition to the common expectation that the panel should possess expertise in the content domain, psychometrics, and pedagogy, special efforts should be made to include intended users of the learning progression assessment on the expert panel. If the learning progression assessment is for formative uses during a course, then instructors of the course should be represented on the item review panel. Similarly, if the learning progression assessment is for summative uses across multiple grades, then teachers who are knowledgeable about the students and curriculums across multiple grades should be part of the expert panel.

Conduct Pilot-Testing/Field-Testing

After items are reviewed and revised, sample items may be tried out with a limited number of students. During this try-out process, primarily qualitative data, such as interview, think aloud, and observations, will be collected. The purpose of item tryout is to obtain perspectives on how students will respond to the measurement instrument. After tryout, the complete instrument is then administered to a small purposefully selected sample as a pilot-test, and the pilot-test data will then be analyzed using a Rasch model. Rasch analysis results will help identify items for improvement and whether or not adding new items is necessary. The revised items and instrument will then be administered to a new sample of students as field-testing.

Considerations for field-testing suggested in previous chapters apply when assessment of learning progression is for summative uses. For formative assessment of learning progression, the field-testing sample should also represent the diversity of teachers. This is because formative assessment of learning progression is integrated into the curriculum units to inform ongoing teaching and learning; how teachers interpret assessment results to plan subsequent teaching and learning activities is an essential component of learning progression formative assessment. Similarly, if formative assessment of learning progression is not specific to a particular sequence of the science curriculum, that is, there is no fixed order among the formative assessment activities, then field-testing should also consider sampling different orders of the formative assessment activities. The ultimate goal for these steps is to collect enough information relevant to the intended assessment contexts of student population, teacher population, and curriculums. Field-testing data will be analyzed using a Rasch model again.

Conduct Rasch Analysis

After data have been collected from pilot-testing/field-testing, analysis of item and test properties through Rasch analysis will begin. Because learning progression is usually multidimensional, multidimensional Rasch analysis is conducted. Chapter 2 reviewed multidimensional Rasch models and popular computer programs. Simultaneous calibration of data from all administrations of assessment should be used whenever possible. Data files for simultaneous calibration can be prepared by creating separate files for each administration first, and then merging them by adding both items and examinees with blanks (i.e., not administered) coded as missing. A typical data structure for a simultaneous calibration of data from different samples of subjects in the situation of summative assessment is shown in Table 6.5.

Similarly, Table 6.6 shows a sample data file structure for a simultaneous calibration of data for formative assessment of learning progression based on a same sample of students during a course.

In the data file represented in Table 6.6, three formative assessment tests of a construct took place at three time points. In addition, two linking tests were also administered. For students at Time 1, the linking test was administered at the same time as Formative Test 1. For the same group of students at Time 2, Linking Test 1 and Linking Test 2 were administered at

TABLE 6.5 A Typical Data Structure for Simultaneous Rasch Calibration of Data From Different Samples of Subjects

<table>
<tr><th>Subjects</th><th colspan="2">Form 1 Items</th><th colspan="2">Form 2 Items</th><th colspan="2">Form 3 Items</th></tr>
<tr><td>Sample 1</td><td colspan="2">x</td><td>x</td><td>missing</td><td>missing</td><td></td></tr>
<tr><td>Sample 2</td><td>missing</td><td></td><td colspan="2">x</td><td>x</td><td>missing</td></tr>
<tr><td>Sample 3</td><td>missing</td><td>x</td><td>missing</td><td></td><td colspan="2">x</td></tr>
</table>

Note: x denotes responses to items

TABLE 6.6 A Typical Data Structure for Simultaneous Calibration of Data From a Same Sample of Repeated Assessment

<table>
<tr><th>Students</th><th>Formative Assessment 1 Items</th><th>Linking Test 1 Items</th><th>Formative Assessment 2 Items</th><th>Linking Test 2 Items</th><th>Formative Assessment 3 Items</th></tr>
<tr><td>S11</td><td>x</td><td>x</td><td colspan="3" rowspan="3">missing</td></tr>
<tr><td>S21</td><td>x</td><td>x</td></tr>
<tr><td>...</td><td>x</td><td>x</td></tr>
<tr><td>S12</td><td rowspan="6">missing</td><td>x</td><td>x</td><td>x</td><td rowspan="3">missing</td></tr>
<tr><td>S22</td><td>x</td><td>x</td><td>x</td></tr>
<tr><td>...</td><td>x</td><td>x</td><td>x</td></tr>
<tr><td>S13</td><td colspan="2" rowspan="3">missing</td><td>x</td><td>x</td></tr>
<tr><td>S23</td><td>x</td><td>x</td></tr>
<tr><td>...</td><td>x</td><td>x</td></tr>
</table>

Note: S11 denotes Student 1 at time 1, S12 Student 1 at time 2, S13 Student 1 at time 3, S21 denotes Student 2 at time 1, S22 Student 2 at time 2, S23 Student 2 at time 3; x denotes responses to items.

the same time as Formative Test 2. Similarly, for the same group of students at Time 3, Linking Test 2 was administered at the same time as Formative Test 3. In the student column, a same student is coded with three different codes corresponding to the three time points (S11, S12, S13, . . .). Thus, the same student will have three ability estimates forming a learning trajectory.

Review Item Fit Statistics, Wright Map, Dimensionality, and Reliability

As described in previous chapters, a variety of fit statistics should be reviewed. Particular attention should be paid to the linking items. Linking item evaluation includes evaluation of these items individually and as a set for measurement of the construct. Evaluating linking items individually follows the same rules as evaluation of fit for any items, which is based on MNSQ, ZSTD, standard errors of measurement, point-measure correlation, category structure, and item characteristic curve. Evaluating link items as a set can be based on two overall statistics: item-within-link fit statistics and the item-between-link statistics (Wright & Bell, 1984). In order to compute the above two statistics, each linking item needs to be coded as two different items within the two linked forms. After submitting all items to a simultaneous Rasch calibration, we will have two difficulty estimates and two sets of fit statistics for each linking item.

Item-within-link fit statistics is defined as follows:

$$IWL = \frac{\sum_{i=1}^{L}(INFIT_{ij} + INFIT_{ik})}{2L}$$

where IWL is the item-within-link fit statistics, i represents a linking item, L is the total number of linking items, $INFIT_{ij}$ is the weighted mean-square residual fit statistics (INFIT MNSQ) for item i within Form J, and $INFIT_{ik}$ is the weighted mean-square residual fit statistics (INFIT MNSQ) for item i within Form K. IWL has an expected value of 1.

Item-between-link statistics is defined as follows:

$$X_{IBL}^2 = \sum_{i=1}^{L}\left(\frac{(d_{ik} - d_{ij})}{\sqrt{SE_{dik}^2 + SE_{dij}^2}}\right)^2$$

where X_{IBL}^2 is a chi-square statistics for item-between-link, L is the total number of linking items, d_{ik} is the item difficulty estimate of linking item i on

Form *K*, and d_{ij} is the item difficulty estimate of linking item *i* on Form *J*, and SE_{dik} is the standard error of measurement for the difficulty estimate of linking item *i* on Form *K*, and SE_{dij} is the standard error of measurement for the difficulty estimate of linking item *i* on Form *J*. X^2_{IBL} has a chi-square distribution with $L-1$ degrees of freedom. A value greater than the critical chi-square value indicates inadequate performance of linking items as a whole. A scatterplot graph between d_{ik} and d_{ij} may also be made to visually examine the overall stability of the estimates of linking items between the two forms.

Table 6.7 presents a hypothetical scenario for an illustration purpose, assuming that there are five linking items Q1 to Q5.

From INFIT statistics, we see that most of the linking items had a good model-data fit except that Q1 may not be fitting well because the INFIT value is beyond the range of 0.7 to 1.3. In order to evaluate the adequacy of linking items as a whole, we can calculate the IWL and IBL statistics as follows:

$$IWL = \frac{\sum_{i=1}^{L}(INFIT_{ij} + INFIT_{ik})}{2L}$$

$$= \frac{(1.35+1.41+1.20+1.19+1.13+1.09+0.98+0.89+0.85+0.90)}{2\times 5}$$

$$= 1.10$$

$$X^2_{IBL} = \sum_{i=1}^{L}\left(\frac{(d_{ik} - d_{ij})}{\sqrt{SE^2_{d_{ik}} + SE^2_{d_{ij}}}}\right)^2$$

$$= \left(\frac{1.25-0.80}{\sqrt{0.34\times0.34+0.33\times0.33}}\right)^2 + \left(\frac{0.55-0.07}{\sqrt{0.12\times0.12+0.23\times0.23}}\right)^2 + \left(\frac{0.51-0.14}{\sqrt{0.19\times0.19+0.16\times0.16}}\right)^2 +$$

$$\left(\frac{0.89-0.34}{\sqrt{0.23\times0.23+0.25\times0.25}}\right)^2 + \left(\frac{1.15-0.68}{\sqrt{0.45\times0.45+0.41\times0.41}}\right)^2$$

$$= 9.76$$

The expected IWL value is 1.0 when linking items function perfectly. Because the obtained IWL is 1.10, which is only 10% deviating from the perfect fit, the linking items overall performed well. In terms of IBL, the critical chi-square value when the degree of freedom is 4 is 9.49, and the obtained chi-square value is 9.76, which is greater than the 9.49, overall the linking items do not seem to function well. When the IWL and IBL statistics suggest different conclusions like the case in this example, a visual display of the scatterplot between the two sets of difficulty estimates may be helpful (see Figure 6.2).

TABLE 6.7 Hypothetical Data for Item-Within-Link and Item-Between-Fit Analysis

	Form 1			Form 2		
Item	**Difficulty**	**SE**	**INFIT MNSQ**	**Difficulty**	**SE**	**INFIT MNSQ**
Q1	1.25	0.34	1.35	0.80	0.33	1.41
Q2	0.55	0.12	1.20	0.07	0.23	1.19
Q3	–0.14	0.19	1.13	–0.51	0.16	1.09
Q4	–0.34	0.23	0.98	–0.89	0.25	0.89
Q5	–0.68	0.45	0.85	–1.15	0.41	0.90

Figure 6.2 shows that the two sets of difficulty estimates are more or less linear, with most observations situated along the straight trend line. Thus, overall we can conclude that the linking items have functioned well.

Reliability evidence of learning progression assessment may be reviewed based on the following: (a) reliability of scoring, (b) reliability of item and examinee estimates, and (c) reliability of the instrument measure. If some of measurement items are performance or constructed-response type, scoring is typically done following a scoring rubric. Inter-rater reliability of scoring by independent raters must be empirically established. After item and person parameters are calibrated, standard errors of measurements for each item and person should also be reviewed. In Rasch

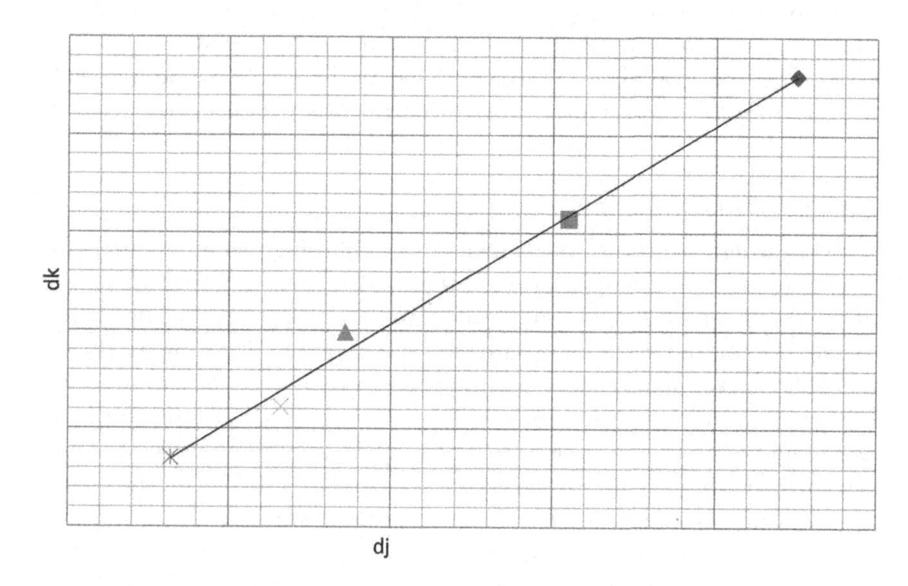

Figure 6.2 Scatterplots for item invariance between linking forms. Overall, item measures from different forms correlate highly, suggesting good invariance.

analysis, there are also two indices produced to indicate the overall item and person reliability. Person separation index and its equivalent Cronbach's alpha should be reviewed for their adequacy of reliability for person measures as a whole. Similarly, item separation index and its equivalent Cronbach's alpha should be reviewed for the adequacy of reliability for item measures as a whole.

Examine Invariance Properties

In order to ensure that item difficulties and student abilities are invariant, DIF and DPF analyses can be performed. Because measurement of learning progression typically involves multiple grades, when a same item is administered to students of different grades, we should expect that the difficulty of a same item remains unchanged from grade to grade if there is a perfect model-data fit. In order to empirically test this hypothesis, when same items have been responded to by students of different grades, difficulties of those items can be estimated separately using data from subsamples of different grades, and the invariance of item difficulties based on different subsamples can be examined both visually (e.g., scatterplots) and statistically (e.g., correlation coefficient). Similarly, measurement of learning progression typically involves multiple linked forms, we should expect that student abilities remain unchanged no matter what form they have responded to if there is a perfect model-data fit. In order to empirically test this hypothesis, student abilities can be estimated separately for different forms, and the invariance of those abilities can be examined both visually (e.g., scatterplots) and statistically (e.g., correlation coefficient).

Establish Validity Claims

Construct validation, which includes establishing evidence related to test content, response processes, internal structure, external structure, and consequences, begins with examination of the construct definition for developing the learning progression measurement instrument. An adequately defined construct of learning progression must be grounded in the research literature, and must present a clear linear progression from one grade to another or from one time point to another. Related to defining constructs of learning progression, the defined test specifications and scoring schemes should also be consistent with the identified behaviors that define the construct of learning progression.

Because learning progression measurement is derived from the defined learning progression, it is necessary to ensure that item and student parameter estimates agree with the intended or hypothesized learning progression. In addition to examining item and person fit statistics, dimensionality, item separation and person separation indices, and adequacy of link items, the Wright map is particularly important for theory-based validity interpretation. If the distribution of both items and examinees is consistent with the defined learning progression, then there is evidence to claim construct validity in terms of internal structure. For example, the Wright map for the measurement of student learning progression on matter through PUM (Liu, 2007) is shown in Figure 6.3.

In Figure 6.3, *cm*, *ph*, *cn*, and *ch* represent four components of matter understanding (composition and structure of matters, physics properties and change, conservation of matter, and chemical properties and changes), and the numerical number after each matter component represents

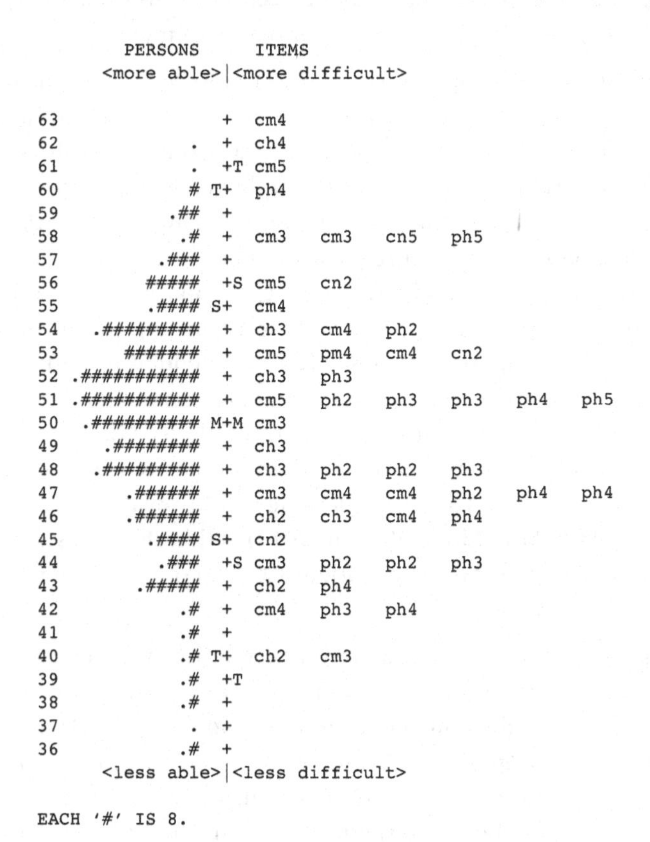

Figure 6.3 Wright map for measurement of understanding of matter. Overall, the distribution of student abilities matches the distribution of items well.

the level of matter understanding (e.g., *ch2* meaning Level 2 of chemical properties and change). The defined learning progression of matter understanding suggests that students from elementary school to high school develop knowledge and understanding of matter through five levels with each level being the result of integrated understanding, like an "overlapping wave," of the four components. The Wright map demonstrates the inter-relations of the four components through "mixing" of items of different components along the vertical line, and the overall progression from Level 2 to Level 5 (Level 1 was not measured by the instrument). Also, students from elementary to high school distributed well on the left hand side of the Wright map. A one-way analysis of variance on student Rasch scale scores using grades as the independent variable showed that there was a statistically significant difference among students of different grades in their knowledge and understanding of matter. All of the above help support the validity of theory-based interpretation of the instrument.

Learning progression assessment can inform various instructional decisions. An important component of validation of learning progression assessment is to collect data on how assessment results have been used to make various decisions. For example, if a learning progression assessment is used for a summative purpose (e.g., awarding student course grades), data on students' future performances (e.g., later course grades) may be collected to see how learning progression assessment may predict student future performances. Similarly, if a learning progression assessment is used for formative purposes (e.g., embedded in an instructional sequence), data may be collected on teachers to understand how learning progression assessment results have informed their planning for ongoing instruction. Both qualitative and quantitative data are useful for this component of validation. As an example, one validity evidence for the CCFA (Liu et al., 2012; Park, Liu, & Waight, 2017) is that there was a statistically significant correlation between student CCFA scores and their state standardized test scores.

While it is important that a variety of validity evidence is established about a learning progression measurement as described above, it is also important to examine the consistency in that evidence. Important questions to ask when evaluating the consistency of the various validity evidences are: (a) clarity, (b) coherence, and (c) plausibility of both inferences and assumptions. A claim about the validity of a learning progression measurement is sound when all types of evidence are clearly stated, coherent, and plausible, and the assumptions underlying the arguments are sound.

Develop Guidelines for Administration, Scoring, and Interpretation of the Test Scores

A final step in developing a learning progression measurement instrument is documentation. Documentation is to facilitate others to use the measurement instrument appropriately and efficiently. In addition to adequate explanation of the purpose of the measurement and constructs assessed as well as instructions to use the instrument, one important purpose of documentation is to facilitate users in interpreting assessment results. Although Rasch analysis plays a crucial role in developing the measurement instrument, we should not expect users to conduct Rasch analysis to estimate item and person parameters, and also we may not expect users to use logits as their chosen measurement scale unit because logits are not intuitively meaningful. Thus, one consideration in documentation is to develop a conversion chart between raw scores and Rasch scale scores. In this way, users will only need to score a student's responses to items to obtain a total raw score, and then obtain its equivalent Rasch scale score by checking the conversion chart. Table 6.8 shows the conversion chart for the PUM measurement instrument (Liu, 2007).

Chapter Summary

Learning progression refers to gradual and systematic development in knowledge and understanding by students over time. It has five characteristics. First, learning progression must pertain to science content as compared to general reasoning or performance skills. Second, it describes different levels of competence, that is, learning progression should map out a hierarchy of student learning outcomes from least competent to most competent. This hierarchy is also called a learning trajectory. Third, a learning progression must include specific student thinking or conceptual change mechanism that enables transition from one performance level to another. Fourth, it is tied to curriculum and instruction. A learning progression represents what is possible for student learning under innovative curriculum and instruction. Finally, it is a model. There are many ways to realize a learning progression in curriculum and instruction and it does not account for all variations in curriculum and instruction. For example, one way is to use a spiral curriculum to organize content for a learning progression. Learning progression may also be realized through hierarchical web-like learning pathways. Assessment is an important component of a learning progression; validation of a learning progression assessment is not sufficient for validation of the learning progression.

TABLE 6.8 Conversion Table Between Raw Scores and Rasch Scale Scores

	Rasch Scale Score		
Raw Score	Elementary School Form	Middle School Form	High School Form
0	31.94	34.31	35.84
1	34.72	35.67	37.64
2	37.13	36.94	39.30
3	39.21	38.13	40.83
4	41.00	39.23	42.26
5	42.53	40.26	43.58
6	43.85	41.22	44.82
7	44.98	42.12	45.98
8	45.96	42.97	47.08
9	46.83	43.76	48.12
10	47.62	44.51	49.13
11	48.38	45.22	50.11
12	49.13	45.90	51.07
13	49.91	46.56	52.02
14	50.76	47.19	52.99
15	51.71	47.81	53.97
16	52.81	48.42	54.99
17	54.08	49.03	56.05
18	55.56	49.65	57.17
19	57.29	50.27	58.35
20	59.30	50.91	59.62
21	61.64	51.57	60.98
22	64.33	52.26	62.44
23	N/A	52.99	64.01
24	N/A	53.75	65.72
25	N/A	54.56	67.56
26	N/A	55.42	69.56
27	N/A	56.34	71.72
28	N/A	57.33	74.06
29	N/A	58.38	76.58
30	N/A	59.51	79.31
31	N/A	60.72	82.25
32	N/A	N/A	85.41
33	N/A	N/A	N/A

Only a few learning progression measurement instruments are currently available. Developing measurement instruments for learning progression starts with defining the primary purposes for which the test scores will be used. Assessment of learning progression can be for either formative or summative, or both purposes. A formative use of test scores of learning progression assessment is to inform teachers to plan for ongoing instruction and for students to reflect on the ongoing progress toward learning goals; and a summative use of learning progression is to identify levels of student achievement along a defined learning progression at different times. Next, the construct to be measured must be defined according to research on learning progression. Based on the defined construct, subject behaviors or performances that represent the construct are then identified. Test specification for learning progression measurement should consider the linkage between and among different assessment forms. After items are reviewed and revised, sample items may be tried out with a limited number of students before the measurement instrument is pilot-tested and field-tested with a larger sample or samples. When conducting Rasch analysis based on pilot-testing/field-testing data, simultaneous calibration of data from all administrations of the measurement instrument should be used whenever possible. Review for model-data fit should pay particular attention to the linking items. Linking item evaluation includes evaluation of these items individually and as a set for measurement of the construct. Evaluating link items as a set can be based on two overall link fit statistics: IWL statistics and IBL fit statistics. Based on Rasch modeling as well as additional sources of data to be collected, the validity and reliability of the learning progression assessment can be claimed through trait interpretation, theory-based interpretation, decision procedures, and qualitative interpretation. Documentation for the learning progression measurement instrument should include adequate explanation of the purpose of the measurement and the assessed construct, instructions to use the instrument, and score reporting.

Exercises

1. Almost all the reported learning progression measurement instruments have been based on Rasch modeling. List reasons why Rasch modeling instead of CTT, is used in developing learning progression measurement instruments.
2. Choose one hypothetical learning progression of a core disciplinary idea or crosscutting concept defined in *Next Generation Science Standards* (NRC, 2013), analyze if the five characteristics described in this chapter are present in the learning progression.

3. Mohan, Chen, and Anderson (2009) reported a learning progression of students' understanding of carbon cycling in socio-ecological systems from upper elementary through high school (*Journal of Research in Science Teaching, 46*[6], 675–698). Based on the learning progression, complete the following steps of developing an instrument for measuring learning progression of carbon-cycling for students from upper elementary through high school: (a) identify student behaviors that represent different levels of the learning progressions, (b) create a table of test specification for three forms (i.e., elementary, junior high school, and high school) of the measurement instrument, and (c) write three sample items for each of the progression levels.

References

American Association for the Advancement of Science. (2001). *Designs for science literacy.* New York, NY: Oxford University Press.

Bruner, J. (1960). *Process of education.* Cambridge, MA: Harvard University Press.

Catley, K., Lehrer, R., & Reiser, B. (2005). *Tracing a prospective learning progression for developing understanding of evolution.* Paper commissioned by the National Academies Committee on Test Design for K–12 Science Achievement. Washington, DC: National Academy of Sciences.

Claesgens, J., Scalise, K., Wilson, M., & Stacy, A. (2009). Mapping student understanding in chemistry: The perspectives of chemists. *Science Education, 93*(1), 56–85.

Duschl, R., Maeng, S., & Sezen, A. (2011). Learning progressions and teaching sequences: A review and analysis. *Studies in Science Education, 47*(2), 123–182.

Lehrer, R., & Schauble, L. (2015). Learning progressions: The whole world is NOT a stage. *Science Education, 99*(3), 432–437.

Liu, X. (2007). Growth in students' understanding of matter during an academic year and from elementary through high school. *Journal of Chemical Education, 84*(11), 1853–1856.

Liu, X., Waight, N., Gregorius, R., Smith, E., & Park, M. (2012). Developing computer model-based assessment of chemical reasoning: A feasibility study. *Journal of Computers in Mathematics and Science Teaching, 31*(3), 259–281.

National Research Council. (2007). *Taking science to school: Learning and teaching science in Grades K–8.* Committee on science learning, Kindergarten through eighth grade. Washington, DC: The National Academies Press.

National Research Council. (2013). *Next generation science standards: For states, by states.* Washington, DC: The National Academies Press.

Neumann, K., Viering, T., Boone, W., & Fischer, H. (2013). Towards a learning progression of energy. *Journal of Research in Science Teaching, 50*(2), 162–188.

Osborne, J. F., Henderson, B., MacPherson, A., Szu, E., Wild, A., & Yao, S.-Y. (2016). The development and validation of a learning progression for argumentation in science. *Journal of Research in Science Teaching, 53*(6), 821–846.

Park. M., & Liu, X. (2016). Assessing understanding of the energy concept in different science disciplines. *Science Education, 100*(3), 483–516.

Park, M., Liu. X., & Waight, N. (2017). Development of the connected chemistry as formative assessment pedagogy for high school chemistry teaching. *Journal of Chemical Education, 94*(3), 273–281.

Romine, W. L., Todd, A. N., & Clark, T. B. (2016). How do undergraduate students conceptualize acid-base chemistry? Measurement of a concept progression. *Science Education, 100*(6), 1150–1183.

Smith, C., & Wiser, M. (2015). On the importance of epistemology-disciplinary core concept interactions in LPs. *Science Education, 99*(3), 417–423.

Testa, I., Capasso, G., Colantonio, A., Galano, S., Marzoli, I., di Uccio, U. S., . . . Zappia, A. (2019). Development and validation of a university students' progression in learning quantum mechanics through exploratory factor analysis and Rasch analysis. *International Journal of Science Education, 41*(3), 388–417.

Todd, A., Romine, W. L., & Whitt, K. C. (2017). Development and validation of the learning progression-based assessment of modern genetics in a high school context. *Science Education, 101*(1), 32–65.

Todd, A., & Romine, W. L. (2016). Validation of the learning progression-based assessment of modern genetics in a college context. *International Journal of Science Education, 38*(10), 1673–1698.

Wilson, M., & Sloane, K. (2000). From principles to practice: An embedded assessment system. *Applied Measurement in Education, 13*(2), 181–208.

Wolfe, E. W. (2004). Equating and item banking with the Rasch model. In E. V. Smith, Jr., & R. M. Smith (Eds.), *Introduction to Rasch measurement* (pp. 366–390). Maple Grove, MN: JAP Press.

Wright, B. D., & Bell, S. R. (1984). Item banks: What, why, how. *Journal of Educational Measurement, 21*, 331–345.

Yao, J., Guo, Y., & Neumann, K. (2017). Refining a learning progression of energy. *International Journal of Science Education, 39*(17), 2361–2381.

Yao, J., & Guo, Y. (2018). Validity evidence for a learning progression of scientific explanation. *Journal of Research in Science Teaching, 55*(2), 299–317.

7

Using and Developing Instruments for Measuring Science Learning Environments

This chapter is concerned with learning environments in science. A learning environment is both an outcome and a predictor variable. Various theoretical frameworks about learning environments are available. This chapter will first review these theoretical frameworks. It will then introduce standardized instruments for measuring various learning environments. Finally, this chapter will describe the process of developing new instruments for measuring learning environments using the Rasch modeling approach.

What Are Science Learning Environments?

Learning environment research in science education is a relatively new field with about 40 years' history. Despite this short history, the field has advanced tremendously, particularly in the development of measurement

Using and Developing Measurement Instruments in Science Education, pages 279–320

instruments. In fact, "few fields of educational research can boast the existence of such a rich array of validated and robust instruments" (Fraser, 2007, p. 105). This large number of measurement instruments have not only contributed to the development of learning environment research as a field of study in science education, but also increased our understanding of the nature and characteristics of learning environments in science.

Fraser (1994) originally defines *learning environments* as "the social-psychological contexts or determinants of learning" (p. 493). Later, he expanded this definition by stating that "learning environment" encompasses "social, physical, psychological, and pedagogical contexts in which learning occurs and which affects student achievement and attitudes" (Fraser, 1998b, p. 1). The above definitions are rooted in an early view of classroom environments as the shared perceptions of people, that is, students and the teacher, in that environment (Moos, 1979). It distinguishes learning environments as perceptions from that as physical settings such as arrangement of classroom and lab furniture, positioning of teacher desks, availability of computers in the classroom, and so forth. Learning environments can be considered as both independent variables and dependent variables. In the former, learning environments are expected to promote or hinder student learning; in the latter, learning environments are considered as outcomes of other factors.

According to Fraser (1994, 1998a, 2007), learning environment research originated from the work by Walberg and Moos during the 1960s and 1970s. The basic premise for learning environment research is that personal goals may be supported or constrained by environmental factors. When personal needs and environmental factors are aligned, student learning outcomes are enhanced. For example, Moos (1979), adopting a social-ecological perspective on educational development, conceptualizes that there is interaction between environmental and personal systems; they influence each other through selection of factors into each system and also through mediating factors. The *environmental system* consists of physical setting, organizational factors, humane aggregate, and social climate; the *personal system* consists of socio-demographic variables, expectations, personality factors, and coping skills. *Mediating factors* include cognitive appraisal, activation or arousal, and efforts at adaption and coping. According to Moos, *determinants of classroom environments* may include aggregate student characteristics (e.g., SES), teacher characteristics (e.g., teacher's gender), organization factors (e.g., small classes), physical and architectural features (e.g., physical arrangement of classrooms), and school and classroom contexts (e.g., type of school).

Specifically related to social climates, Moos (1979) developed a three-dimensional construct of social climates based on data from various social settings (e.g., student living groups, high school classes, families, working groups, hospitals, correctional institutions, military basic training companies, and so on). The three dimensions, which are also called domains, are relationship, personal growth, and system maintenance and change. The *relationship dimension* is concerned with the extent to which people are involved in the setting, the extent to which they support and help one another, and the extent to which they express themselves freely and openly. For example, relationships in the classroom learning environment may pertain to the attentiveness of students to class activities, and their participation in discussion. It may also pertain to the perceptions of affiliation with the class and supports from the teacher. The *personal growth dimension* pertains to the goals of the setting related to personal development and self-enhancement. For example, in the classroom learning environment, personal growth dimension may involve the task orientation, speed and difficulty of tasks, and competition among students. The *system maintenance and change dimension* pertains to the extent to which the environment is orderly and clear in its expectations, and maintains control and responds to change. For example, in the classroom learning environment, system maintenance and change may relate to the order and organization of the classroom activities, clarification of expectations for achievement, and control and innovation of learning processes.

Walberg (1968a) conceptualizes a *classroom climate* to consist of two dimensions: the structural dimension and the affective dimension. The *structural dimension* refers to student roles within the class, such as goal direction and democratic policy. It applies to shared and group-sanctioned classroom behaviors. The *affective dimension* pertains to idiosyncratic personal dispositions to act in a given way to satisfy individual personality needs, such as satisfaction, intimacy and friction in the class. In an evaluation study of the Harvard Project Physics curriculum, Walberg (1968b) found that students gained the most on the physics achievement test when they perceived their classes as being socially homogeneous and intimate groups working on one goal. Students who grew more in science understanding saw classes as being well-organized with little friction between their fellow students, equalitarian and unstratified, and encouraging a greater variety of student interests. In terms of the affective dimension, Walberg found that students who reported greater enjoyment of laboratory work perceived their classes as being unstratified, democratic in policy setting, having clear class goals, and satisfying. Students who gained the most interest in physics saw their classes as being well-organized and unstratified.

There are two common orientations to learning environment research: the outcome-environment interaction orientation and the person-environment fit orientation (Lesniak, 2007). The *outcome-environment orientation* focuses on the association between learning environment variables and student cognitive and affective learning outcomes. That is, learning environments are considered determinants of student learning outcomes. Research over the past three decades has clearly demonstrated that student perceptions of their learning environments are associated with both their achievements in science and their attitudes towards science, and can account for substantial amounts of the variability in student learning. The *person-environment fit* orientation differentiates the *actual* perceived environment from the *preferred* environment. The actual perceived environment is what is happening now, while the preferred learning environment is what students would like to see to happen in an ideal situation. Research has shown that student learning is enhanced when students' actual learning environment is closer to the preferred environment. Research has further demonstrated that teachers' perceived learning environment tends to be more favorable than students'. Thus, student learning outcomes will improve by increasing the match between students' preferred learning environment and the actual learning environment, and between teacher perceived learning environment and student perceived learning environment.

While both outcome-environment and person-environment fit orientations consider learning environment variables as predictors or determinants of learning outcomes, learning environments may also be considered as outcomes. This orientation is particularly pertinent in the context of science education reforms. Educational reforms typically entail certain aspects of learning environments to change. Valid and reliable instruments for measuring these aspects of learning environments can help differentiate innovative science teaching and learning from conventional science teaching and learning.

Research by Fraser, his collaborators and others have also found that it is helpful to differentiate learning environments at the collective level and at the personal level (Fraser, 1994, 1998b; Fraser & Tobin, 1991). The *collective learning environment* is what students perceive the class as a whole; while the *personal learning environment* is what students perceive from their own perspectives. The collective and personal learning environments may not necessarily be the same. Differentiating these two versions of a learning environment can help identity individual differences among students within the same class. Similarly, learning environments may also be differentiated based on units of physical learning settings in which learning takes place. These units can be class, school, family, and so on.

Roth proposes a *"lifeworld" approach* that designates the individual-environment as a unity structure in terms of customary activities and perceptions, conventional uses of tools, materials, language, invariants maintained in it by conventional activities, and so on (Roth, 1999). From this lifeworld perspective, "world (the Other) and individual Self presuppose each other" (Roth, 1999, p. 235). Studying learning environments based on the above lifeworld perspective calls for a different approach, that is, qualitative approach. In fact, Fraser (1994, 1998b, 2007) has consistently pointed out values for combining qualitative and quantitative methods to study science classroom learning environments. This mixed research method approach can produce a much richer understanding of the science classroom learning environments than any one approach alone. For example, Fraser (1994), based on a series of qualitative and quantitative studies of various classroom learning environments, made the following conclusions about science classroom learning environments:

1. In whole-class settings, a small group of target students, typically ranging from three to seven are usually higher ability and/or more active and assertive to volunteer, dominate verbal interactions.
2. Teachers tended to direct higher-level cognitive questions to target students.
3. Target students held more favorable perceptions of the learning environment than nonparticipants.
4. Exemplary teachers used management strategies that facilitate sustained student engagement.
5. Exemplary teachers use strategies designed to increase student understanding of science.
6. Exemplary teachers utilized strategies that encouraged students to participate actively in learning activities.
7. Exemplary teachers maintained favorable classroom learning environments.
8. Teachers conceptualized their roles in terms of metaphors that influenced the way in which they taught.
9. Teacher beliefs had a major impact on the way in which the curriculum was implemented.
10. Knowledge limitations of teachers resulted in an emphasis on students learning facts and completing workbook exercises rather than learning with understanding.
11. The student-perceived learning environment of the classes was related to teachers' knowledge and beliefs.
12. Teacher expectations of and attitudes toward individuals were reflected in individual students' perceptions of the learning environment.

Teacher–student interaction is an important aspect of the science teaching and learning lifeworld. Adopting a communication approach to learning environments, Wubbels and Brekelmans (1998) point out that teacher communication behaviors influence science classroom environments. They specifically recommend that teachers should strive to establish teacher–student relationships characterized by high degrees of leadership, helpfulness/friendliness and understanding behaviors. They also recommend that teachers experiencing undesirable classroom situations should focus on their own behaviors as a means for improvement and should introduce changes in their communication patterns.

Besides the above social-psychological and lifeworld perspectives of learning environments, a *semiotic perspective of learning environments* considers physical settings to be important aspects of learning environment (Shapiro, 1998). "When viewed semiotically, the science classroom can be seen as interweaving sets of sign, symbol and signification systems" (Shapiro, 1998, p. 611). For example, the furniture arrangement as part of the physical environment is a part of an elaborate system of signification. A row and column arrangement of student desks and chairs may suggest to students that they are expected to follow rules and listen to instructions; a circular arrangement of student desks and chairs may suggest to students that they are expected to contribute to the class and to respect others' views; an island arrangement of student desks and chairs may suggest to students that they are expected to take collective initiatives and make collaborative decisions. A semiotic approach to studying learning environments thus first attempts to identify the systems of signs, symbols, and other significations; it then attempts to understand how those systems have been created and are being interpreted by those involved (i.e., students, the teacher).

> Viewing classroom semiotically helps us to see that entire systems of signification within school settings serve as a resource for learners and a means by which they access knowledge. When teachers master the signs and symbols of our culture and become aware of those of others, they know when to break the unspoken rules to become inventive in using new approaches in interaction. (Shapiro, 1998, p. 618)

Specifically in terms of laboratory physical settings, Arzi (1998) suggests that science laboratories should be flexible, multifunctional, and practical.

Learning environments research in science education has benefited from multiple theoretical perspectives and research methodology orientations. As Tobin and Fraser (1998) pointed out that

> learning environments can be described through multiple windows to highlight different issues that are pertinent to the stakeholder goals and extant classroom practices. Any methodology used to explore learning environments will produce a landscape that is incomplete and represents only one of the possible portraits which is likely to be appealing and relevant to different stakeholders. (p. 627)

Thus, researchers need to be aware of limitations of using and developing a particular measurement instrument on science learning environments, and recommend possible ways to address them.

Instruments for Measuring Science Learning Environments

A variety of standardized measurement instruments related to learning environments are available. The following summative descriptions introduce instruments published over the past 40 years in major science education research journals, mostly in *Journal of Research in Science Teaching, Science Education,* and *International Journal of Science education.* The majority of the instruments have been developed based on CTT. They are for various intended uses and based on various theoretical frameworks of learning environments. Each instrument description is information only, not intended to be a critical review of its strength and weakness. Also, descriptions do not follow the same format or even in similar length; how and how much an instrument is described depends on what is reported in the publication. In general, when available each description contains information about the instrument's intended population, purpose, composition (e.g., number and type of items), validation process and key indices of validity and reliability. Instruments are described in their publication years and in alphabetical order of authors' last names.

Descriptions

The Learning Environment Inventory (Anderson & Walberg, 1974; Fraser, Anderson, & Walberg, 1982)

The Learning Environment Inventory (LEI) measures secondary school classroom environments. It contains 15 scales; they are cohesiveness, friction, favoritism, cliqueness, satisfaction, apathy, speed, difficulty, competitiveness, diversity, formality, material environment, goal direction, disorganization, and democracy. Each scale has seven statements about typical school classes; respondents are asked to indicate their degree of agreement to each of the statements by selecting *strongly disagree, disagree, agree,* and *strongly agree.* The LEI originated from the *Classroom Climate Questionnaire* used in the evaluation of the Harvard Project Physics in the 1960s. Scales contained in the final version relate to concepts previously identified as good predictors of learning outcomes, or being relevant to socio-psychological theories. Reported alpha reliability coefficients for the 15 scales ranged from 0.54 to 0.85 (Fraser, 1994).

The Classroom Environment Scale (Moos & Trickett, 1974, 1987; Trickett & Moos, 1973)

The Classroom Environment Scale (CES) focuses on the socio-psychological environment of junior and high school classrooms. It has three parallel forms: the R Form—Real form to assess the actual classroom environment, the I Form—Ideal form to assess the ideal classroom environment, and the E Form—Expectation form to assess expectations about a new classroom. All the forms have 90 items. Students and teachers answer the same items. A complete R form is available as an appendix in Moos (1979).

The CES conceptualizes a classroom environment as a dynamic social system that includes not only teacher behaviors and teacher–student interactions, but also student–student interactions. The conceptual framework for CES comes from relevant literature on educational and organizational psychology. The CES contains three dimensions: relation, system maintenance, and personal growth. The relation dimension involves affective aspects of student–student and teacher–student interactions. System maintenance and change dimension involves aspects of rules and regulations of the classroom and teaching innovations. Personal growth or goal orientation dimension pertains to specific functions of the classroom environment. There are nine subscales related to the above three dimensions.

The KR-20 internal consistency indices for the subscales ranged from 0.67 to 0.86. Inter-correlations among the subscales averaged about 0.25, suggesting good discriminant validity. Factor analysis also conformed to the expected subscales. The CES scores were also found to differ statistically significantly among classrooms with distinct teacher teaching styles. Specifically, based on different CES scores on various dimensions, classroom environments could be classified into the following six types: innovation oriented, structured relationship oriented, supportive task oriented, supportive competition oriented, unstructured competition oriented, and control oriented. There were also statistically significant differences in CES scores on some subscales between teachers and students. Perceptions of real and ideal classroom learning environments were also found to be statistically significant on most subscales.

An updated version of CES measures has 10 scales, each with 10 items. The one additional scale is related to differentiation (Moos & Trickett, 1987). A 24-item short form of CES, with only six scales, each having 4 items, has also been developed (Fraser, 1982; Fraser & Fisher, 1983).

My Classroom Inventory (Fisher & Fraser, 1981)

My Classroom Inventory (MCI) is a simplified version of LEI for elementary and lower junior high school classrooms. It contains only five of LEI's 15 scales, which are cohesiveness, friction, satisfaction, difficulty, and competitiveness. Numbers of items per scale range from 6–9; the total items are 38. Other modifications made on LEI to form MCI include a lower level of reading comprehension, reducing the four-point responses to two point-responses (i.e., yes/no), and presenting both questions and responses on the same sheet of paper. Reported alpha reliability coefficients for the MCI scales ranged from 0.62 to 0.78 (Fraser, 1994). A short form of MCI was created based on statistical analysis of selected items (Fraser, 1982; Fraser & Fisher, 1983). It has 5 scales, each with 5 items. A copy of the short form MCI is available as appendix in Fraser (1994).

College and University Classroom Environment Inventory (Fraser & Treagust, 1986; Fraser, Treagust, & Dennis, 1986)

The College and University Classroom Environment (CUCEI) measures learning environments of small university classes or seminars with up to 30 students. It is not appropriate for university lectures or labs. The CUCEI contains 7 scales—personalization, involvement, student cohesiveness, satisfaction, task orientation, innovation, and individualization. Each scale has 7 items, and each item has four choices of responses (*strongly agree, agree, disagree,* and *strongly disagree*). Items for CUCEI were selected and adapted from LEI, CES, and ICEQ according to their relevance to higher education settings. Validation involved expert review and field-testing. Reported alpha reliability coefficients for the seven scales ranged from 0.70 to 0.90 (Fraser, 1994). The complete instrument of CUCEI is available as an appendix in Fraser (1994).

Individualized Classroom Environment Questionnaire (Fraser, 1990)

The Individualized Classroom Environment Questionnaire (ICEQ) is a 50-item instrument measuring secondary school classrooms in terms of the degree of individualized learning and teaching. It has five dimensions/scales, each containing 10 items. The five dimensions relate to personalization, participation, independence, investigation, and differentiation. Items are presented in a 5-point rating scale ranging from *almost never, seldom, sometimes, often,* and *very often.* The ICEQ originated from the initial long version of the ICEQ instrument (Rentoul & Fraser, 1979). The dimensions included in the instrument met the following criteria: (a) addressing open and inquiry-based classrooms; (b) being considered salient by interviewed teachers and students; (c) meeting expectations of experts, teachers, and students; and (d) meeting item and scale statistical expectations. Reported alpha reliability coefficients for the five scales ranged from 0.68 to 0.79 (Fraser, 1994). Validation for the different forms of ICEQ, that is, student perceived actual learning environment, student preferred learning environment, teacher perceived actual learning environment, and teacher preferred learning environment, was conducted using both individual and class means as units of analysis (Fraser, 1994). A short form of ICEQ was later created based on statistical analysis of selected items (Fraser, 1982; Fraser & Fisher, 1983). This short form has five scales with each scale having five items.

Science Laboratory Environment Inventory (Fraser, McRobbie, & Giddings, 1993)

The Science Laboratory Environment Inventory (SLEI) measures high school and university science laboratory class environments. It has five scales, that is, student cohesiveness, open-endedness, integration, rule clarity, and material environment. Each scale has 6–7 items; each item has five responses to choose from (*almost never, seldom, sometimes, often,* and *very often*). Validation involved 5,447 students in 269 classes from six different countries (USA, Canada, England, Israel, Australia, and Nigeria). Reported alpha reliability coefficients for the five scales ranged from 0.70 to 0.83 (Fraser, 1994). Alpha reliability coefficients for different forms of SLEI, that is, student perceived actual laboratory environment, student preferred laboratory environment, and for different countries were also reported (Fraser, 1994). A personal form (i.e., students' perceptions of themselves as individuals in the class), as compared to the class form (i.e., students' perceptions of the class as a whole) as originally developed, was also created (McRobbie, Fisher, & Wong, 1998).

Questionnaire on Teacher Interaction (Wubbles, Breklmans, & Hooymayers, 1991; Wubbels & Levy, 1993).

The Questionnaire on Teacher Interaction (QTI) measures students' perceptions of the teacher–student interactions in elementary and secondary school classrooms. It has eight scales—helpful/friendly, understanding, dissatisfied, admonishing, leadership, student responsibility and freedom, uncertain, and strict. Each scale has 8–10 items, and each item has 5-point responses ranging from *Never* to *Always*. The QTI is based on the theoretical model of proximity (cooperation–opposition) and influence (dominance–submission). Validation was conducted in the United States, Australia, Singapore, and Brunei. A shorter 48-item version of the QTI for elementary grades has also been created (Goh & Fraser, 1996).

Constructivist Learning Environment Survey (Taylor, Fraser & Fisher, 1997)

The Constructivist Learning Environment Survey (CLES) measures the degree to which a particular secondary classroom learning environment is consistent with a constructivist epistemology. It has 5 scales—personal relevance, uncertainty, critical voice, shared control, and student negotiation. Each scale has seven items. A complete version of CLES is available in the appendix of Fraser (1998a).

Cultural Learning Environment Questionnaire (Fisher & Waldrip, 1997).

The Cultural Learning Environment Questionnaire (CLEQ) measures secondary school students' perceptions of culturally sensitive factors of learning environment. It assumes that at the classroom level, there are distinctions in the preferred learning styles among different high school students. Culture is based on Hofstede's (1984) four dimensions of culture, namely power distance, uncertainty avoidance, individualism, and masculinity/femininity. Initial items for CLEQ were identified from various existing learning environment instruments that were aligned with the above four dimensions. The final CLEQ has 40 items defining eight scales; each scale has 5 items. The eight scales are role differentiation, collaboration, risk involvement, threat of competition, teacher authority, modeling, congruence, and communication. Role differentiation measures the extent to which gender roles are differentiated or overlapped by students; collaboration measures the extent to which students are part of a strong cohesive group; risk involvement measures the extent to which students feel threatened by involvement in class discussion; threat of competition measures the extent to which the students feel threatened by competition from other students; teacher authority measures the extent to which students expect and accept that power is distributed unequally in a school; modeling measures the extent to which the students prefer to learn by a process of modeling; congruence measures the extent to which the students feel threatened by learning in ways that are different from their own cultural pattern; finally, communication measures the extent to which students have more direct forms of communication with the person they are interacting with. All questions are presented in a five-point scale between two extreme alternatives of *Disagree* and *Agree.*

Validation of CLEQ was based on 1,834 students in 95 classrooms of 34 schools. Principal component factor analysis revealed structures of the expected eight scales. Cronbach's alpha internal consistency reliability coefficients for the eight scales ranged from 0.67 to 0.85. Mean correlation coefficients among the scales ranged from 0.08 to 0.22, suggesting the discriminant validity of CLEQ. Further, students' scores on CLEQ were found significantly correlated with students' attitudes and enquiry skills. That is, more favorable student attitudes and greater enquire skills were found to be associated with those who perceived classroom environments with less role differentiation, more collaboration, less teacher authority, less threat of competition, less risk involvement, less modeling, more congruence and more communication.

What Is Happening in This Class Questionnaire (Aldridge & Fraser, 2000)

What Is Happening in This Class (WIHIC) questionnaire is a combination of modified versions of a number of previously published learning environment scales and additional scales on equity and constructivism. It has seven scales; they are student cohesiveness, teacher support, involvement, investigation, task orientation, cooperation, and equity. Each scale has eight items. Items are in a 5-point format with choices of *almost never, seldom, sometimes, often,* and *very often.* Both the personal and class forms of the WIHIC questionnaire are available. The WIHIC questionnaire has been translated into a few languages and validated in such countries as Australia, Canada, South Korea, Singapore, and Taiwan. Confirmatory factor analysis with a sample of 3,980 high school students from Australia, Britain, and Canada supported the 7-scale *a priori* structure of the WIHIC (Dorman, 2003). The items of the WIHIC questionnaire are available in Aldridge, Fraser, and Huang (1999).

The Reformed Teaching Observation Protocol (Sawada et al., 2002)

The Reformed Teaching Observation Protocol (RTOP) is a 25-item classroom observation protocol that is standards based, inquiry oriented, and student centered. It consists of three scales: lesson design and implementation, content, and classroom culture. Items are rated by a scale from *never occurred* (0) to *very descriptive* (4). Reform is defined based on constructivism and follows the framework by the National Science Foundation's

Collaboratives for Excellence in Teacher Preparation. Development of items began with a pool of existing items from various published sources. The items were first piloted by using them to score video-taped lessons taught by student teachers. The final revised set of items were used by trained observers in 141 mathematics and science classrooms in middle schools, high schools, community colleges, and universities, resulting in 287 completed RTOP forms. Face validity was based on analysis of alignment with various science and mathematics reform documents including curriculum standards. Principal components factor analysis was conducted resulting in a three-factor solution. The *R* square for the correlation between two sets of observations was above 0.90. Cronbach's alpha for the entire instrument was 0.97; subscale alphas ranged from 0.80 to 0.93.

Learning Environment Scales (Nolen, 2003)

The Learning Environment Scales (LES) is a 20-item Likert-scale instrument measuring student perceptions of classroom learning environments in terms of teacher goals in mastery, independent thinking and performance, and the classroom competitive and cooperative climate. Validation involved 463 high school students attending one U.S. high school. Principal component factor analysis with varimax rotation resulted in a three-factor solution. The three factors form three scales; they are the science-learning focus scale consisting of items related to teacher goals of mastery and independent thinking, the ability-meritocracy scale consisting of items related to the teacher performance goal as well as competitive climate, and the cooperative climate scale consisting of items related to competitive and cooperative climate. Correlation among the three scales ranged from –0.11 to 0.42. Cronbach's alpha reliability coefficients for the three scales were 0.86, 0.70, and 0.77.

Survey of Instructional and Assessment Strategies (Walczyk & Ramsey, 2003)

The Survey of Instructional and Assessment Strategies (SIAS) is a 51-item survey on college faculty's teaching practices. Among the 51 items, 6 items are directly related to planning for instruction, 15 to delivery of instruction, and 7 to assessment of learning. Each of the above items was responded by faculty on a rating scale from *always* to *never*. Experienced college instructors reviewed the items and judged them to be valid. Principal component factor analysis found that the subscale on planning contained two main factors, the subscale on delivery contained four main factors, and the subscale on assessment contained two factors. Point-biserial correlation between participation in NSF math and science education reform projects and SIAS scores suggested that SIAS discriminated well. The alphas for the above three subscales were 0.67, 0.56, and 0.71.

The Outcome-Based Learning Environment Questionnaire (Aldridge, Laugksch, Seopa, & Fraser, 2006)

The Outcome-Based Learning Environment Questionnaire (OBLEQ) measures students' perceptions of their actual and preferred classroom learning environments in outcome-based learning settings. The development and validation of the questionnaire involved review of literature on outcome-based education, interviews with science curriculum advisors and with Grade 8 science teachers, references to Moos's (1979) three dimensions of social environments (i.e., relation, personal development, and system maintenance) and other relevant learning environment instruments. The OBLEQ consists of seven scales with eight items per scale. The seven scales are:

1. *Involvement*—the extent to which students have attentive interest, participate in discussions, do additional work and enjoy the class.
2. *Investigation*—the extent to which emphasis is placed on the skills and processes of inquiry and their use in problem-solving and investigation.
3. *Cooperation*—the extent to which students cooperate rather than compete with one another on learning tasks.
4. *Equity*—the extent to which students are treated equally and fairly by the teacher.
5. *Differentiation*—the extent to which teachers cater to students differently on the basis of ability, rates of learning, and interests.
6. *Personal Relevance*—the extent to which teachers relate science to students' out-of-school experiences.
7. *Responsibility for Own Learning*—the extent to which students perceive themselves as being in charge of their learning process, motivated by constant feedback and affirmation.

Items are in a 5-point rating scale consisting of *always, often, sometimes, seldom,* and *never.* The actual and preferred response scales of the OBLEQ items are placed side-by-side on a single form of the questionnaire to provide a more economical format. Validation data were collected from 2,638 students in 50 schools. Principal component factor analysis using oblique rotation suggested that some items did not conform to the expected dimensions. The Cronbach alpha reliability coefficients for the scales ranged from 0.66 to 0.84 when the individual was the unit of analysis, and ranged from 0.67 to 0.98 when the class mean was the unit of analysis. For the actual version of the OBLEQ, the discriminant validity (mean correlation of a scale with other scales) ranged from 0.12 to 0.31 when the individual was the unit of analysis, and ranged from 0.13 to 0.42 when the class mean was the unit of analysis. For the preferred version of the OBLEQ, the discriminant validity ranged from 0.18 to 0.43 when the individual was the unit of analysis, and ranged from 0.01 to 0.63 when the class mean was the unit of analysis. The OBLEQ scales also differentiated significantly ($p < 0.01$) between classes.

The Science Teacher School Environment Questionnaire (Huang, 2006)

The Science Teacher School Environment Questionnaire (STSEQ) measures secondary school science teachers' perceptions of the school environment. School environment is defined broadly; it involves science teachers' relationships with students, other teachers, and principals, and other administrators. In addition, school environment also involves science teachers' professional development, and their thoughts about resources, work pressure, gender equity, innovation, and staff freedom.

Items for the STSEQ were based on themes identified from interviews of 34 science and mathematics teachers, and reference to two other related instruments. It contains 45 items distributed equally over 9 scales—teacher–student relation, collegiality, principal leadership, professional interest, gender equity, innovation, staff freedom, resources and equipment, and work pressure. Each item is presented in a 5-point rating scale with choices of *almost always, often, sometimes, seldom,* and *almost never.*

Validation of STSEQ took place in Taiwan involving 900 secondary science teachers from 52 secondary schools. An exploratory factor analysis with promax rotation resulted in a nine-factor structure, with five items in each factor. The factor loadings of the 45 items ranged from 0.35 to 0.87, and the eigenvalues of the nine factors ranged from 1.22 to 10.01. The inter-scale correlation among the nine scales ranged from 0.12 to 0.36 with an overall mean correlation coefficient of 0.28, suggesting that the instrument had adequate discriminant validity. The Cronbach's alpha reliability coefficients of the nine scales, using the individual science teacher as the unit of analysis, ranged from 0.63 to 0.87.

Students' Perception of Assessment Questionnaire (Fisher, Waldrip, & Dorman, 2005; Dhindsa, Omar, & Waldrip, 2007)

The Students' Perception of Assessment Questionnaire (SPAQ) measures secondary school students' perceptions of the assessment process. Assessment refers mainly to tests and homework. It has 5 scales: congruence with planned learning, assessment of applied learning, student consultation on assessment types, transparency in assessment, and accommodation of students' diversity in assessment procedures. Congruence with planned learning measures the extent to which assessment covers the students' learning experience; assessment of applied learning measures the extent to which assessment evaluates the application of students' learning to daily life; student consultation on assessment measures the extent to which students are consulted in deciding the assessment tasks; transparency in assessment measures the extent to which students are informed about the assessment procedures; diversity in assessment measures the extent to which assessment accounts for individual differences. Each scale has 6 items presented in a 4-point rating scale: *almost always* (4), *often* (3), *sometimes* (2), and *almost never* (1).

Validation of the instrument took place in Brunei. The questionnaire was administered to 1,028 upper secondary science students 15–16 years. The three university lecturers who read the original instrument judged that the language of the content and construct of the instrument was valid for upper secondary students but less appropriate for the lower secondary students who were less familiar with the English language. For the original 30-item instrument, the Flesch Reading Ease (FRE) and Flesch–Kincaid Grade Level (FKGL) coefficients were 56.7 and 7.5, respectively. Principal component factor analysis using varimax rotation found that 24 of the original 30 items loaded highly on the five expected dimensions; the five factors reported in this study accounted for 50.6% variance. For the 24-item version, the FRE and FKGL coefficients were 68.6 and 6.4. The mean correlation among the five scales ranged from 0.15 to 0.20, indicating that the SPAQ measures distinct, although somewhat overlapping aspects of dimensions of assessment. Statistical significance was also found among different classes, suggesting discriminant validity. Cronbach's alpha reliability coefficients for the 5 scales of the 30-item instrument ranged from 0.64 to 0.77. Qualitative data from interviews and classroom observations were also used to cross-validate the instrument.

Science Lesson Plan Analysis Instrument (Jacobs, Martin, & Otieno, 2008)

The Science Lesson Plan Analysis Instrument (SLPAI) is a 20-item rating scale of science teachers' lesson plans. The 20 items are grouped into the following categories: (a) alignment with endorsed practices (2 items), (b) lesson design and implementation—cognitive and metacognitive issues (7 items), (c) lesson design and implementation—sociocultural and affective issues (6 items), and (d) portrayal and use of the practices of science (5 items). Each item is rated as *exemplary* (2 points), *making progress* (1 point), and *needs improvement* (0 points). A weight from 1 to 3 is also assigned to an item. A total weighted score, standardized as a percentage, is derived as the measure of a lesson plan. Development of the items were based on some published lesson plan evaluation and science class observation instruments, and informed by current reform documents and research syntheses. Validation of the instrument was conducted through comparing the results with that obtained from two other instruments, one based on teacher self-reporting (Standards-Based Teaching Practices Questionnaire, SBTPQ), and another based on classroom observation (Reformed Teaching Observation Protocol, RTOP). Agreement was found between conclusions made based on SLPAI and on SBTPQ. However, no consistent statistically significant correlations in scores between relevant SLPAI and RTOP items were found, likely due to the fact that the lessons evaluated using SLPAI were not the same lessons taught and evaluated using RTOP. Inter-rater agreement between two developers of the instrument was 96%, and between one developer of the instrument and another new researcher was 89%.

Student Actions Coding Sheet (Erdogan, Campbell, & Abd-Hamid, 2011)

The Student Actions Coding Sheet (SACS) was created to measure the extent to which student-centered actions are occurring in science classrooms. The instrument was developed through the following five stages: (a) student action identification, (b) use of both national and international content experts to establish content validity, (c) refinement of the item pool based on reviewer comments, (d) pilot testing of the instrument, and (e) statistical reliability and item analysis leading to additional refinement and finalization of the instrument. The final version of SACS instrument consisted of 24 items. Exploratory factor analysis suggested a three-factor solution accounting for 64.96% of the total variance. All but two items loaded above the 0.30 level with the remaining 24 items with loading values ranging from 0.343 to 0.951. The three factors were: student actions indicative of lower cognitive domains (remembering and understanding), student actions indicative of the medium cognitive domain level (higher stages of understanding and applying), and student actions indicative of the highest cognitive domain level (analyzing, evaluating, and creating). The finalized instrument was found to be internally consistent. Inter-rater intraclass correlation reliability coefficients among different groups of raters was statistically significant ($p < 0.01$) with the average ICC for all 22 groups combined to be 0.74. Correlation between SACS and RTOP scores was 0.65.

The Classroom Observation Protocol for Undergraduate STEM (Smith, Jones, Gilbert, & Wieman, 2013)

The Classroom Observation Protocol for Undergraduate STEM (COPUS) is a coding scheme documenting what the instructor and the students are doing during a class period. Instructor practices are recoded by 12 codes such as lecturing (L), real-time writing on the board (RrW), asking a clicker question (CQ), and so forth; student activities are recorded by 13 codes, such as listening (L), working in groups (WG), student asks questions (SQ), and so forth. Levels of student engagement may also be recorded: *low* when only 10–20% of students obviously are engaged, *medium* when 20–80% students are obviously engaged, and *high* when more than 80% students are obviously engaged. Teaching practices and student activities are coded continuously in 2-minute intervals. Training needed to use the protocol can be as little as 1.5 hours. The development of initial codes was informed by TDOP codes and refinement of the codes was facilitated by many rounds of uses in a variety of university STEM classes. Validation of the codes was conducted in 31 classes of 11 departments at two universities; majority of the classes were introductory courses. Content validity was based on researchers' and science education associates' judgement of the appropriateness of the codes for STEM teaching. Coding agreement based on Jaccard similarity scores across all pairs of observers ranged from 0.80 to 0.99. The averaged Kappa for all codes was above 0.83; the Kappa for student codes was 0.87 and for instructor codes was above 0.79.

The Teaching Practices Inventory (Wieman & Gilbert, 2014)

The Teaching Practices Inventory (TPI) is an instructor's self report of teaching practices in university and college mathematics and science courses. Instructors can complete the inventory in 10 minutes or less. TPI includes 72 items with a scoring rubric for a total possible score of 67. Items are a checklist of specific practices either used or not used during the course. The teaching practices include the following categories: (a) course information provided, (b) supporting materials provided, (c) in-class features and activities, (d) assignments, (e) feedback and testing, (f) other (diagnostics, pre-post testing, etc.), (g) training and guidance of TAs, and (h) collaboration or sharing in teaching. The development of TPI underwent two major iterations and one final round of minor revisions involving a large number of diverse mathematics and science faculty. A total of 179 instructors from five math and science departments at one large institution completed Version 2 of the inventory. No practices were identified that were used by more than two instructors and not captured by the inventory. The accuracy of item interpretation was tested by the department director interviews and the reviews of 179 instructor responses. The TPI responses for seven team-taught courses in which two instructors provided responses for the same course showed that the differences in TPI scores of a same course were small (0–2 points). The scoring rubric was based on the extent of use of a research-based teaching practice (ETP) by assigning points to each practice for which there is research showing that the practice improves learning.

The Advancing Science by Enhancing Learning in the Laboratory Student Laboratory Experience Survey (Barrie et al., 2015)

The Advancing Science by Enhancing Learning in the Laboratory Student Laboratory Experience Survey (ASLES) is an 11-item Likert scale instrument. The development of the ASLES started with open-ended questions for students to evaluate their laboratory experiences. Student responses to open-ended questions were then converted into Likert-scale items. Teams of experienced educators in chemistry, biology, and physics reviewed the items. Validation data were from 56 studies in 19 institutions of three countries. Most of the experiments were at the first year level and 12 at upper-levels. A total of 3,153 students responded to the survey. Exploratory factor analysis showed a three factor solution with six, three and two items, respectively. The first factor was about student motivators, the second factor was about assessment, and the third factor was about instructor resources. Cronbach's alphas for the three factors were 0.86, 0.71 and 0.69.

Teaching Dimensions Observation Protocol (Hora, 2015)

Teaching Dimensions Observation Protocol (TDOP) is a system of codes for describing teaching and learning activities taking place in an undergraduate classroom; it is not intended to measure any constructs related to teaching or learning. The TDOP covers five key aspects of classroom dynamics: teaching methods, pedagogical strategies, student–instructor interactions, types of cognitive engagement, and use of instructional technology. Observers code what is happening in 2-minute intervals; the frequencies of codes representing different activities are then computed. Inter-rater reliability between two coders ranged from 0.80 to 0.90 for the category of codes on teaching methods, 0.73 to 0.85 for the category of codes on pedagogical moves, 0.72 to 0.83 for the category of codes on interactions, 0.71 to 0.78 for the category of codes on cognitive engagement, and 0.89 to 0.94 for the category of codes on instructional technology.

The Behavioral Engagement Related to Instruction (Lane & Harris, 2015)

The Behavioral Engagement Related to Instruction (BERI) measures university students' behavioral engagement in large enrolled courses. The development of the observation protocol began with a definition of student in-class behaviors that indicate engagement and student in-class behaviors that indicate disengagement. Student in-class behaviors indicating engagement are listening, writing, reading, engaged computer use, engaged student interaction, and engaged interaction with instructor. Student in-class behavior indicating disengagement are settling in/packing up, being unresponsive, off-task, disengaged computer use, disengaged student interaction, and distracted by another student. When using the protocol, the observer randomly chooses a place to sit in the classroom where 10 students can be clearly seen. An observation point is taken for every page of instructor notes, for any major change in activity or content, or at 2-minute intervals depending on which time interval is shorter. After only one practice session, new observers were able to produce reliable codes compared with their trained partner. The averaged inter-rater agreement was calculated to be 96.5%. Findings from simultaneous observations of the class by three trained observers suggest that although the overall level of engagement varied between different groups of 10 students, the general trends over time in student engagement remained consistent. However, the average level of student engagement was significantly less at the back of the class than in the front of the class. The BERI was able to document the difference in student engagement between different teaching methods.

The Laboratory Observation Protocol for Undergraduate STEM (Velasco et al., 2016).

The Laboratory Observation Protocol for Undergraduate STEM (LOPUS) is a segmented protocol that characterizes (a) instructional behaviors without a priori criteria of quality, (b) students' and TA's behaviors, (c) the extent and initiator of verbal interactions between students and TA, and (d) the nature of the content being discussed during these verbal interactions. The LOPUS includes three categories of codes: TA's instructional behaviors, students' behaviors, and verbal interactions between the TA and students. The LOPUS is based on a theoretical framework of instructional capacity involving interactions between instructors, students, and instructional materials. The development of initial codes was informed by COPUS and new codes were added based on literature on university laboratory instruction. Validation data were collected from 15 TAs' chemistry labs at one large research university. The labs were at the first and second semester first year general chemistry level. Face validity of LOPUS was established through interviews with chemistry TAs ($n = 3$), chemistry faculty ($n = 3$), and STEM laboratory coordinators ($n = 4$). Median inter-rater agreement between independent coders using Krippendorff's alpha was 0.82.

Social Media and Science Learning Survey (Moll & Nielsen, 2017)

Social Media and Science Learning Survey is a descriptive instrument to collect information of secondary and university students' uses of social media for science learning. The development of the survey followed an inductive approach. Focus-group interviews of secondary and post-secondary physics students were used to identify the most frequently used social media tools and common behaviors when participants used social media tools to support their science learning. The survey has three main sections: demographics and everyday social media use, social media use in the science subject of interest, and social media use in high-school science learning. Within each science learning context (e.g., high school science or university physics), respondents are asked to self-report a level of proficiency with the listed social media tool. The initial version was piloted in three post-secondary learning contexts in three countries. Validation interviews ($n = 4$) were conducted with preservice teachers who took the survey. The survey was revised based on the pilot test and validation interviews. The revised survey was administered again to another group of preservice teachers and another round of validation interviews ($n = 6$) was conducted.

Instructional Practices Log in Science (Adams et al., 2017)

The Instructional Practices Log in Science (IPL-S) is a daily teacher log developed for K–5 teachers to self-report their science instruction. It measures five dimensions of science instruction: low-level sense making, high-level sense-making, communication, integrated practices, and basic practices. The IPL-S is formatted as a survey for teachers to complete shortly after a lesson. Teachers are asked to log only about the behaviors in which at least half of the students in the class engaged during the science lesson. There are 12 items about science content, 1 item about time in minutes spent teaching science, 1 item on teacher use of tools, 4 items on instructional science goals, and 35 items on student activities and behaviors. Student activities and behaviors are rated on the following scales: *not today, little, moderate,* and *considerable.* A total of 206 elementary teachers completed their daily log entries. The development of IPL-S was informed by a theoretical framework of science instruction consisting of sense-making and science practices. The IPL-S item development went through several recursive stages. The large-scale implementation of the IPL-S included 136 second-year teachers and the experienced teacher pilot included a sample of 58 teachers who have been teaching for at least 4 years. Validation of the scale included cognitive interviews about teacher response processes and descriptive statistical analysis of teacher responses to items. Also, a total of 28 science lessons from 11 teachers were observed by one of seven trained IPL-S raters, and their logs were compared with the teachers' logs. On average across the 28 lessons, teachers and raters had exact agreement on 66% of the items, with a minimum of 46% and a maximum of 80%. A two-level confirmatory factor analysis showed a good model-data fit. Within-level Cronbach's alphas were all above 0.60 with the exception of low-level sense-making scale which was 0.144; between-level Cronbach's alphas ranged from 0.756 to 0.925.

Dimensions of Success (Shah, Wylie, & Gitomer, 2018)

Dimensions of Success (DoS) is an instrument to measure the quality of out-of-school STEM programing. It includes 12 dimensions categorized in four domains: features of the learning environment (3 dimensions), activity engagement (3 dimensions), STEM knowledge and practices (3 dimensions), and youth development in STEM (3 dimensions). A scoring rubric was developed for each dimension based on a 4-point scale: *compelling evidence* (level 4), *reasonable evidence* (level 3), *inconsistent evidence* (level 2), and *evidence absent* (level 1). Observers not only assign a numerical rating, but also they write detailed evidence justifying the score. A study involving 284 observations was used for validation. Validity evidence was established through scoring inference and generalization. Weighted Kappa suggests that here was a moderate agreement for four dimensions, fair agreement for six dimensions, and only slight agreement for the remaining two dimensions. A follow-up study with revised training and scoring guide resulted in higher inter-rater reliability. Generalizability study indicated that multiple observations are needed to get a stable measure of quality.

Developing Instruments for Measuring Science Learning Environments

State the Purpose and Intended Population

Although there are dynamic interactions between individuals and their environments, we could conceptualize learning environments to be both determinants of student learning outcomes and outcomes of student learning. Therefore, there can be two main purposes for measuring learning environments; one is to study how a particular learning environment is related to or can predict a certain student learning outcome, and the other is to study how a learning environment changes as the result of an intervention or differs from other learning environments due to different types of learning and teaching. For the first purpose, units of analysis can be either individual students or classes; but for the second purpose, units of analysis should be classes. For example, measurement of students' perceptions of the learning environment can help understand how different perceptions of a learning environment may be related to different levels of science achievement. In this case, units of analysis are individual student scores on a learning environment scale. As another example involving program evaluation of a science curriculum innovation, the measurement of learning environments will help understand if the intended learning environment has been created by implementation of a science curriculum innovation. In this case, units of analysis can be both individual student scores and class means on the learning environment scale. For both the above types of purposes, the intended population of a learning environment measurement instrument should be defined based on units of analysis, such as classes and individuals, and their identifiable characteristics such as geographical locations, types of classes or schools, individual demographics, and so on.

Besides the above two main purposes, one other historical purpose of measuring learning environments is to study the fit between the actual learning environment and the preferred or ideal learning environment, the so-called person-environment fit study (Fraser, 1996, 1998a; Moos, 1979). The basic premise for this approach is that personal goals may be supported or constrained by environmental interactions. When personal needs and environmental factors are aligned, student learning outcomes will be enhanced. Measurement of learning environments for this purpose typically involves developing parallel forms of a learning environment scale and administering them simultaneously to students. Units of analysis are typically the different scores between two forms. Person-environment fit studies have played and will continue to play an important role in understanding the context of science teaching and learning. Once again, the intended population of the measurement instrument for this purpose should also be clearly defined, because no learning environment is ideal for all types of populations of students.

Define the Construct to Be Measured

Learning environments have been commonly considered as the social-psychological contexts in which learning takes place. This socio-psychological approach naturally focuses on individuals' perceptions instead of external observations of a physical environment. Because different perceptions of the environment are results of different interactions between individuals and the environment, learning environments can be defined based on different components and patterns of the human-environment interactions. Moos (1979), based on studies of various social settings such as student living groups, hospitals, correctional institutions, military training camps, and school classrooms, identified three fundamental dimensions of a learning environment: relationship, personal growth, and system maintenance and change. The relationship dimension is concerned with the extent to which people are involved in the setting, the extent to which they support and help one another, and the extent to which they express themselves freely and openly. Personal growth dimension pertains to the goals of the setting related to personal development and self-enhancement. Finally, the system maintenance and change dimension pertains to the extent to which the environment is orderly and clear in its expectations, maintains control, and responds to change. Majority of learning environment instruments described earlier are based on this conceptualization of learning environments.

Adopting a different socio-psychological conception of learning environments, Walberg (1968a) conceptualizes learning environments to consist of two dimensions: the structural dimension and the affective dimension. The structural dimension refers to student roles within the class such as goal direction and democratic policy. It applies to shared, group-sanctioned classroom behaviors. The "affective" dimension pertains to idiosyncratic personal dispositions to act in a given way to satisfy individual personality needs, such as satisfaction, intimacy, and friction in the class.

Although physical settings have not been typically conceptualized as part of learning environments in the past 4 decades, the semiotic approach to learning environments distinguishes itself by explicitly focusing on signs in the environment (Shapiro, 1998). If a "learning environment" is defined broadly to encompass "social, physical, psychological, and pedagogical contexts in which learning occurs and which affects student achievement and attitudes" (Fraser, 1998b, p. 1), then, separation between the socio-psychological aspects from the physical settings of a learning environment is difficult. No matter what theoretical approach is to be adopted to define learning environments, one thing is clear: learning environment is not a unidimensional construct, but it is multi-dimensional. Different theories may focus on different dimensions of a learning environment, and some dimensions may pertain to physical settings. Further, learning environment is not a generic construct; it is specific to settings. For example, an elementary science classroom learning environment may be defined differently from that of a secondary school science classroom. Similarly, an online learning environment may be defined differently from that of a classroom learning environment. Defining the construct of learning environments to be measured must take into consideration not only the theoretical framework in which learning environments are conceptualized, but also the nature of the learning setting and characteristics of the students.

Identify Performances of the Defined Construct

Once the construct to be measured is defined, specific behaviors, that is, the domain of a learning environment, must be identified. Depending on the purpose of the measurement of the learning environment, the identified behaviors should reflect the hypothesized or inferred differences in measured learning environment. That is, if the purpose of measuring a learning environment is to study the relationship between the learning environment and student learning outcomes, then what perceptions of the learning environment are associated with higher student learning outcomes and what perceptions are associated with lower student learning outcomes should be derived from the defined construct so that a variation or a range of student perceptions of the learning environment may be anticipated. For example, if the purpose of measuring a learning environment is to study the relationship between student conceptual understanding and the inquiry learning environment, then it is necessary to identify desirable student perceptions of the inquiry learning environment, such as students being encouraged to change their research questions during the inquiry, which may lead to higher student conceptual understanding; it is also possible to identify less desirable student perceptions of the inquiry learning environment, such as students only studying the problems given by the teacher, which may lead to a lower level of conceptual understanding.

Similarly, if the purpose of measuring a learning environment is to study how different learning environments are created by different curriculum innovations, then different student perceptions of the learning environment at different levels of implementation of the curriculum innovation should be derived. For example, if the interest is to study the difference between an inquiry science learning environment and a didactic science learning environment, what student perceptions of the learning environment are associated with inquiry learning and what student perceptions are associated with didactic learning should be inferred from the theory.

The above rationale for identifying student behaviors is also applicable to person-environment fit studies. In these situations, no external references are available to guide the inferences of student perceptions corresponding to different degrees of learning environments. No matter what purpose measurement of a learning environment serves, the learning environment must be considered as a continuum that can be quantified by a continuous variable based on a measurement instrument. Based on this continuous variable, the learning environment measured can be differentiated into various degrees that are thought to be associated with different student perceptions of the learning environment. This process must rely on a valid theory. Without a valid theory, it is difficult to conceptualize various degrees of a learning environment, and in turn it is difficult to identify different student behaviors or perceptions.

Once student behaviors are identified, a test specification for a learning environment measurement instrument that involves deciding how many items are needed for each construct may be created. Although the exact number of items necessary for the construct can only be known later based on statistical and Rasch analyses, basic consideration is that there should be enough items of sufficient variation to cover the anticipated range in student perceptions of the measured learning environment. Thus, a more heterogeneous student population will require more items to measure the construct. It is always a good practice to develop more items than what may be needed in the final instrument allowing for more flexibility in the revision stage.

Once the number of items needed is tentatively decided, the next step is to create a pool of preliminary items. Because measurement of learning environments is primarily based on students' and teachers' perceptions, types of items appropriate for measurement of learning environments are the same as that for measurement of affective variables. Likert scale and rating scale are commonly used; other formats, such as semantic differential, inventory, and checklist, may also be used.

The following are sample items from the Constructivist Learning Environment Survey (CLES; Taylor, Fraser, & Fisher, 1997):

In this class...

It is OK for me to ask the teacher "Why do I have to learn this?"

Almost never (1) *Sometimes* (3) *Seldom* (2) *Often* (4) *Almost always* (5)

It's OK for me to question the way I'm being taught.

Almost never (1) *Sometimes* (3) *Seldom* (2) *Often* (4) *Almost always* (5)

It's OK for me to complain about teaching activities that are confusing.

Almost never (1) *Sometimes* (3) *Seldom* (2) *Often* (4) *Almost always* (5)

The above items are in a rating scale format. Students are asked to rate how often each of the described teaching practices actually takes place in the classroom. The above items are for the actual form of CLES. If they are for the preferred form, then the instruction should be that students are asked to rate how often they PREFER each of the described teaching practices to take place in the classroom. Thus, although the statements and choices may be the same, students will be asked to rate based on different perspectives, actual or preferred.

Conduct Pilot-Test/Field-Test

After a pool of items has been created, it is necessary to have items reviewed by a panel of experts consisting of both content experts and methodologists. After items have been reviewed by a panel and necessary revisions made to the items, the items should then be tried out by a small number of respondents—the pilot-test. The respondents will not only respond to the items, but also state their impressions or comment on the clarity of the items. Interviews with a few respondents may also be helpful. Pilot-test data may be analyzed using a Rasch model and results will inform item revision and deletion/addition.

The revised items from pilot-test will then go for a field-testing. The sample chosen for the field testing should represent the intended population in terms of the range along the construct to be measured. Depending on the item format, a minimum of 25 observations per category are preferred. For example, if items are written in a checklist format, thus each item has five categories (e.g., the Likert scale), the preferred sample size for field-testing should be minimally 125. The selected sample should also possess sufficient variation in student perceptions of the measured learning environment construct similar to that of the intended population.

Conduct Rasch Analysis

Data collected from pilot-testing/field-testing are used for Rasch analysis. Because item formats used in learning environment measurement instruments are the same as that used in affective measurement instruments, data file setup and control file commands are the same as those in analyzing affective measurement data.

Figure 7.1 shows a sample data file for a university student engagement survey. This survey contains three sections on student cognitive engagement, curriculum emphasis, and instructional practices. Item formats include three types of rating scales (e.g., *never* [1], *one to two times* [2], *three to five times* [3], and *more than five times* [4]); although students responded to 40 items, only the first 34 items in the above three sections were analyzed. Student responses were coded in numbers starting from column 31; all codes are numerical.

Figure 7.2 shows the Winsteps control file for analyzing the data shown in Figure 7.1. It can be seen that this control file includes a few additional commands. When an instrument contains a few different groups of items following different response category formats (e.g., one item set uses a 5-category Likert scale format, another set of items uses a 4-category frequency

```
File  Edit  Format  View  Help
 1 Online Fall 2017 BIO200    4422444224421421123144441424243444223441
 2 Online Fall 2017 BIO200    1113344121142211142242431434242123222211
 3 Online Fall 2017 BIO200    1111131111221211111233231334142143323442
 4 Online Fall 2017 BIO200    3211141442442411114442341434144443431343
 5 Online Fall 2017 BIO200    1111111111121211112332221423142134232332
 6 Online Fall 2017 BIO200    1211141121431131111111111144234312432134 3
 7 Online Fall 2017 BIO200    2213233121442111133333222433243223322443
 8 Online Fall 2017 BIO200    1111144424444311444444441324244444422431
 9 Online Fall 2017 BIO200    1322144422311421113222221332142223341341
10 Online Fall 2017 BIO200    1222141441111311113233312
11 Online Fall 2017 BIO200    1111131111221111111222221434143213221332
12 Online Fall 2017 BIO200    1311111441421311112433221414243432222341
13 Online Fall 2017 BIO200    2322332232322423222423423233132322222242
14 Online Fall 2017 BIO200    3232323234232323232223233333212433421142
15 Online Fall 2017 BIO200    1111141121211111111443344142314312333134 1
16 Online Fall 2017 BIO200    1121244422432211224444441423143333441241
17 Online Fall 2017 BIO200    2313344241443311213444441343244324431431
18 Online Fall 2017 BIO200    1111144411441311141211111444242424443444
19 Online Fall 2017 BIO200    1122233131122311122332331322142123413344
20 Online Fall 2017 BIO200    2234434434444424333232434424143324332342
21 Online Fall 2017 BIO200    2111241111111311111132223
22 Online Fall 2017 BIO200    3422424423442411113432441344244413332341
23 Online Fall 2017 BIO200    1412244121121111141333311323142123323444
24 Online Fall 2017 BIO200    2222232222212311112233331322243434333441
25 Online Fall 2017 BIO200    3211143333431311112333331333142424433341
26 Online Fall 2017 BIO200    3333313334344411444444441214142113213334
27 Online Fall 2017 BIO200    2111141211132411112444341424143233342341
28 Online Fall 2017 BIO200    1133341141122211223433331324244124412142
29 Online Fall 2017 BIO200    2112143322231311123443321424143214433341
30 Online Fall 2017 BIO200    2411144212131211124222331344143214442241
31 Online Fall 2017 BIO200    1111121111411111122444441412143312333233
32 Online Fall 2017 BIO200    3422444322132311123344441423143213423441
33 Online Fall 2017 BIO200    2313342332343321323433331423243123333433
34 Online Fall 2017 BIO200    3411244111122211133444111333132223322342
35 Online Fall 2017 BIO200    3112243333411311112423431233144444212341
36 Online Fall 2017 BIO200    1211214441421411221444441413131313313142
37 Online Fall 2017 BIO200    1122122211312221223222311233133112323232
38 Online Fall 2017 BIO200    1112143124211311111431111322143233443344
39 Online Fall 2017 BIO200    2211124111222111112222211321132112313444
40 Online Fall 2017 BIO200    4422143242343421233344 14
```

Figure 7.1 Shows a data file for measuring student classroom engagement. Student responses to questions start from Column 31, and student identity information is contained in Columns 1 through Column 30.

format), the command "isgroup" is used. Also, to reversely code student responses to an original item, the command group of IREFER, IVALUEA, IVALUEB, and so forth is used. In this example, items represented by Code B are reversely coded, and by A are kept unchanged. Data file can be integrated into the control file as one file.

When developing instruments for learning environment measurement, one common issue is to equate measures between parallel forms. Consider two parallel forms of a learning environment instrument, the actual form and the preferred form. It is necessary to directly compare students' perceptions on the two forms by calculating the difference between the two scores. We need to obtain two sets of scale scores based on Rasch analysis: one for the actual form and another for the preferred form. Given that two parallel forms measure two different constructs, in order to make the scales of the two constructs directly comparable, we need to equate them

```
File  Edit  Format  View  Help
&INST
Title= "Student Engagement Survey"
ITEM1 = 31 ; Starting column of item responses
NI = 34 ; Number of items
NAME1 = 1 ; Starting column for person label in data record
NAMLEN = 30 ; Length of person label
XWIDE = 1 ; Matches the widest data value observed
isgroups = 1111111111111111111222223456789xyz
IREFER=AAAAAAAAAAAAAAAAAAAAAAAABAAAAAAAAA
CODES = "1234 " ; matches the data
IVALUEA=1234
IVALUEB=4321
&END ; Item labels follow: columns in label
1 ; Item 1 : 1-1
2 ; Item 2 : 2-2
3 ; Item 3 : 3-3
4 ; Item 4 : 4-4
5 ; Item 5 : 5-5
6 ; Item 6 : 6-6
7 ; Item 7 : 7-7
8 ; Item 8 : 8-8
9 ; Item 9 : 9-9
10 ; Item 10 : 10-10
11 ; Item 11 : 11-11
12 ; Item 12 : 12-12
13 ; Item 13 : 13-13
14 ; Item 14 : 14-14
15 ; Item 15 : 15-15
16 ; Item 16 : 16-16
17 ; Item 17 : 17-17
18 ; Item 18 : 18-18
19 ; Item 19 : 19-19
20 ; Item 20 : 20-20
21 ; Item 21 : 21-21
22 ; Item 22 : 22-22
23 ; Item 23 : 23-23
24 ; Item 24 : 24-24
25 ; Item 25 : 25-25
26 ; Item 26 : 26-26
27 ; Item 27 : 27-27
28 ; Item 28 : 28-28
29 ; Item 29 : 29-29
30 ; Item 30 : 30-30
31 ; Item 31 : 31-31
32 ; Item 32 : 32-32
33 ; Item 33 : 33-33
34 ; Item 34 : 34-34
END NAMES
  1 Online Fall 2017 BIO200    4422444224421421123144441424243444223441
  2 Online Fall 2017 BIO200    1113344121142211142242431434242123222211
```

Figure 7.2 Shows a Winsteps control file for analyzing student engagement survey data shown in Figure 7.1. The command "isgroup" indicates which items have the same response category format; the "irefer" command specifies which items need to be reversely coded.

so that they have the same measurement units, that is, means and standard deviations. The ideal way to accomplish this is to conduct multidimensional Rasch modeling using such computer programs as ConQuest (Wu, Adams, & Wilson, 1997). In multidimensional Rasch modeling, items for the actual form measure one construct, and items for the preferred form measure another construct. Using a simultaneous calibration, we will obtain two scale scores for each student, one for each construct. We will also know the correlation between the two constructs.

If multidimensional scaling is not feasible for whatever reasons, we may still use unidimensional Rasch scaling. Because simultaneous calibration is not appropriate due to lack of linkage between the two forms and more importantly, violation of the unidimensionality requirement, we have to conduct two separate Rasch analysis sessions, one for the actual form and another for the preferred form. In order to place scale scores from the two separate calibrations on the same scale, we can use a set of scaling commands, such as UMEAN and USCALE in Winsteps, to force one scale onto another. That is, after Rasch analysis for one form (e.g., actual form), the mean and standard deviation of the item measures for this form are noted. When conducting Rasch analysis for the second form, we will then use commands UMEAN and USCALE to force the mean and standard deviation of item measures of the second form to be equal to that obtained from the previous form. For example, if the mean and standard deviation of item difficulties on the actual form are 0.0, and 1.1, then the commands for Rasch analysis for the preferred form should be UMEAN = 0.0 and USCALE = 1.1.

Review Item Fit Statistics, the Wright Map, Dimensionality, and Reliability

Fit statistics are reviewed in the same way as in developing measurement instruments for affective variables. Item category structure is also examined. For example, Figure 7.3 shows that the progression from Category 1 to Category 4 is quite orderly, indicating that the four-step structure of the rating scale is appropriate.

In the situation of learning environment measurement, item difficulty and person ability parameters have different meanings. Item difficulty may be interpreted as the tendency for an individual to endorse a category of an item (e.g., selecting *agree, always,* or *yes* to a positively phrased statement). Consider the following example:

1. It is OK for me to ask the teacher "Why do I have to learn this?"

Almost never (1) *Sometimes* (3) *Seldom* (2) *Often* (4) *Almost always* (5)

The difficulty for the above item reflects the tendency for an individual to select *almost always.* Because item difficulty is on the interval scale without absolute zero, an item with a difficulty of 0.5 is more likely for individuals to endorse as *almost always* than an item with a difficulty of 0.2, but less likely for individuals to endorse as *almost always* than an item with a difficulty of 1.0. Similarly, an individual's ability may be interpreted as the individual's overall level of positive perception of the learning environment. Thus, a

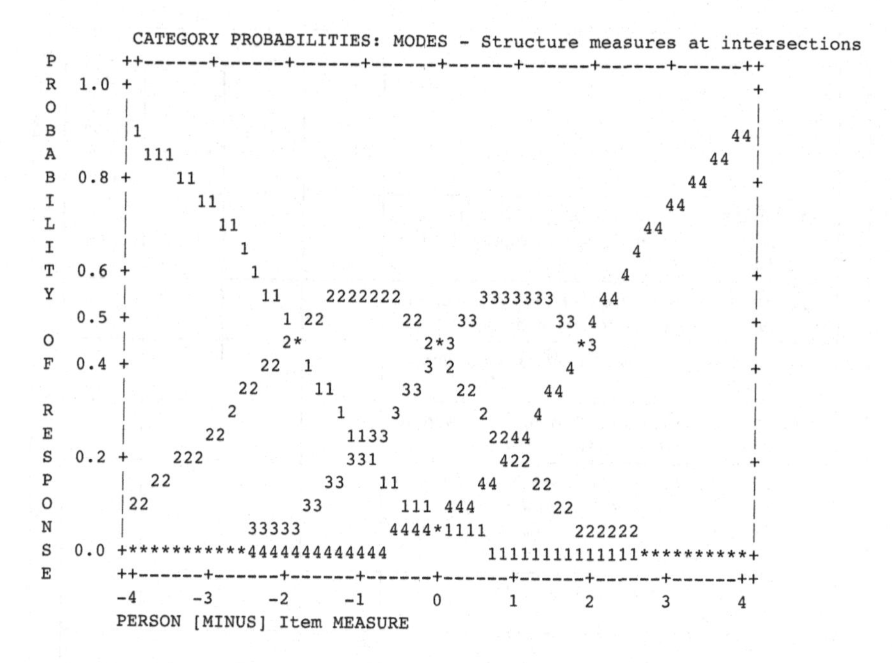

Figure 7.3 Rating scale structure for PISA Inquiry Science Learning Environment scale. All categories have distinct characteristic curves, suggesting a good item category structure.

person with an ability level of 0.5 is overall more positive about the learning environment than a person with an ability level of 0.2, but less positive than a person with an ability level of 1.0

The following items are extracted from the 2006 PISA Student Survey Questionnaire (OECD, 2006); they may be conceptualized as measuring students' perceptions of inquiry science learning environment:

When learning school science topics at school, how often do the following activities occur? (Please tick only one box in each row)

	In all lessons (4)	*In most lessons (3)*	*In some lessons (2)*	*Never or hardly ever (1)*
Q1. *Students are given opportunities to explain.*				
Q2. *Students spend time in the laboratory doing practical experiments.*				

	In all lessons (4)	*In most lessons (3)*	*In some lessons (2)*	*Never or hardly ever (1)*
Q3. *Students are required to design how a school science question could be investigated in the laboratory.*				
Q4. *The students are asked to apply a school science concept to an everyday problem.*				
Q5. *The lessons involve students' opinions about the topics.*				
Q6. *Students are asked to draw conclusions from an experiment they have conducted.*				
Q7. *The teacher explains how a school science idea can be applied to a number of different phenomena (e.g., the movement of objects, substances with similar properties).*				
Q8. *Students are allowed to design their own experiments.*				
Q9. *There is a class debate or discussion.*				
Q10. *Experiments are done by the teacher as demonstrations.*				
Q11. *Students are given the chance to choose their own investigations.*				
Q12. *The teacher uses school science to help students understand the world outside school.*				
Q13. *Students have discussions about the topics.*				
Q14. *Students do experiments by following the instructions of the teacher.*				
Q15. *The teacher clearly explains the relevance of broad science concepts to our lives.*				
Q16. *Students are asked to do an investigation to test out their own ideas.*				
Q17. *The teacher uses examples of technological application to show how school science is relevant to society.*				

Figure 7.4 presents the Wright map for the above set of items based on a 5% random subsample ($n = 278$) of the USA sample.

From Figure 7.4, we can see that Q8 and Q11 are the most difficult for students to endorse, while Q1 and Q14 are the easiest for students to endorse. Q8 is "Students are allowed to design their own experiments" and

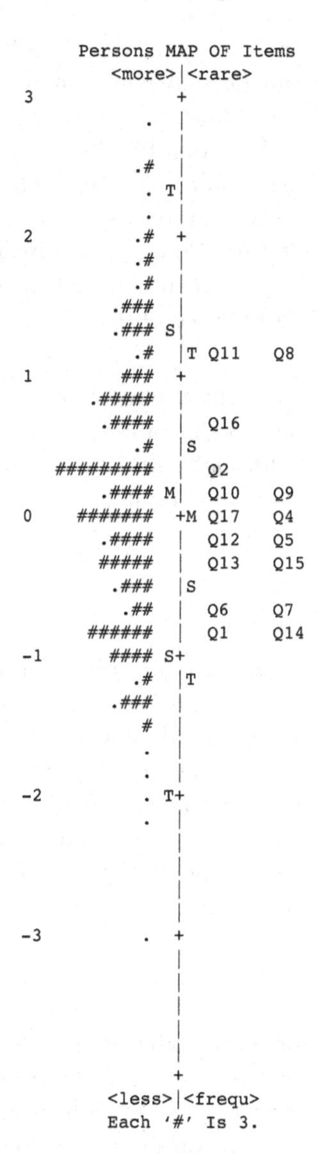

Figure 7.4 Wright map of PISA Inquiry Science Learning Environment scale. There are noticeable gaps in the distribution of items.

Q11 is "Students are given the chance to choose their own investigations." These two practices are characteristics of open-ended inquiry. Although they are essential for authentic science inquiry, they are not common in science classrooms, which is why they are the most difficult for students to endorse. On the other hand, Q1 is "Students are given opportunities to explain" and Q14 is "Students do experiments by following the instructions of the teacher." These two practices are common in science teaching, which is why they are the easiest for students to endorse. Figure 7.4 also shows that the 17 items do not have a sufficient difficulty range to match the range of students' perceptions of the inquiry science teaching environment, thus more items on both ends, that is, more easy to endorse and more difficult to endorse, are needed. In terms of the continuum among the items, there are three noticeable gaps, one between Q11/Q8 and Q16, another between Q16 and Q2/Q3, and the last between Q13/Q15 and Q6/Q7. More items are needed to fill the gaps.

Examination of dimensionality also indicates that, overall, the 17 items are unidimensional. The person separation index is 2.54, and person reliability coefficient is 0.87. Although both indices are acceptable, there is room for further improvement, and adding more items on both ends to fill the gaps should help achieve this.

Examine Invariance Properties

After examination of model-data fit of items, the Wright map, and dimensionality as well as reliability, some items may be revised or deleted, and a new set of items may be created. This new set of items will form a revised version of the measurement instrument. The revised measurement instrument will then go through another cycle of field-testing and Rasch analysis until all items and the measurement instrument as a whole meet the quality expectations. Item invariance property such as DIF and person invariance proper may then be examined at this stage.

Establish Validity Claims

Construct validation includes establishing evidence related to test content, response processes, internal structure, external structure and consequences. Construct validation may begin with examination of the definition of the learning environment construct measured. Important questions to ask about the construct definition may include: (a) "Is the definition based on commonly accepted theories?"; (b) "Does the defined construct have an underlying linear progression for the construct?"; and (c) "Are the specific behaviors described by the items clearly related to the construct?" Answers to the above questions should all be positive; a negative answer undermines validity claims.

Because Rasch modeling is a theory-based approach, when there is good model-data fit of items and model-data fit of persons, the measurement instrument which produces the data can be considered valid in terms of internal structure. The Wright map and dimensionality also provide important evidence for the validity related to the internal structure of the instrument. That is, when the Wright map shows that items are evenly distributed and target the population well, then the measurement instrument possesses an adequate internal structure.

Also, the agreement between the actual order of items and the predicted order of items based on valid theories can also provide evidence for validity related to external structure. For example, the order of items for the PISA inquiry learning environment scale described earlier according to the obtained Rasch difficulties is presented in Figure 7.5. Items are

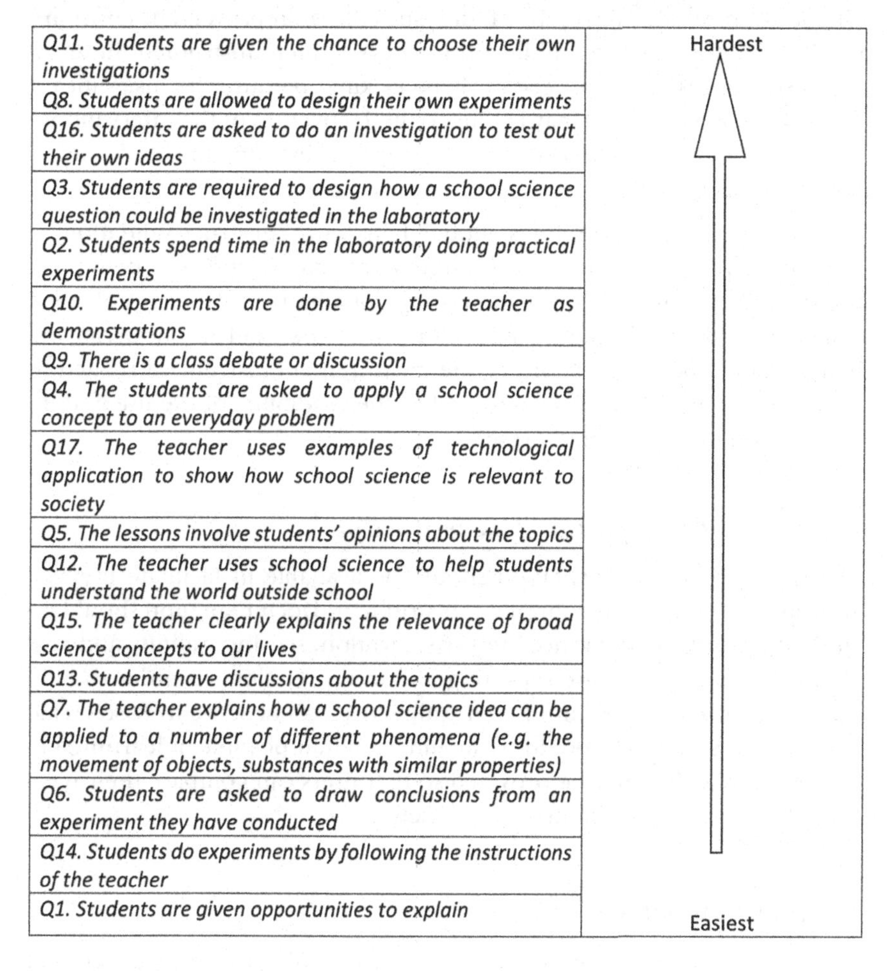

Figure 7.5 Item order of the PISA Inquiry Learning Environment items. Teaching practices noted on the top are harder to implement than that noted at the bottom.

presented from the most difficult item (on the top) to the easiest item (on the bottom). Is the order reasonable? For example, is Q11 more indicative of inquiry science teaching than Q8? Is Q1 easiest to implement in inquiry science teaching? Besides comparing the orders of items between the theoretical prediction and the actual one, we may also ask experts to predict the order of items and compare it to the obtained one.

Results from learning environment measurement are typically used for developing instructional interventions. Validation related to consequences is thus about the effectiveness of such interventions for improving student learning. Results from program evaluation of instructional improvement efforts should provide evidence on the validity related to consequences. Of course, insignificant results of an instructional improvement effort are possible. In this case, the validity of the measurement instrument in terms of consequences may not necessarily be lacking, because the insignificant instruction improvement may be due to ineffectiveness of the intervention. However, a re-examination of the measurement instrument should nevertheless be necessary.

Finally, qualitative findings should help with construct validation in terms of response processes. In addition to qualitative studies of item functioning during item tryout/pilot-testing, qualitative evidence collected from respondents through observation, interview, and artifacts can enhance claims about the validity of the measurement instrument related to response processes, which in turn add to the credibility of the use and interpretation of measurement results.

Develop Guidelines for Use of the Instrument

Appropriate documentation should be available to facilitate users to appropriately use the measurement instrument. Documentation should include the construct definition, test specification, scoring, validity and reliability evidence, and score reporting and interpretation. In order to help users convert raw scores into Rasch scale scores, a raw score to Rasch scale score conversion table should be included. When possible, a learning environment baseline may be provided so that users can compare their measured learning environments to the baseline.

Chapter Summary

Learning environments are social, physical, psychological, and pedagogical contexts in which learning occurs and that affect student learning outcomes in science. The basic premise for learning environment research is that personal goals may be supported or constrained by environmental factors. When personal needs and environmental factors are aligned, student learning outcomes are enhanced. Learning environments include three dimensions or domains; they are relationship, personal growth, and system maintenance and change. The relationship dimension is concerned with the extent to which people are involved in the setting, the extent to which they support and help one another, and the extent to which they express themselves freely and openly; the personal growth dimension pertains to the goals of the setting related to personal development and self-enhancement; and the system maintenance and change dimension pertains to the extent to which the environment is orderly and clear in its expectations, maintains control, and responds to change.

There can be three orientations to learning environment research: the outcome-environment orientation, the person-environment fit orientation, and the environment-outcome orientation. The outcome-environment orientation focuses on the association between learning environment variables and student cognitive and affective learning outcomes; the person-environment fit orientation differentiates various forms of the learning environment, such as the *actual* perceived environment and the *preferred* environment; and the environment-outcome orientation is concerned with how a learning environment may change as the result of changes in other factors of a science teaching and learning system. A wide variety of measurement instruments are currently available; they possess various degrees of validity and reliability.

Developing instruments for measuring learning environments begins with stating the purposes of the measurement in terms of the three orientations and clearly defining the construct of the learning environment to be measured. Items to be constructed for the measurement instrument should define a domain to provide adequate coverage of the anticipated variation in the measured learning environment of a particular population. Rasch analysis facilitates development of the measurement instrument by providing various statistics and measures related to model-data fit, the person-item match, dimensionality, person and item separation and reliability, and scale equating. Measurement instruments developed through Rasch analysis produce interval scale scores that are directly comparable between parallel forms.

Exercises

1. The *Constructivist Learning Environment Survey* (CLES; Taylor, Fraser, & Fisher, 1997), is one of the popular science learning environment scales of science education research. A complete ver-

sion of CLES is available in the appendix of Fraser (1998a). Apply Standard 1 Validity (Joint Committee of the AERA, NCME and APA, 2014) to critique this instrument. What are the strengths and weaknesses of this instrument? What do you recommend on how to use this instrument?

2. Most measurement instruments introduced in this chapter have been developed using the CTT approach. From the perspectives of Rasch modeling, what are common limitations of these measurement instruments? What do we need to consider when using these measurement instruments? How may Rasch modeling help address the limitations?
3. Identify one specific science learning environment for which a measurement instrument is currently not available, complete the following initial steps of developing a measurement instrument for this science learning environment: (a) define the science learning environment construct and support the definition with pertinent theories, (b) identify student or teacher performances that represent the defined construct, and (c) write five sample items for each of the levels of the defined construct.

References

Adams, E. L., Carrier, S. J., Minogue, J., Porter, S. R., McEachin, A., Walkowiak, T. A., & Zulli, R. A. (2017). The development and validation of the instructional practices log in science: A measure of K–5 science instruction. *International Journal of Science Education, 39*(3), 335–357.

Aldridge, J. M., & Fraser, B. J. (2000). A cross-cultural study of classroom learning environments in Australia and Taiwan. *Learning Environment Research: An International Journal, 3,* 101–134.

Aldridge, J. M., Fraser, B. J., & Huang, T.-C. I. (1999). Investigating classroom environments in Taiwan and Australia with multiple research methods. *Journal of Educational Research, 93,* 48–62.

Aldridge, J. M., Laugksch, R. C., Seopa, M. A., & Fraser, B. J. (2006). Development and validation of an instrument to monitor the implementation of outcomes-based learning environments in science classrooms in South Africa. *International Journal of Science Education, 28*(1), 45–70.

Anderson, G. J., & Walberg, H. J. (1974). Learning environments. In H. J. Walberg (Ed.), *Evaluating educational performance: A sourcebook of methods, instruments, and examples.* Berkeley, CA: McCutchan.

Arzi, H. J. (1998). Enhancing science education through laboratory environment: More than walks, benches and widgets. In B. J. Fraser & K. G. Tobin (Eds.), *International handbook of science education* (pp. 595–608). Dordrecht, The Netherlands: Kluwer.

Barrie, S. C., Bucat, R. B., Buntine, M. A., da Silva, K. B., Crisp, G. T., George, A. V., . . . Yeung, A. (2015). Development, evaluation and use of a student experience survey in undergraduate science laboratories: The advancing science by enhancing learning in the laboratory student laboratory learning experience survey. *International Journal of Science Education, 37*(11), 1795–1814.

Dhindsa, H. S., Omar, K., & Waldrip, B. (2007). Upper secondary Bruneian science students' perceptions of assessment. *International Journal of Science Education, 29*(10), 1261–1280.

Dorman, J. P. (2003). Cross-national validation of the *What Is Happening in This Class?* (WIHIC) questionnaire using confirmatory factor analysis. *Learning Environments Research: An International Journal, 6,* 231–245.

Erdogan, I., Campbell, T., & Abd-Hamid, N. H. (2011). The student actions coding sheet (SACS): An instrument for illuminating the shifts toward student-centered science classrooms. *International Journal of Science Education, 33*(10), 1313–1336.

Fisher, D. L., & Fraser, B. J. (1981). Validity and use of my class inventory. *Science Education, 65,* 145–156.

Fisher, D. L., & Waldrip, B. G. (1997). Assessing culturally sensitive factors in the learning environment of science classrooms. *Research in Science Education, 27,* 41–49.

Fisher, D. L., Waldrip, B. G., & Dorman, J. (2005, April). *Student perceptions of assessment: Development and validation of a questionnaire.* Paper presented at the Annual Meeting of the American Educational Research Association, Montreal, Canada.

Fraser, B. J. (1982). Development of short-forms of several classroom environment scale. *Journal of Educational Measurement, 19,* 221–227.

Fraser, B. J. (1990). *Individualized classroom environment questionnaire.* Melbourne, Australia: Australian Council for Educational Research.

Fraser, B. J. (1994). Research on classroom and school climate. In D. Gable (Ed.), *Handbook of research on science teaching and learning* (pp. 493–541). New York, NY: Macmillan.

Fraser, B. J. (1998a). Science learning environments: Assessment, effects and determinants. In B. J. Fraser & K. G. Tobin (Eds.), *International handbook of science education* (pp. 527–564). Dordrecht, The Netherlands: Kluwer.

Fraser, B. J. (1998b). The birth of a new journal: Editor's introduction. *Learning Environment Research: An International Journal, 1,* 1–5.

Fraser, B. J. (2007). Classroom learning environments. In S. K. Abell & N. G. Lederman (Eds.), *Handbook of research on science education* (pp. 103–124). New York, NY: Routledge.

Fraser, B. J., Anderson, G. J., & Walberg, H. J. (1982). *Assessment of learning environments: Manuals for learning environment inventory (LEI) and my class inventory (MCI) (third version).* Perth, Australia: Western Australian Institute of Technology.

Fraser, B. J., & Fisher, D. L. (1983). Development and validation of short forms of some instruments measuring student perceptions of actual and preferred classroom learning environment. *Science Education, 67*, 115–131.

Fraser, B. J., McRobbie, C. J., & Giddings, G. J. (1993). Development and cross-national validation of a laboratory classroom environment instrument for senior high school science. *Science Education, 77*, 1–24.

Fraser, B. J., & Tobin, K. (1991). Combining qualitative and quantitative methods in classroom environment research. In B. J. Fraser & H. J. Walberg (Eds.), *Educational environments: Evaluation, antecedents, consequences* (pp. 271–292). London, England: Pergamon.

Fraser, B. J., & Treagust, D. F., (1986). Validity and use of an instrument for assessing classroom psychosocial environment in higher education. *Higher Education, 15*, 37–57.

Fraser, B. J., Treagust, D. F., & Dennis, N. C. (1986). Development of an instrument for assessing classroom psychosocial environment at universities and colleges. *Studies in Higher Education, 11*, 43–54.

Goh, S. C., & Fraser, B. J. (1996). Validation of an elementary school version of the questionnaire on teacher interaction. *Psychological Reports, 79*, 512–522.

Hofstede, G. (1984). *Culture's consequences.* Newbury Park, CA: SAGE.

Hora, M. T. (2015). Toward a descriptive science of teaching: How the TDOP illuminates the multidimensional nature of active learning in postsecondary classrooms. *Science Education, 99*(5), 783–818.

Huang, S. L. (2006). An assessment of science teachers' perceptions of secondary school environments in Taiwan. *International Journal of Science Education, 8*(1), 25–44.

Jacobs, C. L., Martin, S. N., & Otieno, T. C. (2008). A science lesson plan analysis instrument for formative and summative program evaluation of a teacher education program. *Science Education, 92*(6), 1096–1126.

Joint Committee of American Educational Research Association, American Psychological Association, & National Council on Measurement in Education. (2014). *Standards for educational and psychological testing.* Washington, DC: American Psychological Association.

Lane, E. S., & Harris, S. E. (2015). A new tool for measuring student behavioral engagement in large university classes. *Journal of College Science Teaching, 44*(6), 83–91.

Lesniak, K. (2007). Positive classroom and laboratory environments for science learning: Barry J. Fraser's contributions to science education. In X. Liu (2007). *Great ideas in science education* (pp. 15–30). Rotterdam, The Netherlands: Sense.

McRobbie, C. J., Fisher, D. L., & Wong, A. F. L. (1998). Personal and class forms of classroom environment instruments. In B. J. Fraser & K. G. Tobin (Eds.), *International handbook of science education* (pp. 581–594). Dordrecht, The Netherlands: Kluwer.

Moll, R., & Nielsen, W. (2017). Development and validation of a social media and science learning survey. *International Journal of Science Education, Part B, 7*(1), 14–30.

Moos, R. H. (1979). *Evaluating educational environments.* San Francisco, CA: Jossey-Bass.

Moos, R. H., & Trickett, E. J. (1974). *Classroom environment scale manual* (1st ed.). Palo Alto, CA: Consulting Psychologists Press.

Moos, R. H., & Trickett, E. J. (1987). *Classroom environment scale manual* (2nd ed.). Palo Alto, CA: Consulting Psychologists Press.

Nolen, S. B. (2003). Learning environment, motivation, and achievement in high school science. *Journal of Research in Science Teaching, 40*(4), 347–368.

OECD. (2006). *Assessing scientific, reading and mathematical literacy: A framework for PISA 2006.* Paris, France: Author.

Rentoul, A. J., & Fraser, B. J. (1979). Conceptualization of enquiry-based or open classrooms learning environments. *Journal of Curriculum Studies, 11,* 233–245.

Roth, W.-M. (1999). Learning environment research, life world analysis, and solidarity in practice. *Learning Environment Research: An International Journal, 3,* 225–247.

Sawada, D., Piburn, M. D., Judson, E., Turley, J., Falconer, K., Benford, R., & Bloom, I. (2002). Measuring reformed practices in science and mathematics classrooms: The Reformed Teaching Observation Protocol. *School Science and Mathematics, 102*(6), 245–253.

Shah, A. M., Wylie, C., & Gitomer, D. (2018). Improving STEM program quality in out-of-school-time: Tool development and validation. *Science Education, 102,* 238–259.

Shapiro, B. (1998). Reading the furniture: The semiotic interpretation of science learning environments. In B. J. Fraser & K. G. Tobin (Eds.), *International handbook of science education* (pp. 609–621). Dordrecht, The Netherlands: Kluwer.

Smith, M. K., Jones, F. H. M., Gilbert, S., & Wieman, C. E. (2013). The Classroom Observation Protocol for Undergraduate STEM (COPUS): A new instrument to characterize university STEM classroom practices. *CBE-Life Sciences Education, 12,* 618–627.

Taylor, P. C., Fraser, B. J., & Fisher, D. L. (1997). Monitoring constructivist classroom learning environments. *International Journal of Educational Research, 27,* 293–302.

Trickett, E. J., & Moos, R. H. (1973). The social environment of junior high and high school classrooms. *Journal of Educational Psychology, 65,* 93–102.

Tobin, K. G., & Fraser, B. J. (1998). Qualitative and quantitative landscapes of classroom learning environment. In B. J. Fraser & K. G. Tobin (Eds.), *International handbook of science education* (pp. 623–640). Dordrecht, The Netherlands: Kluwer.

Velasco, J. B., Knedeisen, A., Xue, D., Vickrey, T. L., Abebe, M., & Stains, M. (2016). Characterizing instructional practices in the laboratory: The

Laboratory Observation Protocol for Undergraduate STEM. *Journal of Chemical Education, 93*(7), 1191–1203.

Walberg, H. J. (1968a). Structural and affective aspects of classroom climate. *Psychology in schools, 5*, 247–253.

Walberg, H. J. (1968b). Classroom climate and individual learning. *Journal of Educational Psychology, 59*(6), 414–419.

Walczyk, J. J., & Ramsey, L. L. (2003). Use of learner-centered instruction in college science and mathematics classrooms. *Journal of Research in Science Teaching, 40*(6), 566–584.

Wieman, C., & Gilbert, S. (2014). The teaching practices inventory: A new tool for characterizing college and university teaching in mathematics and science. *CBE-Life Sciences Education, 13*, 552–569.

Wu, M. L., Adams, R. J., & Wilson, M. (1997). *Conquest: Generalized item response modeling software—Manual.* Melbourne, Australia: Australian Council for Educational Research.

Wubbels, T., & Brekelmans, M. (1998). The teacher factor in the social climate of the classroom. In B. J. Fraser & K. G. Tobin (Eds.), *International handbook of science education* (pp. 565–580). Dordrecht, The Netherlands: Kluwer.

Wubbles, T., Breklmans, M., & Hooymayers, H. (1991). Interpersonal teacher behavior in the classroom. In B. J. Fraser & H. J. Walberg (Eds.), *Educational environments: Evaluation, antecedents and consequences* (pp. 141–160). London, England: Pergamon.

Wubbles, T., & Levy, J. (Eds.). (1993). *Do you know what you look like: Interpersonal relationships in education.* London, England: Falmer Press.

Author Index

A

B

Using and Developing Measurement Instruments in Science Education, pages 321–330

C

D

E

F

G

H

M

N

O

P

R

S

T

U

V

W

X

Y

Z

Subject Index

A

C

D

E

F

I

Using and Developing Measurement Instruments in Science Education, pages 331–333

S

T

U

V

W